Qualitative Insights Through Applied Cognitive Task Analysis

Rafiza Abdul Razak
Universiti Malaya, Malaysia

Nurul Fitriah Alias
Universiti Teknologi MARA, Malaysia

Aizal Yusrina Idris
International Islamic University Malaysia, Malaysia

IGI Global
Scientific Publishing
Publishing Tomorrow's Research Today

Published in the United States of America by
IGI Global Scientific Publishing
701 East Chocolate Avenue
Hershey, PA, 17033, USA
Tel: 717-533-8845
Fax: 717-533-8661
Website: https://www.igi-global.com E-mail: cust@igi-global.com

Library of Congress Cataloging-in-Publication Data

Names: Abdul Razak, Rafiza, 1974- editor. | Alias, Nurul Fitriah, 1988- editor. | Idris, Aizal Yusrina, 1974- editor.
Title: Qualitative insights through applied cognitive task analysis / edited by Rafiza Razak, Nurul Alias, Aizal Yusrina Idris.
Description: Hershey, PA : IGI Global Scientific Publishing, [2025] | Includes bibliographical references and index. | Summary: "This book delves into the advanced methodologies of Applied Cognitive Task Analysis (ACTA) with a focus on qualitative data. The book aims to elucidate how ACTA can be effectively utilized to decode complex mental processes and decision-making strategies in various fields, ranging from education to high-stakes environments like healthcare and military operations"-- Provided by publisher.
Identifiers: LCCN 2024059925 (print) | LCCN 2024059926 (ebook) | ISBN 9798369377338 (hardcover) | ISBN 9798369377345 (paperback) | ISBN 9798369377352 (ebook)
Subjects: LCSH: Qualitative research--Data processing. | Qualitative research--Methodology. | Decision making.
Classification: LCC H62 .Q3476 2025 (print) | LCC H62 (ebook) | DDC 001.4/2--dc23/eng/20250212
LC record available at https://lccn.loc.gov/2024059925
LC ebook record available at https://lccn.loc.gov/2024059926

British Cataloguing in Publication Data
A Cataloguing in Publication record for this book is available from the British Library.

Table of Contents

Preface

The field of cognitive task analysis (CTA) has long been instrumental in understanding the intricacies of human cognition in complex tasks. Applied Cognitive Task Analysis (ACTA) is a streamlined method that enhances traditional cognitive task analysis by focusing on practical applications and flexibility. It's particularly valuable in designing training systems and interfaces for various professional settings. This book, Qualitative Insights through Applied Cognitive Task Analysis, delves deeply into ACTA, offering a nuanced perspective on its methodologies, theoretical underpinnings, and practical applications.

As technological and educational landscapes evolve, the need for robust analytical tools to decipher cognitive processes becomes more pressing. ACTA serves as a bridge between qualitative insights and actionable strategies, enabling researchers, educators, and industry practitioners to design interventions that align with cognitive realities.

Structured across six parts, this book offers a detailed guide to ACTA, from foundational principles to advanced applications. Part 1 introduces the theoretical and practical evolution of ACTA, while Part 2 provides a systematic breakdown of methodologies, covering data collection techniques like structured interviews, direct observation, and work sample analysis. Triangulation methods are discussed as a means to enhance data validity, ensuring comprehensive insights. ACTA dissects the cognitive processes underlying task performance, bridging theoretical knowledge with practical implementation. Its foundations are supported by key theories like Cognitive Load Theory and Situated Cognition, which emphasize optimizing cognitive resources and integrating real-world contexts. Research by Sharma et al. (2023) highlights the role of social network analysis in online collaborative learning, revealing how peer interactions and discourse patterns affect cognitive load and engagement. This reinforces the importance of understanding the social dynamics that shape cognitive performance, further validating ACTA's focus on qualitative insights.

In Part 3, the use of Applied Cognitive Task Analysis (ACTA) in educational settings is explored through tools such as task diagrams, knowledge audits, simulations, and cognitive demand tables. These methods are showcased in two case studies that highlight ACTA's transformative impact on learning outcomes—one focusing on electronic training programs and the other on microlearning pedagogical practices.

As education shifts toward digital and lifelong learning environments, there is an increasing need to support self-directed learning, particularly in informal settings. The rise of intelligent devices and advancements in web technologies has profoundly changed the way learners access, interact with, and generate knowledge (Haleem, 2022).

Parts 4 and 5 explore innovative approaches, including the integration of artificial intelligence and virtual reality (VR), interdisciplinary contributions, and the ethical challenges of ACTA. The integration of VR with ACTA principles is particularly powerful because it allows for a nuanced exploration of both the physical and cognitive aspects of task performance. ACTA focuses on dissecting the cognitive processes involved in complex tasks, such as decision-making, problem-solving, and skill execution. VR environments can be tailored to replicate these tasks with high fidelity, providing a platform to study cognitive functions in contexts that closely resemble real-world scenarios.

For instance, research by Zhou et al. (2023) introduced an innovative method of evaluating cognitive processing speed using an immersive virtual street environment embedded with a Stroop task, known as VR-Street. This study illustrates the synergy between VR and ACTA by engaging participants in a dual-task scenario where they navigate a virtual street while performing a Stroop task. The VR-Street environment allowed researchers to capture comprehensive behavioral and performance data, offering insights into how participants manage cognitive resources under realistic conditions.

By addressing these topics, the book equips readers to navigate the complexities of applying ACTA in dynamic, high-stakes environments.

Finally, Part 6 examines practical implications and evaluates ACTA's impact through case studies and metrics. Success stories across sectors highlight the methodology's ability to optimize performance, foster innovation, and improve organizational outcomes. The long-term benefits of ACTA-based interventions extend well beyond immediate improvements, significantly enhancing organizational decision-making, training efficiency, adaptability, and sustained learning. For instance, ACTA improves decision-making in high-stakes fields, as demonstrated by Militello et al. (2023) and Ahmad et al. (2020), who showed its value in improving cognitive processes among anaesthesiologists and novice surgeons. Furthermore, Simon and Richards (2022) highlighted how ACTA enhanced training efficiency for Olympic coaches by refining cognitive task analysis, achieving notable outcomes like medals and records. Adaptability was another key outcome, exemplified by Stanik et al. (2022), whose ACTA-based method facilitated innovative device designs for emergency care teams. Lastly, ACTA fosters a culture of continuous learning, as Taylor et al. (2023) and Marshall and Finlayson (2022) emphasized in studies on coaching and nursing, respectively, demonstrating how updating training to address cognitive demands supports ongoing professional development and organizational performance.

CASE STUDIES

Case Study 1: Electronic Training for Professional Development

This case study examines the development of TaSPoD (e-Training System for Professional Development), an electronic training platform designed to enhance professional development for educators in higher education institutions. The study identifies key barriers to effective digital training, including cultural, technological, and pedagogical challenges.

To address these issues, the Design-Based Research (DBR) approach was employed, incorporating iterative phases: Analysis, Design, Validation, Development, Evaluation, and Innovation. The research highlights critical gaps in existing electronic training systems, such as insufficient instructional support, limited interactivity, and a lack of multimedia integration. To refine the TaSPoD prototype, ACTA techniques including Task Diagrams, Knowledge Audit Tables, Simulation Interviews, and Cognitive Demand Tables; were applied.

Findings from this study emphasize the importance of user-friendly interfaces, interactive learning strategies, and structured feedback mechanisms in improving digital professional training. Hoq (2020) and Saintika et al. (2021) pointed out that learners and educators are familiar with online communication and utilizing internet for communication purposes. The research underscores that effective electronic training must incorporate diverse instructional methods, enhance engagement, and provide comprehensive support for educators. Rafique & Aziz (2019) asserted that exist significant relationship between pedagogical content knowledge and pedagogical design.

Case Study 2: Developing Pedagogical Practices of Microlearning in Higher Education

This case study investigates the implementation of microlearning in higher education, focusing on short, targeted learning units designed to enhance student engagement and knowledge retention. Research has demonstrated that microlearning can significantly boost students' confidence, interest, self-efficacy, learning abilities, and overall academic performance (Fitzgerald & Tisdell, 2019). The study introduces the AMPLIFY framework (Advancing Microlearning Practices for Enhanced Learning Experiences), which consists of eight key elements, including personalized learning activities, interactive experiences, and social integration.

The research follows a structured, four-phase process—Analysis, Design, Development, and Evaluation—while employing ACTA techniques to assess the cognitive processes involved in microlearning. The findings highlight the growing preference for flexible, on-demand learning solutions, particularly among Generation Z learners, who thrive in digital and mobile-friendly educational environments.

Despite its advantages, the study identifies several challenges in microlearning adoption, such as unclear implementation guidelines and limited educator preparedness. The research underscores the need for structured training programs to help educators effectively integrate microlearning strategies into their curriculum.

By leveraging ACTA techniques, educators and institutions can develop more efficient, engaging, and accessible educational experiences, bridging the gap between traditional and digital learning environments.

ORGANIZATION OF THE BOOK

The book is organized into 16 chapters. A brief description of each of the chapters follows:

Chapter 1 introduces Applied Cognitive Task Analysis (ACTA), a crucial methodology for understanding and optimizing human cognition across various domains. ACTA dissects cognitive processes underlying task performance, bridging theoretical knowledge with practical applications. The chapter highlights the importance of social science research and qualitative methods in revealing the motivations and behaviors shaping cognition. It provides a foundation for exploring how ACTA improves task performance in fields such as aviation, healthcare, organizational behavior, and education. The evolution from traditional Cognitive Task Analysis (CTA) to ACTA reflects a shift toward dynamic, context-sensitive tools. The chapter also addresses challenges in cognitive task analysis, including capturing tacit knowledge, task complexity, and ecological validity. By leveraging innovative methods and interdisciplinary approaches, ACTA advances our understanding of cognitive processes and enhances human performance in professional settings.

Chapter 2 explores the foundational theories behind Applied Cognitive Task Analysis (ACTA), focusing on Cognitive Load Theory (CLT) and Situated Cognition. CLT is crucial for managing cognitive resources during learning and task execution by optimizing intrinsic, extraneous, and germane loads. This helps design effective instructional materials and training programs for better information processing and retention. Situated Cognition emphasizes the role of context and social interactions in learning, aligning with ACTA's focus on practical knowledge application in real-world settings. The chapter examines how these theories enhance understanding of cognitive processes and stress the need for personalized, context-rich learning experiences. By integrating qualitative insights, it highlights the value of capturing expert performance nuances often missed by quantitative methods. This theoretical foundation supports the application of ACTA across various fields, from education to industry, where improving cognitive performance is essential.

Chapter 3 provides an overview of the Applied Cognitive Task Analysis (ACTA) process, outlining its key steps from start to finish. It begins with the planning phase, which involves defining the scope, identifying tasks, and selecting appropriate tools. The data collection stage follows, utilizing methods like cognitive interviews and task observations to gather detailed information on cognitive demands. The data analysis phase focuses on identifying cognitive bottlenecks and task complexities, providing insights for improving task design, training, and protocols. The chapter concludes with the reporting stage, emphasizing clear, actionable communication of findings to stakeholders. By systematically guiding readers through each step, this chapter offers a practical resource for those applying ACTA in education, healthcare, industry, and military settings.

Chapter 4 explores the methodologies used in the Applied Cognitive Task Analysis (ACTA) process, detailing how data is gathered to understand cognitive tasks. It covers structured interviews, direct observation, and work sample analysis, providing a step-by-step guide on applying each technique. The chapter discusses the strengths and limitations of these methods and how to combine them for a comprehensive view of cognitive processes. It highlights the importance of aligning data collection strategies with research objectives, offering practical examples such as using structured interviews for verbal explanations and direct observation for real-time task performance. Work sample analysis is emphasized for its ability to provide concrete evidence of task execution and cognitive demands. This chapter equips researchers and practitioners with the tools needed to conduct effective cognitive task analysis, providing rich, actionable data for task design, training, and system optimization.

Chapter 5 explores the data analysis techniques used in the Applied Cognitive Task Analysis (ACTA) process, detailing how the cognitive data is being analysed. It covers structured interviews, direct observation, and work sample analysis, providing a step-by-step guide to analysed all data gathered from each technique in ACTA. The chapter discusses the strengths and limitations of these methods and how to combine them for a comprehensive view of cognitive processes. It highlights the importance of aligning data analysis strategies with research objectives, offering practical examples such as using structured interviews for verbal explanations and direct observation for real-time task performance. Work sample analysis is emphasized for its ability to provide concrete evidence of task execution and cognitive demands. This chapter equips researchers and practitioners with the tools needed to conduct effective cognitive task analysis, providing rich, actionable data for task design, training, and system optimization.

Chapter 6 explores the application of Applied Cognitive Task Analysis (ACTA) within educational settings, focusing on two distinct case studies that demonstrate its versatility and effectiveness. The first case study examines Electronic Training for Professional Development, highlighting how ACTA can enhance the design and delivery of digital training programs, ensuring they meet the evolving needs of professionals in various fields. The second case study delves into Developing Pedagogical Practices of Microlearning Educational Experience in Higher Education. It showcases how ACTA can inform the development of innovative microlearning strategies, leading to more personalized, engaging, and effective learning experiences for students in higher education. By analyzing these case studies, the chapter underscores the potential of ACTA to revolutionize educational practices, providing educators and trainers with robust tools to optimize learning outcomes and foster continuous improvement in educational environments.

Chapter 7 discusses the use of task diagrams as a key tool in Applied Cognitive Task Analysis (ACTA) within educational settings, divided into two sections. The first, "Task Diagrams for Electronic Training for Professional Development," explains how task diagrams break down complex tasks into manageable components, optimizing the design of electronic training programs. These diagrams visually represent cognitive processes, helping to address challenges and enhance professional development. The second section, "Task Diagrams for Designing Microlearning Pedagogical Practices," shows how task diagrams aid in developing microlearning strategies in higher education, offering insights into structuring content, personalizing learning, and aligning pedagogy with educational goals. Together, these sections demonstrate how task diagrams streamline the design process and improve educational outcomes.

Chapter 8 explores conducting a knowledge audit within Applied Cognitive Task Analysis (ACTA) in educational settings. It is divided into two sections. The first, "Knowledge Audit for Electronic Training for Professional Development," focuses on assessing existing knowledge and identifying gaps in electronic training programs, ensuring content aligns with professional development needs. The second section, "Knowledge Audit for Developing Microlearning Pedagogical Practices," discusses how knowledge audits can help design microlearning strategies in higher education, tailoring experiences to learners' needs and improving engagement. The chapter emphasizes the role of knowledge audits in refining educational practices and optimizing content delivery for both training and microlearning.

Chapter 9 examines the use of simulation techniques in Applied Cognitive Task Analysis (ACTA) to improve educational practices. It is divided into two sections. The first, "Simulation of Electronic Training for Professional Development," explores how simulation models real-world scenarios to enhance electronic training, offering hands-on learning that improves skill acquisition and retention. This section shows how simulation-based training provides an immersive experience that bridges theory and practice for professional development. The second section, "Simulation for Microlearning Pedagogical Practices," discusses how simulations create dynamic, adaptive learning environments in higher education. These simulations help deliver personalized, relevant content, fostering engagement and better learning outcomes. The chapter underscores simulation's importance in optimizing educational strategies for both electronic training and microlearning.

Chapter 10 discusses the development and use of Cognitive Demand Tables in Applied Cognitive Task Analysis (ACTA) to assess and optimize cognitive load in educational settings. It is divided into two sections. The first, "Cognitive Demand Table of Electronic Training for Professional Development," examines how cognitive demands in electronic training are identified, analyzed, and balanced to prevent overload, ensuring effective and accessible professional development. The second, "Cognitive Demand Table of Microlearning Pedagogical Practices," focuses on optimizing the cognitive load of microlearning strategies in higher education, enhancing understanding and retention. This chapter highlights the importance of managing cognitive demands to create effective, learner-centered instructional strategies.

Chapter 11 explores advancements in Applied Cognitive Task Analysis (ACTA) driven by emerging technologies and methodologies. It examines how artificial intelligence (AI) and virtual reality (VR) are transforming ACTA practices by automating data analysis, optimizing task performance assessments, and personalizing learning experiences. The chapter highlights AI's role in enhancing ACTA and VR's application in simulating complex cognitive tasks and providing immersive training environments. It also looks at future trends and tools that could further advance ACTA. By providing insights into these innovations, the chapter discusses their implications for educational and professional settings, emphasizing how ACTA can evolve in a technology-driven world.

Chapter 12 explores the interdisciplinary dimensions of Applied Cognitive Task Analysis (ACTA), highlighting its integration with cognitive science, psychology, and organizational studies. It examines how ACTA benefits from advancements in these fields, showcasing its versatile applications across healthcare, education, and technology. By exploring these cross-sectoral applications, the chapter illustrates the mutual benefits of ACTA's collaboration with diverse disciplines, enhancing its effectiveness and impact. Specific examples demonstrate how ACTA methodologies address complex problems, contributing to scientific knowledge while gaining insights from various research perspectives. This comprehensive overview emphasizes how interdisciplinary contributions enrich ACTA and offers insights into future research and practical applications.

Chapter 13 examines the methodological and practical challenges of applying Applied Cognitive Task Analysis (ACTA) across complex domains. It highlights the difficulties in capturing and modeling cognitive processes in high-pressure environments, such as aviation and healthcare, where expert decision-making is critical. Additionally, it discusses the need for scalable and adaptable ACTA methodologies to accommodate varying expertise levels, as well as the integration of emerging technologies like AI and VR. The chapter also offers solutions, including hybrid methodologies and interdisciplinary collaboration, to enhance ACTA's effectiveness in real-world applications and ensure actionable insights.

Chapter 14 addresses the ethical considerations essential to Applied Cognitive Task Analysis (ACTA). Key topics include the importance of obtaining informed consent, minimizing harm during data collection, and maintaining confidentiality and data integrity, especially in sensitive environments like healthcare and military settings. It also examines the risks of dual-use dilemmas, where cognitive insights could be misused, and stresses the need for responsible dissemination of findings. By adopting ethical practices, including robust informed consent, distress protocols, and cybersecurity measures, ACTA researchers can ensure their work adheres to high ethical standards while fostering trust and safeguarding participant welfare.

Chapter 15 explores the practical application of Applied Cognitive Task Analysis (ACTA) findings across various sectors. It outlines strategies for translating cognitive insights into real-world outcomes, including tailoring findings to stakeholder needs, developing training programs, designing user-centered interfaces, and integrating ACTA findings into policies. Emphasizing continuous evaluation and iteration, the chapter showcases success stories from industries like healthcare, military, and education. By applying ACTA, organizations can improve performance, safety, and efficiency while fostering a culture of continuous improvement and innovation.

Chapter 16 examines methods for evaluating the effectiveness of Applied Cognitive Task Analysis (ACTA)-based interventions, focusing on both short-term and long-term impacts. Key assessment strategies include pre- and post-intervention evaluations, longitudinal studies, behavioral observations, self-reported measures, and performance metrics. These methods provide comprehensive insights into how ACTA improves decision-making, training efficiency, adaptability, and organizational learning. By assessing cognitive processes and task performance over time, the chapter highlights the value of ACTA in optimizing training, enhancing operational efficiency, and supporting sustained organizational development across various fields.

This book serves as a valuable resource for researchers, educators, and practitioners. It combines theoretical rigor with actionable strategies, inspiring its readers to harness the power of ACTA in enhancing human performance and understanding. By bridging the gap between theory and practice, this work underscores the potential of ACTA to address complex cognitive challenges and drive meaningful progress in diverse fields.

Rafiza Abdul Razak
Universiti Malaya, Malaysia

Nurul Fitriah Alias
Universiti Teknologi MARA, Malaysia

Aizal Yusrina Idris
International Islamic University Malaysia, Malaysia

REFERENCES

Ahmad, K., Bhattacharyya, R., & Gupte, C. (2020). Using cognitive task analysis to train orthopaedic surgeons: Is it time to think differently? A systematic review. *Annals of Medicine and Surgery (London)*, 59, 131–137. DOI: 10.1016/j.amsu.2020.09.031

Fitzgerald, J. H., & Tisdell, C. C. (2019, June). The impact of educational microcontent on the student learning experience. In *Proceedings of the 2019 International Conference on Mathematics, Science and Technology Teaching and Learning* (pp. 17-22). Association for Computing Machinery DOI: 10.1145/3348400.3348412

Haleem, A., Javaid, M., Qadri, M. A., & Suman, R. (2022). Understanding the role of digital technologies in education: A review. *Sustainable Operations and Computers*, 3, 275–285. DOI: 10.1016/j.susoc.2022.05.004

Hoq, M. Z. (2020). E-Learning During the Period of Pandemic (COVID-19) in the Kingdom of Saudi Arabia: An Empirical Study. *American Journal of Educational Research*, 8(7), 457–464. http://pubs.sciepub.com/education/8/7/2

Marshall, D., & Finlayson, M. (2022). Applied cognitive task analysis methodology: Fundamental cognitive skills surgical nurses require to manage patient deterioration. *Nursing Praxis in Aotearoa New Zealand*, 38(1), 25–37. DOI: 10.36951/27034542.2022.04

Militello, L. G., Salwei, M. E., Reale, C., Sushereba, C., Slagle, J. M., Gaba, D., Weinger, M. B., Rask, J., Faiman, J., Andreae, M., Burden, A. R., & Anders, S. (2023). Adapting cognitive task analysis methods for use in a large sample simulation study of high-risk healthcare events. *Journal of Cognitive Engineering and Decision Making*, 17(4), 315–331. DOI: 10.1177/15553434231192283

Rafique, S., & Aziz, S. (2019). Relationship of Pedagogical Skills and Content Knowledge with Pedagogical Design Capacity in Higher Education of Pakistan. *Global Regional Review*, IV(IV), 85–94. DOI: 10.31703/grr.2019(iv-iv).10

Saintika, Y., Astiti, S., Kusuma, D. J. A., & Muhammad, A. W. (2021). *Analysis of E-learning readiness level of public and private universities in central Java, Indonesia*. Register. *Jurnal Ilmiah Teknologi Sistem Informasi*, 7(1), 15–30. DOI: 10.26594/register.v7i1.2042

Sharma, P., Akgun, M., & Li, Q. (2023). Understanding student interaction and cognitive engagement in online discussions using social network and discourse analyses. *Educational Technology Research and Development*. Advance online publication. DOI: 10.1007/s11423-023-10261-w

Simon, S., & Richards, P. (2022). Investigating cognitive observational difficulties and expert skills in elite Olympic and Paralympic coaches using an applied cognitive task analysis. *Journal of Expertise*.

Taylor, J., Ashford, M., & Jefferson, M. (2023). High performance coach cognition in the wild: Using applied cognitive task analysis for practical insights—Cognitive challenges and curriculum knowledge. *Frontiers in Psychology*, 14. Advance online publication. DOI: 10.3389/fpsyg.2023.1154168

Zhou, Y., Zhao, Y., Xiang, Z., Yan, Z., Shu, L., Xu, X., Zhang, L., & Tian, X. (2023). A dual-task-embedded virtual reality system for intelligent quantitative assessment of cognitive processing speed. *Frontiers in Human Neuroscience*, 17. Advance online publication. DOI: 10.3389/fnhum.2023.1158650

Introduction

PREAMBLE: QUALITATIVE INSIGHTS THROUGH APPLIED COGNITIVE TASK ANALYSIS

Aim and Purpose of the Book

The book, *Qualitative Insights through Applied Cognitive Task Analysis (ACTA)*, seeks to serve as a comprehensive guide to understanding and leveraging ACTA as a powerful tool for exploring and improving cognitive processes in various domains. It bridges the gap between theoretical constructs and practical applications, offering insights that are grounded in qualitative methodologies and enriched by real-world case studies. The aim is to provide readers with a clear understanding of ACTA's potential to enhance performance, design better training programs, and inform decision-making in diverse professional and educational settings.

Designed for researchers, practitioners, and educators, this book highlights the importance of qualitative insights in ACTA, emphasizing their role in uncovering hidden cognitive demands and facilitating practical interventions. It offers an accessible yet in-depth exploration of ACTA's frameworks, methods, and applications, ensuring relevance for both newcomers and experienced professionals.

Organisation and Structure of the Book

The book is structured into six parts, each focusing on a specific aspect of ACTA to guide readers through its foundational principles, methodologies, and practical implications:

1. **Foundation and Theoretical Background**

Chapters 1 and 2 introduce readers to ACTA, detailing its evolution, distinguishing features, and theoretical underpinnings. This section establishes the groundwork for understanding the cognitive and qualitative aspects of ACTA.

2. **Methodologies and Techniques**

Chapters 3 to 5 delve into the processes and techniques central to ACTA, from data collection methods such as structured interviews and work sample analysis to advanced analytical strategies like triangulation. These chapters provide step-by-step guidance for conducting ACTA effectively.

3. **ACTA in Educational Settings**

Chapters 6 to 10 explore the application of ACTA in educational environments, with specific emphasis on electronic training and microlearning pedagogical practices. Key tools like task diagrams, knowledge audits, simulations, and cognitive demand tables are discussed in detail.

4. **Advanced Topics and Future Directions**

Chapters 11 and 12 address cutting-edge developments in ACTA, including the integration of technologies like artificial intelligence and virtual reality. This section also explores interdisciplinary contributions and cross-sectoral applications, showcasing ACTA's versatility.

5. **Challenges and Ethical Considerations**

Chapters 13 and 14 identify and address methodological, practical, and ethical challenges in ACTA. Solutions and best practices are provided to ensure integrity and efficacy in research and implementation.

6. **Practical Implications and Case Studies**

Chapters 15 and 16 focus on translating ACTA insights into actionable strategies and evaluating their impact. Real-world case studies illustrate the successful application of ACTA in diverse sectors, offering valuable lessons for readers.

Focus and Contribution

By weaving together theoretical foundations, detailed methodologies, practical applications, and forward-looking perspectives, this book provides a holistic view of ACTA. Readers will gain not only a robust understanding of the framework but also actionable insights to apply ACTA in real-world contexts. Through a blend of qualitative rigor and practical utility, the book underscores ACTA's potential to drive innovation and enhance cognitive understanding across disciplines.

Section 1
Foundation and Theoretical Background

Chapter 1
An Introduction to Applied Cognitive Task Analysis

ABSTRACT

This chapter introduces Applied Cognitive Task Analysis (ACTA), a crucial methodology for understanding and optimizing human cognition across various domains. ACTA dissects cognitive processes underlying task performance, bridging theoretical knowledge with practical applications. The chapter highlights the importance of social science research and qualitative methods in revealing the motivations and behaviors shaping cognition. It provides a foundation for exploring how ACTA improves task performance in fields such as aviation, healthcare, organizational behavior, and education. The evolution from traditional Cognitive Task Analysis (CTA) to ACTA reflects a shift toward dynamic, context-sensitive tools. The chapter also addresses challenges in cognitive task analysis, including capturing tacit knowledge, task complexity, and ecological validity. By leveraging innovative methods and interdisciplinary approaches, ACTA advances our understanding of cognitive processes and enhances human performance in professional settings.

DOI: 10.4018/979-8-3693-7733-8.ch001

INTRODUCTION

In the vast landscape of understanding human cognition, Applied Cognitive Task Analysis (ACTA) stands as a beacon illuminating the intricate workings of the mind. Within these pages, we embark on a journey to decode the mysteries of human thought processes and behaviours, guided by the principles of ACTA.

At the heart of our exploration lies the recognition of the profound relevance of social science research in unravelling the complexities of cognitive phenomena. Through the lens of social science, we gain invaluable insights into the interconnectedness of individual cognition and the broader societal context in which it unfolds.

Central to our approach is the utilisation of qualitative research methods, a powerful toolset for delving deep into the nuances of human experience. By embracing qualitative inquiry, we seek not only to uncover surface-level behaviours but also to illuminate the underlying meanings, motivations, and intricacies that shape cognitive processes.

In this introductory chapter, we lay the foundation for our journey ahead, providing a roadmap for how we will navigate the terrain of human cognition through the lens of ACTA. With a firm grasp of the relevance of social science research and the qualitative research approach, we are poised to embark on a transformative exploration into the depths of the human mind.

1.1. Understanding Cognitive Processes

Cognitive processes serve as the building blocks of human thought, perception, and action, shaping the very essence of our existence. These processes, alongside unique cognitive abilities like logical reasoning and problem-solving, allow humans to understand their environment and find practical solutions (Thakur, 2023). At their core, cognitive processes encompass a wide array of mental activities, including perception, attention, memory, language, reasoning, and decision-making (Dhakal & Bobrin, 2023). They underpin virtually every aspect of human behaviour, from the mundane routines of daily life to the most complex feats of innovation and creativity.

The significance of cognitive processes in human behaviour cannot be overstated. They serve as the cognitive machinery through which we navigate the world, make sense of our surroundings, and interact with others. At the heart of decision-making, problem-solving, and task performance lie a myriad of cognitive processes working in concert to guide our actions and inform our choices (Nielsen & Minda, 2019).

Consider, for instance, the process of decision-making. Whether choosing between two options at a grocery store or making high-stakes decisions in a boardroom, cognitive processes play a pivotal role in evaluating alternatives, weighing risks and rewards, and arriving at a course of action. Similarly, in problem-solving endeavours, cognitive processes enable us to identify obstacles, generate potential solutions, and evaluate their efficacy in achieving desired outcomes (Nielsen & Minda, 2019).

Central to our exploration is the concept of decoding the activity of the mind through cognitive task analysis. By dissecting the cognitive processes underlying task performance, we gain invaluable insights into the mechanisms driving human behaviour. Cognitive task analysis offers a systematic framework for uncovering the cognitive demands of a task, identifying the knowledge, skills, and strategies required for successful performance, and pinpointing areas for improvement (Militello & Hutton, 1998).

As we delve deeper into the realm of cognitive processes, we embark on a journey of discovery, unravelling the intricate workings of the mind and gaining a deeper understanding of what it means to be human. Through the lens of cognitive task analysis, we illuminate the pathways through which cognition shapes behaviour, paving the way for enhanced performance, innovation, and growth.

1.2 Applied Cognitive Task Analysis (ACTA)

1.2.1 Definitions and Distinctions from Traditional CTA

Applied Cognitive Task Analysis (ACTA) is a streamlined method that enhances traditional cognitive task analysis by focusing on practical applications and flexibility. It's particularly valuable in designing training systems and interfaces for various professional settings. ACTA helps extract information about the cognitive demands and skills required for a task, allowing practitioners to represent this information in a format that translates more directly into applied products, such as improved training scenarios or interface recommendations (Militello & Hutton, 1998). An evaluation study found that ACTA techniques are easy to use, flexible, and provide clear output, making them important for training purposes (Gore, Banks, McDowall, 2015).

Applied Cognitive Task Analysis (ACTA) is a comprehensive approach aimed at understanding and optimising cognitive processes underlying task performance in various domains, including psychology, human factors, and organisational behaviour. At its core, ACTA seeks to bridge the gap between theoretical knowledge of cognition and practical application in real-world settings, making it an invaluable tool for researchers and practitioners alike (Militello & Hutton, 1998).

Central to the importance of ACTA is its multidisciplinary nature, drawing upon insights from psychology, cognitive science, and applied fields to address complex challenges in human performance. By dissecting the cognitive demands of tasks, ACTA provides a systematic framework for identifying cognitive bottlenecks, optimising workflow efficiency, and enhancing overall task performance (Militello & Hutton, 1998).

The methodologies employed in Applied Cognitive Task Analysis (ACTA) encompass a diverse array of tools and techniques tailored to the specific context of the task under investigation. These methodologies include task diagrams, which provide a visual representation of task structure, clarifying sequences of actions and identifying areas of cognitive demand. A knowledge audit systematically examines the expertise and knowledge required for effective task performance, identifying key cognitive skills and problem-solving strategies. Simulation involves creating realistic scenarios or virtual environments to immerse participants in task contexts, enabling observation and analysis of cognitive processes in action. Additionally, the Cognitive Demand Table offers a structured framework for categorising and analysing cognitive demands at different task stages, identifying areas of cognitive workload and potential performance bottlenecks. These adaptable methodologies ensure a comprehensive analysis of cognitive processes and task performance, facilitating the optimization of cognitive demands for improved task proficiency.

1.2.2 Evolution and Scope of ACTA

The evolution of ACTA from traditional CTA represents a shift towards more dynamic and context-sensitive tools in cognitive analysis. This evolution is driven by the need to address real-world complexities where traditional models of task analysis may fall short. ACTA's methodologies are designed to capture intricate judgments, critical cues, and strategies used by professionals across fields, offering a more granular and applicable insight into cognitive tasks (Gore, Banks, & McDowall, 2018).

Examples abound of how ACTA has been applied across diverse domains to understand and improve cognitive performance. In the realm of aviation, ACTA has been used to analyse cockpit procedures and decision-making processes, leading to the development of more efficient training programs and enhanced safety protocols (Seamster & Redding, 2017). In healthcare settings, ACTA has been instrumental in identifying cognitive errors and improving clinical decision-making among healthcare providers (Graham et al., 2023; Russ-Jara et al., 2023). Similarly, in the realm of organisational behaviour, ACTA has been applied to optimise workflow design, enhance team coordination, and improve overall productivity (Knisely et al., 2021; Papautsky et al., 2020). In the education field, ACTA has been utilised to analyse teaching methodologies, identify cognitive demands in learning tasks, and develop more effective instructional strategies (Wahid et al., 2023; Idris et al., 2022). Through its diverse applications, ACTA continues to demonstrate its versatility and effectiveness in addressing cognitive challenges across various fields, highlighting its significance in enhancing performance and decision-making processes.

Through its emphasis on practical application and real-world relevance, ACTA offers a powerful lens to understand and optimise cognitive performance across diverse domains. Leveraging insights from cognitive psychology and applied research, ACTA drives innovation, improves decision-making, and enhances human performance in various contexts. These developments underscore ACTA's potential as both a research tool and a practical framework for performance enhancement and training in professional environments. The expansion of ACTA's methodologies highlights its growing importance in cognitive research, emphasising the nuances of expert behaviour and decision-making.

1.3 Challenges in Cognitive Task Analysis

Conducting cognitive task analysis (CTA) presents researchers and practitioners with a multitude of challenges, each requiring careful consideration and innovative solutions to overcome. These challenges range from capturing elusive tacit knowledge to navigating the complexities of real-world tasks while ensuring ecological validity.

One significant challenge in CTA is the capture and interpretation of tacit knowledge, the implicit knowledge and expertise that individuals possess but may struggle to articulate (Militello & Hutton, 1998). Tacit knowledge is often deeply ingrained in expert practitioners and can be challenging to elicit through traditional research methods. Without access to this valuable reservoir of knowledge, researchers risk overlooking critical insights into cognitive processes and decision-making strategies.

Dealing with the inherent complexity of real-world tasks is another formidable challenge in CTA. Tasks in naturalistic settings are often multifaceted, dynamic, and unpredictable, posing significant challenges for researchers attempting to isolate and analyse specific cognitive processes (Clark et al., 2008). Moreover, the interplay between cognitive, social, and environmental factors further complicates the task analysis process, requiring researchers to adopt a holistic perspective.

Ensuring ecological validity, or the extent to which findings from CTA reflect real-world conditions, is yet another challenge that researchers must contend with. This involves overcoming issues such as incomplete descriptions and inaccuracies in the content, which can lead to misrepresentations of the conceptual knowledge required for tasks. Additionally, the difficulty participants have in recalling problem-solving strategies, especially when many processes have become automatic and lack conscious awareness, further complicates the accurate depiction of real-world scenarios. Moreover, identifying important cues and decision points that experts use in practice can be elusive, and the variation in CTA methods adds another layer of complexity. Ensuring that CTA findings are both cost-effective and beneficial, without conflicting with other design theories, while also integrating unconscious expert knowledge, requires meticulous attention to maintaining ecological validity throughout the analysis process (Clark et al., 2008).

Laboratory-based studies offer the advantage of controlled and precise environments, which facilitate the isolation and manipulation of variables. However, this control often comes at the expense of ecological validity, the extent to which study findings can be generalised to real-world settings. Research indicates that laboratory tasks, due to their artificial nature, may fail to capture the complexity and nuances of tasks performed in authentic environments (Holleman et al., 2020), thus limiting the practical utility of CTA findings.

To address the challenges of enhancing the effectiveness of Cognitive Task Analysis (CTA) methodologies, researchers have been leveraging innovative technologies, such as simulation platforms and virtual environments. Recent studies indicate that these immersive technologies create realistic task scenarios that closely mirror real-world conditions, thereby improving the ecological validity of CTA findings. For instance, virtual reality (VR) and immersive environments have been shown to provide effective and engaging training and assessment platforms, making them valuable tools in both educational and professional settings (Rubio-Tamayo et al., 2017). These technologies allow for the creation of dynamic and interactive simulations that can replicate the complexities and nuances of real-world tasks, offering a more practical and applicable approach to CTA. By integrating such technologies, researchers can design more accurate and representative task analyses, enhancing both the reliability and utility of their findings (Rubio-Tamayo et al., 2017).

Additionally, researchers have explored novel data collection methods, such as eye-tracking technology and physiological measures, to supplement traditional observational and interview-based approaches (Krejtz et al., 2018; Ayres et al., 2021). These advanced techniques provide deeper insights into cognitive processes by capturing real-time data on visual attention and physiological responses, thereby enriching the overall analysis and interpretation of task performance. By integrating multiple data sources, researchers can triangulate findings and gain a more comprehensive understanding of cognitive processes underlying task performance.

Recent research supports the notion that collaborative and interdisciplinary approaches significantly enhance the effectiveness of Cognitive Task Analysis (CTA). These approaches bring together diverse expertise from fields such as psychology, human factors, and computer science, fostering innovation through a richer understanding of complex problems. One study illustrates how interdisciplinary collaboration in science teams can yield substantial outcomes, particularly through the co-design and co-production of solution-oriented knowledge. This process enhances the development of new scientific insights by integrating diverse disciplinary perspectives, which is critical for addressing complex research questions effectively (Specht & Crowston, 2022).

CONCLUSION

In conclusion, while cognitive task analysis presents formidable challenges, innovative approaches and interdisciplinary collaborations hold promise for overcoming these obstacles and advancing our understanding of human cognition in real-world settings. By addressing the challenges of tacit knowledge capture, task complexity, and ecological validity, researchers can unlock new insights into cognitive processes and pave the way for more effective interventions and solutions.

REFERENCES

Clark, R. E., Feldon, D., van Merriënboer, J. J. G., Yates, K., & Early, S. (2008). Cognitive task analysis. In Spector, J. M., Merrill, M. D., van Merriënboer, J. J. G., & Driscoll, M. P. (Eds.), *Handbook of research on educational communications and technology* (3rd ed., pp. 577–593). Macmillan/Gale.

Dhakal, A., & Bobrin, B. D. (2023). *Cognitive Deficits*. Europe PMC.

Graham, L. A., Gray, C., Wagner, T. H., Illarmo, S., Hawn, M. T., Wren, S. M., Iannuzzi, J., & Harris, A. H. S. (2023). Applying cognitive task analysis to health services research. *Health Services Research*, 58(2), 415–422. DOI: 10.1111/1475-6773.14106

Holleman, G. A., Hooge, I. T. C., Kemner, C., & Hessels, R. S. (2020). The 'Real-World Approach' and Its Problems: A Critique of the Term Ecological Validity. *Frontiers in Psychology*, 11. Advance online publication. DOI: 10.3389/fpsyg.2020.00721

Idris, A. Y., Razak, R. A., & Rahman, S. S. A. (2022). Investigating knowledge task difficulties in an e-training system for professional development design. *Docens Series in Education*, 2, 51–65.

Knisely, B. M., Joyner, J. S., & Vaughn-Cooke, M. (2021). Cognitive task analysis and workload classification. *MethodsX*, 8, 101235. DOI: 10.1016/j.mex.2021.101235

Krejtz, K., Duchowski, A. T., Niedzielska, A., Biele, C., & Krejtz, I. (2018). Eye tracking cognitive load using pupil diameter and microsaccades with fixed gaze. *PLoS One*, 13(9), e02036291. DOI: 10.1371/journal.pone.0203629

Militello, L. G., & Hutton, R. J. B. (1998). Applied cognitive task analysis (acta): A practitioner's toolkit for understanding cognitive task demands. *Ergonomics*, 41(11), 1618–1641. DOI: 10.1080/001401398186108

Nielsen, E. G., & Minda, J. P. (2019). Problem Solving and Decision Making. *Psychology (Irvine, Calif.)*. Advance online publication. DOI: 10.1093/obo/9780199828340-0246

Papautsky, E. L., Strouse, R., & Dominguez, C. (2020). Combining cognitive task analysis and participatory design methods to elicit and represent task flows. *Journal of Cognitive Engineering and Decision Making*, 14(4), 288–301. DOI: 10.1177/1555343420976014

Rubio-Tamayo, J. L., Gertrudix Barrio, M., & García García, F. (2017). Immersive environments and virtual reality: Systematic review and advances in communication, interaction and simulation. *Multimodal Technologies and Interaction*, 1(4), 21. DOI: 10.3390/mti1040021

Russ-Jara, A. L., Elkhadragy, N., Arthur, K. J., DiIulio, J. B., Militello, L. G., Ifeachor, A. P., Glassman, P. A., Zillich, A. J., & Weiner, M. (2023). Cognitive task analysis of clinicians' drug–drug interaction management during patient care and implications for alert design. *BMJ Open*, 13, e075512. DOI: 10.1136/bmjopen-2023-075512

Seamster, T. L., & Redding, R. E. (2017). *Applied Cognitive Task Analysis in Aviation*. Routledge. DOI: 10.4324/9781315262321

Specht, A., & Crowston, K. (2022). Interdisciplinary collaboration from diverse science teams can produce significant outcomes. *PLoS One*, 17(11), e0278043. Advance online publication. DOI: 10.1371/journal.pone.0278043

Thakur, S. N. (2023). The human brain, mind, and consciousness: Unveiling the enigma. *Philosophia*, 20(2), 163–179. DOI: 10.13140/RG.2.2.29992.14082

Wahid, S. A. N., Razak, R. A., & Halili, S. H. (2023). Investigating knowledge task difficulties in designing professional ethics instructional framework for technical educators. *Jurnal Penyelidikan Pendidikan*, 41(1), 45–57.

Chapter 2
Theoretical Frameworks in ACTA

ABSTRACT

This chapter explores the foundational theories behind Applied Cognitive Task Analysis (ACTA), focusing on Cognitive Load Theory (CLT) and Situated Cognition. CLT is crucial for managing cognitive resources during learning and task execution by optimizing intrinsic, extraneous, and germane loads. This helps design effective instructional materials and training programs for better information processing and retention. Situated Cognition emphasizes the role of context and social interactions in learning, aligning with ACTA's focus on practical knowledge application in real-world settings. The chapter examines how these theories enhance understanding of cognitive processes and stress the need for personalized, context-rich learning experiences. By integrating qualitative insights, it highlights the value of capturing expert performance nuances often missed by quantitative methods. This theoretical foundation supports the application of ACTA across various fields, from education to industry, where improving cognitive performance is essential.

1.0 INTRODUCTION

In exploring the depths of Applied Cognitive Task Analysis (ACTA), it is crucial to understand the theoretical foundations that guide its methodologies and practices. ACTA, as a methodological tool, draws heavily from cognitive science to map out the thought processes and decision-making strategies of experts across various domains. These insights are essential for developing instructional designs and training programs that align cognitively with the complexities of real-world tasks. By

DOI: 10.4018/979-8-3693-7733-8.ch002

leveraging these frameworks, ACTA ensures that training and task analysis reflect the mental demands placed on individuals in high-stakes environments.

To achieve these outcomes, ACTA is built upon well-established cognitive theories that explain how individuals process, retain, and apply knowledge. This chapter delves into the cognitive theories that underpin ACTA's approach, offering a comprehensive understanding of the foundational concepts that enhance its application across diverse professional settings. Specifically, we will focus on two key theories—Cognitive Load Theory (CLT) and Situated Cognition—both of which play a pivotal role in shaping the principles and practices of ACTA, guiding how it captures and models complex cognitive tasks.

2.0 COGNITIVE THEORIES SUPPORTING ACTA

Applied Cognitive Task Analysis (ACTA) leverages cognitive theories to improve task performance and learning outcomes. By dissecting complex tasks into their cognitive components, ACTA helps to uncover the implicit knowledge that experts use to perform efficiently. These insights are then used to develop training programs and instructional designs that align with the way humans naturally think and learn. Understanding these theories provides a solid foundation for effectively applying ACTA across various professional domains. This section explores how Cognitive Load Theory (CLT) and Situated Cognition underpin ACTA methodologies, ensuring that tasks are structured to optimize cognitive resources and contextual understanding.

2.1 Cognitive Load Theory (CLT)

Cognitive Load Theory is central to understanding how ACTA enhances learning and performance by optimising the use of working memory. Developed by Sweller (2011), CLT posits that instructional designs must account for the cognitive load imposed on learners, balancing it to avoid overwhelming their limited working memory capacity. This balancing act is crucial in ACTA, where tasks often demand significant cognitive resources. In practical applications, such as nursing education, CLT principles help structure learning experiences to maximize the retention and application of knowledge in complex, high-stakes environments.

2.1.1 Key Types of Cognitive Load

Cognitive Load Theory differentiates between three types of cognitive load: intrinsic, extraneous, and germane load. Each of these plays a significant role in how ACTA practitioners design and analyse tasks.

- **Intrinsic Cognitive Load**: This type of cognitive load is inherent to the complexity of the task itself. In ACTA, understanding the intrinsic load is crucial for breaking down complex tasks into smaller, more manageable components. For example, in medical training, the intrinsic cognitive load of diagnosing a patient may involve analysing symptoms, assessing medical history, and applying clinical knowledge. ACTA helps to map out these steps and optimize the sequence in which they are taught to reduce overload.
- **Extraneous Cognitive Load**: This refers to the unnecessary cognitive load introduced by poorly designed instructional materials or irrelevant information. In ACTA, reducing extraneous load is critical for enabling learners to focus on essential task-related information. For instance, removing unnecessary technical jargon or overcomplicated instructions from a training manual can help learners focus on the core task without being distracted by irrelevant details.
- **Germane Cognitive Load**: This load relates to the cognitive resources dedicated to processing and understanding information, contributing to the construction of schemas for long-term memory retention. ACTA leverages germane load by creating learning environments that encourage deeper understanding, allowing learners to internalize knowledge and develop mental models that can be applied in real-world situations. In aviation training, for instance, ACTA might include simulations that foster critical decision-making skills, which are crucial for pilots under high-pressure scenarios.

2.1.2 CLT in Instructional Design for ACTA

Cognitive Load Theory emphasizes that our working memory has limited capacity and that instructional designs must avoid overloading it. This limitation is a pivotal consideration in ACTA when designing educational materials and learning experiences.

- **Task Breakdown and Sequencing**: ACTA applies CLT by breaking down complex tasks into smaller, manageable parts to avoid overwhelming learners. This aligns with the findings of Tabatabaee et al. (2024), who demonstrated how managing cognitive load in workplace-based learning directly impacts the decision-making skills of nursing students. In their study, the relationship between extraneous cognitive load and decision-making skills was found to be particularly strong, underscoring the importance of reducing unnecessary cognitive burden in clinical training environments. By breaking down complex tasks such as patient diagnosis into stages, educators can manage both intrinsic and extraneous cognitive load, improving learning outcomes. This systematic task sequencing not only optimizes the learner's ability to retain information but also enhances their problem-solving capabilities, as supported by the path analysis findings in the clinical setting. This approach prevents cognitive overload by presenting information in a structured, sequential manner that aligns with the learner's ability to process each component before moving on to the next.

- **Use of Multimedia and Technology:** Recent research has highlighted the benefits of using multimedia in instructional design to enhance cognitive engagement and improve retention (Liu, 2024). Visuals, simulations, and animations can be used to optimize germane load, enabling learners to focus on essential information. However, ACTA ensures that such tools are designed in a way that minimizes extraneous load. For example, interactive simulations in healthcare training, when properly segmented and learner-controlled, enhance comprehension by allowing learners to actively engage with the material while avoiding overwhelming their working memory. This approach aligns with Graham et al. (2023), who introduced Cognitive Task Analysis (CTA) to the health services research field. By adapting CTA methods to better understand the cognitive processes underlying clinical decision-making, they provided a systematic framework that can inform the design of decision-support tools. These tools, when developed with a clear understanding of task-related cognitive processes, help manage cognitive load effectively, improving both the learning experience and the quality of care. Such an adaptation is critical in designing health technologies that reduce cognitive burden, allowing healthcare professionals to perform tasks more efficiently without being overloaded by unnecessary information.

2.1.3 Measurement of Cognitive Load in ACTA

Accurately measuring cognitive load has become increasingly important in educational research, particularly within the context of task analysis. The ability to measure cognitive load is essential for ensuring that instructional designs are effective in reducing unnecessary cognitive strain and enhancing learning outcomes.

- **Subjective and Objective Measures:** Subjective measures, such as learner self-reports, are often used in combination with objective physiological measures, like eye-tracking or EEG, to gauge cognitive load (Krell et al., 2022). In high-cognitive-demand environments such as aviation and engineering, this combination of measures provides deeper insights into how learners are processing information and how instructional designs can be adjusted to improve performance. ACTA leverages these insights to refine its task analysis methods, ensuring that instructional materials are cognitively aligned with the task's complexity and the learner's abilities.

2.1.4 Application of CLT Across Domains

Cognitive Load Theory is highly adaptable and has been applied across a variety of professional settings, including second language acquisition (SLA), healthcare, education, and aviation. In each of these domains, ACTA uses CLT to ensure that tasks are structured in a way that optimizes learning without overwhelming learners.

- **Second Language Acquisition (SLA):** In the context of SLA, multimodal instructional designs that combine visual, auditory, and written inputs have been found to reduce intrinsic load while enhancing germane load (Liu, 2024). ACTA practitioners apply these principles by incorporating multimodal inputs and feedback mechanisms that help learners process information more efficiently, thereby improving language acquisition outcomes.
- **Healthcare:** In clinical settings, the application of CLT has helped to design more effective training programs by reducing extraneous cognitive load, particularly in fast-paced, high-stress environments (Venkat et al., 2020). For example, ACTA practitioners ensure that medical trainees are not inundated with irrelevant information during simulations, allowing them to focus on critical decision-making and patient care tasks.
- **Aviation:** In aviation, where the cognitive demands of tasks such as flight planning and emergency response are high, ACTA applies CLT to ensure that pilots are trained in a structured manner that avoids cognitive overload. By using simulations that reflect real-world scenarios, ACTA enables learners to practice and refine their decision-making skills in a safe, controlled environment.

2.1.5 Optimizing Digital Learning Environments

CLT is particularly relevant in the design of digital learning environments, which are often used in ACTA to simulate complex tasks. Research by Skulmowski and Xu (2022) highlights that while interactive elements in digital learning can sometimes increase extraneous load, they can also promote motivation and engagement when carefully designed.

- **Interactive Learning Tools:** ACTA practitioners use CLT to guide the design of digital tools, ensuring that they minimize extraneous load while maximizing engagement. For example, virtual reality (VR) simulations in industrial training environments allow learners to engage in complex tasks within a controlled virtual setting, reducing the risk of real-world consequences while still allowing for deep learning.

- **User Experience and Usability:** The usability of digital tools is closely tied to cognitive load. When tools are intuitive and easy to navigate, learners experience reduced cognitive strain, allowing them to focus on the task at hand (Skulmowski & Xu, 2022). ACTA practitioners consider these usability factors when designing training platforms, ensuring that learners are not distracted by unnecessarily complicated interfaces.

2.1.6 Integrating CLT Principles in ACTA

The integration of Cognitive Load Theory into ACTA is essential for designing professional training programs where cognitive tasks are complex and require precision and decision-making.

- **Targeted Learning Outcomes:** By leveraging CLT principles, ACTA practitioners can develop instructional materials that target specific learning outcomes without overwhelming the learner. This is particularly important in fields such as medicine, aviation, and industrial training, where errors can have significant consequences.
- **Real-World Applications:** The integration of CLT principles within ACTA enables learners to engage in real-world scenarios while managing cognitive load effectively. This ensures that learners not only understand the tasks they are performing but also retain the knowledge and skills needed to apply them in real-world settings.

2.1.7 Review of Cognitive Load Theory (CLT) in Shaping Applied Cognitive Task Analysis (ACTA)

Cognitive Load Theory (CLT) plays a crucial role in shaping how ACTA enhances learning and task performance by optimizing cognitive resources, particularly working memory. By distinguishing between intrinsic, extraneous, and germane cognitive load, CLT provides a structured framework for designing educational and professional training programs that align with human cognitive limitations. Through the careful management of these cognitive loads, ACTA ensures that learners can engage with complex tasks without becoming overwhelmed, thereby promoting deeper understanding, retention, and the ability to apply knowledge in real-world contexts.

The integration of CLT into ACTA is particularly beneficial in domains requiring high-stakes decision-making, such as healthcare, aviation, and industrial training. By breaking tasks into manageable components, leveraging multimedia tools, and accurately measuring cognitive load, ACTA practitioners can design learning environments that not only minimize unnecessary cognitive strain but also enhance the learner's ability to develop expertise. Furthermore, the application of CLT in digital learning environments ensures that instructional materials are both engaging and effective, optimizing the learner's experience.

In conclusion, Cognitive Load Theory provides a foundational framework for ACTA, guiding the development of instructional strategies that are cognitively aligned and contextually relevant. This synergy between CLT and ACTA ensures that learners are equipped with the tools and knowledge to perform complex tasks effectively, thereby improving overall competence and performance in professional settings.

2.2 Situated Cognition Theory

Situated Cognition, another foundational theory in ACTA, emphasizes that knowledge is inseparable from the context in which it is used. This theory posits that learning is not a detached cognitive process but one deeply embedded in interactions with specific environments, tasks, and social practices (Wilson & Myers, 2000). Learning, according to Situated Cognition, occurs through active engagement in meaningful activities within real-world contexts, where the learner interacts with the environment and constructs knowledge based on those experiences. This theory diverges from traditional views of learning as a process of absorbing abstract knowledge; instead, it emphasizes the importance of real-life situations in shaping cognition. Situated Cognition aligns closely with ACTA's practical approach, where the focus is on real-world applications and understanding cognitive processes as they naturally occur in specific professional settings (Militello & Hutton, 1998).

ACTA, through Cognitive Task Analysis (CTA) tools, allows investigators to capture the cognitive processes individuals engage in when performing tasks, generating data that reflects practitioner knowledge and reasoning (Klein & Militello, 2001). Such data is invaluable in designing training programs that mirror the authentic contexts in which experts operate. These insights are particularly valuable in complex professional settings where tasks require not just technical skills, but also tacit knowledge gained through experience (Hoffman & Militello, 2009). By focusing on the real-world cognitive demands of tasks, ACTA provides a versatile methodology applicable across domains such as healthcare, aviation, and engineering (Hogenboom et al., 2021), enabling the exploration of knowledge management practices essential to professional competence (Gore, 2013).

Situated Cognition Theory further posits that learning occurs through active engagement in both real-world and virtual activities within specific social, cultural, and physical contexts. This perspective highlights that perception and action are co-determined by the interaction between individuals and their environments, with learning being deeply embedded in the situations where it takes place. Rather than separating knowledge from its context, Situated Cognition recognizes that what and how people learn are shaped by their specific activities and environments (Ataizi, 2012). In this way, learning is not simply the acquisition of abstract concepts but is constructed and developed through hands-on experience, making it more robust and applicable to real-world tasks. This perspective is integral to ACTA's approach, which seeks to analyse and design learning experiences that are embedded in authentic professional practice contexts (Militello & Hutton, 1998).

For instance, in ACTA, methodologies are designed to dissect how experts use both explicit and tacit knowledge to perform tasks in their actual work environments. By understanding the knowledge, thought processes, and decision-making strategies used by experts, ACTA aims to create training and educational interventions that are not only grounded in theory but are also directly applicable to the tasks learners need to perform. This ensures that learning is contextually relevant, practically grounded, and effective in fostering competence in professional settings.

2.2.1 Key Principles of Situated Cognition and ACTA

Both Situated Cognition and ACTA emphasize the importance of context in learning. Below are key principles that align these two approaches:

- **Contextual Learning:** Situated Cognition's focus on learning within real-world contexts aligns with ACTA's goal of embedding learning experiences within the authentic settings where tasks are performed. By mirroring real-world conditions, ACTA ensures that learners not only grasp abstract principles but also develop the capacity to apply these principles in context-specific situations.
- **Perception and Action:** The co-determination of perception and action in Situated Cognition mirrors ACTA's approach to analysing how individuals interact with their environment to perform tasks effectively. This dynamic relationship between an individual and their environment is critical to successful task completion, especially in high-stakes professions such as emergency response, aviation, and healthcare.

- **Practical Application:** Both approaches recognize that effective learning involves engaging with the specific social, cultural, and physical contexts relevant to the learner's goals and activities. For instance, ACTA ensures that learners develop not just technical skills, but also the situational awareness necessary to adapt to complex and dynamic environments.

2.2.2 Situated Cognition in Professional Practice

Situated Cognition suggests that knowledge is constructed within and linked to the activity, context, and culture in which it is developed and used. This is particularly relevant in fields that require deep expertise, such as healthcare, education, and engineering, where learning must be practical, hands-on, and aligned with professional contexts. For example, in a study by Choi and Hannafin (1995), the authors highlight how learners engaged in authentic tasks within meaningful contexts develop deeper understanding and more effective problem-solving skills. This approach contrasts with more traditional educational models that focus on theoretical knowledge divorced from real-world applications.

Previous research supports the idea that learning environments that closely mimic real-world settings significantly enhance the transfer of skills and knowledge. Lave and Wenger's (1991) work on communities of practice demonstrates that learning occurs most effectively when novices participate in social and professional communities, engaging with experienced practitioners. Through this social engagement, learners not only acquire explicit knowledge but also gain tacit knowledge that is critical for proficiency in professional tasks. In professional settings such as healthcare, Situated Cognition plays a crucial role. For instance, a scoping review by O'Brien and Battista (2020) on situated learning theory (SLT) in health professions education found that understanding social learning processes and designing interventions aligned with communities of practice are vital for effective learning in healthcare. This review highlights how situated learning, through hands-on experience in clinical settings, fosters the development of both clinical reasoning and interpersonal skills, which are crucial for success in healthcare professions.

Similarly, in the field of education, the work of Mahmood, Aziz, and Bibi (2023) emphasizes the importance of teaching practice in professional development. Their study shows that the ability of prospective teachers to apply theoretical knowledge in practical classroom settings significantly enhances their preparedness for real-world teaching. By engaging in authentic teaching practices, teachers develop critical skills for managing classrooms and addressing pedagogical challenges. This research further underscores how situated learning experiences that incorporate real-world contexts not only enhance professional competence but also ensure that theoretical knowledge is effectively transferred into practice.

2.2.3 The Role of Situated Cognition in Enhancing ACTA

By incorporating the principles of Situated Cognition, ACTA ensures that learning experiences are relevant, practical, and grounded in the specific contexts where they will be applied. This contextual relevance not only aids in the acquisition of new knowledge but also enhances the application of that knowledge in professional settings, leading to improved performance and competence. Whether in healthcare, military, or industrial training, ACTA leverages Situated Cognition to ensure that learners are equipped with the skills and knowledge necessary to perform effectively in their respective fields.

For instance, in military training, ACTA methodologies are used to break down complex mission scenarios. According to Cannon-Bowers et al. (2013), the pressures to improve the efficacy of military medical training have driven the use of simulation technologies. They describe a cognitive task analysis technique specifically designed for military medical tasks, which generates training requirements, performance metrics, and scenario and simulator needs. By having expert instructors perform critical tasks like cricothyroidotomy and haemorrhage control, their scenario-based approach breaks down these complex tasks into manageable components. This aligns with ACTA's goal of simplifying complex mission scenarios to enhance learning outcomes.

Furthermore, this approach reflects the principles of situated cognition, which emphasizes that learning and skill acquisition are most effective when they occur within the context of real-world situations. By embedding trainees in realistic, scenario-based simulations, they gain an understanding of how to apply their knowledge in environments that closely mirror the complexities of actual military operations. This situated approach allows learners to develop decision-making and procedural skills within a context that replicates the challenges they will face in the field. The technique not only informs the development of effective simulation-based training systems but also ensures that trainees focus on critical aspects without becoming overwhelmed by cognitive overload, much like ACTA's objective to reduce extraneous load in learning environments. Through this immersive, context-driven approach, both ACTA and situated cognition contribute to more effective learning and skill retention in military training.

In summary, Situated Cognition, by emphasizing the role of context, social interaction, and real-world engagement in learning, aligns perfectly with ACTA's goal of analysing and improving task performance in complex, professional environments. By ensuring that learning is both contextually grounded and practically applicable, ACTA enhances learners' ability to perform effectively in their professional roles, contributing to overall competence and expertise across various domains.

2.2.4 Review of Situated Cognition Theory in Shaping Applied Cognitive Task Analysis (ACTA)

Situated Cognition Theory is integral to how ACTA enhances learning and task performance by embedding knowledge acquisition within real-world contexts. This theory highlights the inseparability of knowledge from the environment in which it is applied, underscoring the importance of authentic settings for meaningful learning. Through active engagement in practical, context-specific tasks, learners develop a deeper understanding of both explicit and tacit knowledge, which is essential for effective performance in professional environments.

By leveraging Situated Cognition, ACTA ensures that learning is not merely theoretical but grounded in the practical realities of professional tasks. The use of real-world simulations and scenario-based learning helps learners develop critical skills that go beyond technical expertise, encompassing decision-making, problem-solving, and adaptability within dynamic environments. Whether in healthcare, military training, or industrial contexts, ACTA's focus on situated learning enables learners to interact with their environment in ways that reflect actual professional demands, making their training more relevant and impactful.

The alignment of Situated Cognition with ACTA's task analysis methodologies supports the development of training programs that are highly contextualized, engaging, and practically grounded. By focusing on the interactions between learners and their environments, ACTA provides a powerful approach for designing educational experiences that foster not only knowledge acquisition but also real-world application. This immersive, context-driven learning model ensures that learners are better prepared to apply their skills effectively in complex professional tasks.

In conclusion, Situated Cognition Theory enhances the effectiveness of ACTA by ensuring that learning experiences are contextualized, interactive, and directly applicable to the tasks learners will encounter in professional practice. This synergy between Situated Cognition and ACTA promotes a deeper level of engagement and competence, enabling professionals to perform complex tasks with greater confidence and expertise across various domains.

2.3 Role of Qualitative Insights in Understanding Cognitive Processes

Qualitative research plays a crucial role in comprehending cognitive processes by providing deep, nuanced insights that quantitative methods may overlook. This approach allows researchers to explore the intricate details of how individuals think, learn, and solve problems, offering a comprehensive understanding of cognitive functions in various contexts.

2.3.1 Depth and Richness of Data

Qualitative methods such as interviews, focus groups, and observations enable researchers to gather rich, detailed data about cognitive processes. These methods allow participants to express their thoughts, feelings, and experiences in their own words, providing a deeper understanding of their cognitive strategies and mechanisms.

- **Interviews**: Interviews are a cornerstone of qualitative research, enabling researchers to delve deeply into participants' cognitive processes. Through semi-structured or unstructured formats, interviews facilitate a detailed exploration of participants' thoughts and experiences. This method is particularly effective in understanding how individuals process information and make decisions, as it allows for in-depth probing and follow-up questions that can clarify and expand upon initial responses (Gill, Stewart, Treasure & Chadwick, 2008).
- **Focus Groups:** Focus groups gather multiple participants to discuss a topic, fostering interactive discussions that can reveal shared cognitive strategies and group dynamics. This method captures a range of perspectives and can highlight commonalities and differences in cognitive processes among participants. Focus groups are particularly useful in examining how social interactions and group contexts influence cognitive mechanisms (Gill & Baillie, 2018).
- **Observations:** Observational methods involve watching participants in their natural settings, providing contextual insights into their cognitive behaviours. This method allows researchers to see firsthand how individuals apply their cognitive strategies in real-world scenarios, offering a more comprehensive understanding of cognitive processes that might not be fully articulated through interviews or focus groups alone (Kelemen et al., 2023).

These qualitative methods collectively enable a richer, more nuanced understanding of cognitive processes, as they capture the complexity of human thought and behaviour in a way that quantitative methods often cannot. They provide a platform for participants to express their experiences in their own words, thereby uncovering the depth and intricacy of cognitive strategies and mechanisms.

2.3.2 Contextual Understanding

Understanding cognitive processes requires considering the context in which these processes occur, and recent research underscores the importance of qualitative methods in capturing these contextual factors. Qualitative research excels at exploring how environmental, social, and cultural contexts influence cognition. For example, studies have shown that analysing contextual factors is crucial for understanding how interventions are implemented and their success in real-world settings. This includes examining how various levels of context, such as organisational leadership and external motivators, impact the effectiveness of quality improvement initiatives (Mielke et al., 2022; Reed et al., 2018).

In the field of education, qualitative research has been particularly valuable in revealing how students interact with learning materials and their peers. This contextual understanding helps inform the design of educational interventions. By capturing the intricate details of the learning environment, researchers can better understand the cognitive processes involved and develop strategies that enhance learning outcomes. For instance, examining how students' engagement with STEM education is influenced by the contextual factors of their school environment provides insights that can improve teaching practices and student learning experiences (Fairhurst, Koul & Sheffield, 2023).

These findings highlight the vital role of qualitative research in providing a comprehensive understanding of cognitive processes within their contextual frameworks, which is essential for designing effective educational and intervention strategies.

2.3.3 Integration with Cognitive Theories

Qualitative research plays a vital role in bridging the gap between theoretical cognitive frameworks and practical applications. Cognitive theories such as Cognitive Load Theory (CLT), Situated Cognition, and Schema Theory often present abstract models of how people learn and process information. However, these models can benefit from real-world validation, which qualitative research provides. Through interviews, observations, and cognitive task analyses, qualitative research uncovers how these cognitive processes manifest in practice, often revealing insights that can refine or expand upon the original theories.

For instance, recent research by Su and Chen (2023) demonstrates how qualitative interviews were used to explore learners' cognitive strategies in STEM education, highlighting the challenges students face in processing complex information in high-load environments. Their study revealed that learners often struggled with cognitive overload when dealing with hands-on STEM tasks, yet qualitative feedback provided valuable insights into how students utilized problem-solving and critical thinking to navigate these challenges. This study also aligns with the principles of Cognitive Load Theory (CLT), showing how instructional materials can be better designed to accommodate learners' cognitive limitations. By aligning theoretical models with practical experiences, Su and Chen's research underscores the importance of qualitative inquiry in ensuring cognitive theories remain grounded in the realities of learners and educators.

2.3.4 Uncovering Hidden Mechanisms

Qualitative research can uncover hidden cognitive mechanisms that are not easily accessible through quantitative methods. Techniques like applied cognitive task analysis (ACTA) enable researchers to explore the underlying thought processes involved in complex tasks. For example, a study by Taylor, Ashford, and Jefferson (2023) used ACTA to identify the cognitive strategies employed by experts in high-stakes environments. The study revealed how these experts adaptively use their knowledge and skills to navigate complex situations, offering valuable insights that can inform the development of training programs and improve performance in similar high-pressure contexts. Such qualitative approaches provide a richer understanding of cognitive processes, highlighting the importance of context and the flexibility of expert thinking, which are often overlooked in traditional quantitative research.

2.3.5 Personalised Insights

Qualitative research provides personalised insights into cognitive processes by focusing on individual experiences and perspectives. This personalised approach is essential for understanding how different people perceive, process, and respond to information. Recent research in various fields supports this perspective.

Qualitative research in psychology and health research utilises methods such as in-depth interviews, focus groups, and ethnography to gain deep insights into individual beliefs, values, and motivations. This personalised approach helps uncover the cognitive processes behind behaviours and decision-making. For example, qualitative research in health studies has been instrumental in understanding how patients and healthcare providers perceive and manage health-related issues, thereby informing the design of effective interventions and healthcare policies (Rana, Poudel, & Chimoriya, 2023).

Moreover, the personalised approach of qualitative research is particularly valuable in educational settings. Studies have highlighted how qualitative insights into students' interactions with learning materials and their peers can reveal important contextual factors that affect learning. This understanding is crucial for developing educational interventions that are tailored to the specific needs and experiences of students, thereby enhancing their learning outcomes. For instance, understanding how students engage with learning materials, collaborate with peers, and navigate their learning environment can inform the design of tailored educational interventions. These personalised approaches enhance learning outcomes by addressing specific student needs and experiences. As Fairhurst, Koul, and Sheffield (2023) emphasise in their study, qualitative insights play a crucial role in shaping effective teaching practices and promoting student success.

Overall, the personalised insights gained through qualitative research are indispensable for understanding the complex and individualised nature of cognitive processes across different contexts.

2.3.6 Methodological Triangulation

Combining qualitative insights with quantitative data, a practice known as methodological triangulation, can enhance the validity and reliability of cognitive research. This mixed-methods approach allows researchers to validate findings across different methodologies, providing a more comprehensive picture of cognitive processes. For example, using qualitative interviews to supplement quantitative surveys can help explain the reasons behind statistical trends, offering a deeper understanding of the data (Creswell & Clark, 2017).

2.3.7 Practical Applications of Qualitative Cognitive Research

Qualitative research into cognitive processes offers practical applications across a variety of fields, including education, healthcare, human-computer interaction (HCI), and organizational decision-making. By uncovering the nuanced ways people perceive, process, and respond to complex information, qualitative insights can directly inform the design of training programs, interfaces, and decision-support tools.

In HCI, for example, a study by Christou (2023) used qualitative interviews and observations to understand how users interact with AI-powered educational software. The findings revealed key cognitive challenges that users face in navigating automated feedback systems, such as difficulties in interpreting feedback and integrating it into their learning process. These insights informed significant improvements to the software's user interface and design, leading to a more intuitive and user-friendly experience. By addressing issues like information overload and real-time processing demands, the study showed how AI systems can better support users' cognitive capabilities, ensuring that the technology is aligned with how users think and learn. This research highlights the value of qualitative approaches in capturing nuanced user experiences that quantitative data alone may overlook. Furthermore, it underscores the role of qualitative insights in refining AI tools to enhance both usability and learning outcomes, demonstrating how qualitative research contributes to the practical application of AI in educational settings.

In the context of education, a study by Eslit (2023) explores the adaptations and experiences of college-level language learners in the post-pandemic era, providing a rich example of the practical applications of qualitative cognitive research. By employing qualitative methods such as interviews and document analysis, the study delves into the cognitive strategies learners use to navigate self-directed learning, technology integration, and social connections in language acquisition. These methods allow researchers to capture the nuanced experiences of students as they adapt to the shifting educational landscape, shedding light on the cognitive and emotional challenges they face.

The use of qualitative research in this context offers practical implications for educators and policymakers, highlighting the significance of socio-cultural dynamics in language learning. Eslit (2023) emphasizes how qualitative research provides insights into learners' self-regulation, problem-solving techniques, and how they interact with new technological tools in the post-pandemic world. This is especially valuable for refining pedagogical practices to support student success in increasingly digital and self-directed learning environments.

The triangulation of data sources and the use of member checking in the study underscore the importance of rigor in qualitative research, ensuring the reliability and validity of the findings. By investigating real-world language learning challenges, this research highlights the broader applications of qualitative cognitive research in understanding how learners adapt to changing educational conditions. In turn, these insights can guide the development of more effective, learner-centred approaches to language instruction in the post-pandemic era.

2.3.7 Summary

Qualitative research is indispensable for understanding cognitive processes, offering a depth of insight that quantitative methods often fail to capture. Through methods such as interviews, focus groups, and observations, qualitative research explores the intricacies of how individuals think, learn, and adapt in real-world settings. These methods provide a nuanced view of the cognitive strategies employed by learners and professionals across various contexts, uncovering hidden mechanisms and offering personalized insights that are vital for designing effective interventions.

Moreover, the integration of qualitative research with cognitive theories, such as Cognitive Load Theory (CLT) and Situated Cognition, ensures that these frameworks are not just theoretical but grounded in the realities of learners' and practitioners' experiences. By capturing contextual factors and the interaction between individuals and their environments, qualitative approaches highlight the importance of socio-cultural dynamics and personal experiences in shaping cognitive processes. These insights are crucial for fields like education, healthcare, and human-computer interaction, where the application of knowledge and decision-making in complex, dynamic settings is key to success.

Additionally, the practice of methodological triangulation, which combines qualitative and quantitative data, further strengthens the reliability and validity of cognitive research. This approach offers a more comprehensive understanding of how people perceive, process, and act upon information, ensuring that the findings are robust and applicable in diverse professional and educational contexts. Practical applications of these insights are vast, informing everything from educational program design to AI system usability and organizational decision-making.

3.0 CONCLUSION

In conclusion, qualitative insights into cognitive processes are critical for advancing both theoretical understanding and practical applications. By offering a rich, contextualized, and personalized perspective on how individuals think and learn, qualitative research contributes significantly to the development of more effective and adaptive learning environments, interventions, and tools across multiple domains.

REFERENCES

Ataizi, M. (2012). Situated Cognition. In Seel, N. M. (Ed.), *Encyclopedia of the Sciences of Learning*. Springer. DOI: 10.1007/978-1-4419-1428-6_16

Cannon-Bowers, J., Bowers, C., Stout, R., Ricci, K., & Hildabrand, A. (2013). Using cognitive task analysis to develop simulation-based training for medical tasks. *Military Medicine*, 178(Suppl_10), 15–21. DOI: 10.7205/MILMED-D-13-00211

Choi, J. I., & Hannafin, M. (1995). Situated cognition and learning environments: Roles, structures, and implications for design. *Educational Technology Research and Development*, 43(2), 53–69. DOI: 10.1007/BF02300472

Christou, P. A. (2023). The use of artificial intelligence (AI) in qualitative research for theory development. *The Qualitative Report*, 28(9), 2739–2755. DOI: 10.46743/2160-3715/2023.6536

Creswell, J. W., & Clark, V. L. P. (2017). *Designing and conducting mixed methods research* (3rd ed.). SAGE Publications.

Eslit, E. R. (2023). *Unveiling the cognitive depths: Delving into language learning and acquisition amidst the post-pandemic era* (ED628255). ERIC. https://files.eric.ed.gov/fulltext/ED628255.pdf

Fairhurst, N., Koul, R., & Sheffield, R. (2023). Students' perceptions of their STEM learning environment. *Learning Environments Research*, 26(3), 977–998. DOI: 10.1007/s10984-023-09463-z

Gill, P., & Baillie, J. (2018). Interviews and focus groups in qualitative research: An update for the digital age. *British Dental Journal*. Advance online publication. DOI: 10.1038/sj.bdj.2018.815

Gill, P., Stewart, K., Treasure, E., & Chadwick, B. (2008). Methods of data collection in qualitative research: Interviews and focus groups. *British Dental Journal*, 204(6), 291–295. DOI: 10.1038/bdj.2008.192

Gore, J. (2013). *Transforming knowledge: Capturing engineers' cognitive expertise. Paper presented at the 11th International Conference on Naturalistic Decision Making*, Marseilles, France.

Graham, L. A., Gray, C., Wagner, T. H., Illarmo, S., Hawn, M. T., Wren, S. M., Iannuzzi, J., & Harris, A. H. S. (2023). Applying cognitive task analysis to health services research. *Health Services Research*, 58(2), 415–422. DOI: 10.1111/1475-6773.14106

Hoffman, R. R., & Militello, L. (2009). *Perspectives on Cognitive Task Analysis: Historical Origins and Modern Communities of Practice*. Psychology Press.

Kelemen, W. L., Jhangiani, R. S., Chiang, I.-C. A., Cuttler, C., & Leighton, D. C. (2023). *Research methods in psychology*. The Saylor Foundation. http://www.saylor.org/site/textbooks/

Klein, G., & Militello, L. (2001). Some guidelines for conducting a cognitive task analysis. In Salas, E. (Ed.), *Advances in human performance and cognitive engineering research* (pp. 163–199). Elsevier Science/JAI Press. DOI: 10.1016/S1479-3601(01)01006-2

Krell, M., Xu, K. M., Rey, G. D., & Paas, F. (2021). Editorial: Recent approaches for assessing cognitive load from a validity perspective. *Frontiers in Education*, 6, 838422. Advance online publication. DOI: 10.3389/feduc.2021.838422

Lave, J., & Wenger, E. (1991). *Situated learning: Legitimate peripheral participation*. Cambridge University Press. DOI: 10.1017/CBO9780511815355

Liu, D. (2024). The effects of segmentation on cognitive load, vocabulary learning, retention, and reading comprehension in a multimedia learning environment. *BMC Psychology*, 12(4). Advance online publication. DOI: 10.1186/s40359-023-01489-5

Mahmood, M., Aziz, S., & Bibi, M. (2023). Effects of teaching practice in shaping prospective teachers' professional identities: Comparative analysis. *Journal of Education and Educational Development*, 10(1), 72–94. DOI: 10.22555/joeed.v10i1.731

Mielke, J., Brunkert, T., & Zúñiga, F.. (2022). Methodological approaches to study context in intervention implementation studies: An evidence gap map. *BMC Medical Research Methodology*, 22, 320. DOI: 10.1186/s12874-022-01772-w

Militello, L. G., & Hutton, R. J. B. (1998). Applied cognitive task analysis (acta): A practitioner's toolkit for understanding cognitive task demands. *Ergonomics*, 41(11), 1618–1641. DOI: 10.1080/001401398186108

O'Brien, B. C., & Battista, A. (2020). Situated learning theory in health professions education research: A scoping review. *Advances in Health Sciences Education : Theory and Practice*, 25, 483–509. DOI: 10.1007/s10459-019-09900-w

Rana, K., Poudel, P., & Chimoriya, R. (2023). Qualitative methodology in translational health research: Current practices and future directions. *Health Care*, 11(2665). Advance online publication. DOI: 10.3390/healthcare11192665

Reed, J. E., Kaplan, H. C., & Ismail, S. A. (2018). A new typology for understanding context: Qualitative exploration of the model for understanding success in quality (MUSIQ). *BMC Health Services Research*, 18, 584. DOI: 10.1186/s12913-018-3348-7

Skulmowski, A., & Xu, K. M. (2022). Understanding cognitive load in digital and online learning: A new perspective on extraneous cognitive load. *Educational Psychology Review*, 34(1), 171–196. DOI: 10.1007/s10648-021-09624-7

Su, K.-D., & Chen, H.-Y. (2023). Exploring the learning efficacy of students' STEM education from the process of hands-on practical experience. In *Innovative technologies and learning* (pp. 541–550). Springer. DOI: 10.1007/978-3-031-40113-8_41

Sweller, J. (2011). Cognitive load theory. In Mestre, J. P., & Ross, B. H. (Eds.), *The psychology of learning and motivation: Cognition in education* (pp. 37–76). Elsevier Academic Press. DOI: 10.1016/B978-0-12-387691-1.00002-8

Tabatabaee, S. S., Jambarsang, S., & Keshmiri, F. (2024). Cognitive load theory in workplace-based learning from the viewpoint of nursing students: Application of a path analysis. *BMC Medical Education*, 24, 678. DOI: 10.1186/s12909-024-05664-z

Taylor, J., Ashford, M., & Jefferson, M. (2023). High performance coach cognition in the wild: Using applied cognitive task analysis for practical insights–cognitive challenges and curriculum knowledge. *Frontiers in Psychology*, 14, 1154168. DOI: 10.3389/fpsyg.2023.1154168

Venkat, M. V., O'Sullivan, P. S., Young, J. Q., & Sewell, J. L. (2020). Using Cognitive Load Theory to improve teaching in the clinical workplace. *MedEdPORTAL: the Journal of Teaching and Learning Resources*, 16, 10983. DOI: 10.15766/mep_2374-8265.10983

Wilson, B. G., & Myers, K. M. (2000). Situated cognition in theoretical and practical context. In Jonassen, D. H., & Land, S. M. (Eds.), *Theoretical foundations of learning environments* (pp. 57–88). Erlbaum.

Section 2
Methodologies and Techniques

Chapter 3
Key Steps in the ACTA Process

ABSTRACT

This chapter provides an overview of the Applied Cognitive Task Analysis (ACTA) process, outlining its key steps from start to finish. It begins with the planning phase, which involves defining the scope, identifying tasks, and selecting appropriate tools. The data collection stage follows, utilizing methods like cognitive interviews and task observations to gather detailed information on cognitive demands. The data analysis phase focuses on identifying cognitive bottlenecks and task complexities, providing insights for improving task design, training, and protocols. The chapter concludes with the reporting stage, emphasizing clear, actionable communication of findings to stakeholders. By systematically guiding readers through each step, this chapter offers a practical resource for those applying ACTA in education, healthcare, industry, and military settings.

INTRODUCTION

This chapter introduces the key steps in the Applied Cognitive Task Analysis (ACTA) process, a structured methodology designed to uncover and analyse the cognitive components of complex tasks. ACTA simplifies cognitive task analysis, making it accessible to practitioners without specialized training in cognitive science. It is particularly valuable in domains where observational data is hard to obtain, enabling the identification of decision-making strategies and tacit knowledge critical to expert performance.

The ACTA process is organized into three data collection techniques—task diagram, knowledge audit, and simulation interviews—followed by an analytical stage where findings are synthesized into a cognitive demands table. These steps provide a comprehensive framework for analysing cognitive skills, facilitating the

DOI: 10.4018/979-8-3693-7733-8.ch003

design of systems, training programs, and performance improvements aligned with real-world challenges.

This chapter explores each step in detail, illustrating how ACTA's tools can be applied to document task complexities, elicit expert insights, and transform those findings into actionable recommendations. By bridging cognitive science with practical application, ACTA serves as a powerful tool for optimizing performance and supporting innovation across diverse fields.

3.1 Understanding ACTA Processes

Applied Cognitive Task Analysis (ACTA) is a set of streamlined cognitive task analysis tools that have been developed specifically for use by professionals who have not been trained in cognitive psychology, but who do develop applications that can benefit from the use of cognitive task analysis. ACTA methodology is suitable for job domains where observational data are difficult to obtain and can be used for identification and codification of complex skills on the workplace. The ACTA methodology requires less training, resources and time for its application and provides tools for extracting knowledge and coding complex decision-making skills.

The ACTA provides the opportunity for general domain application and flexibility, which allows for the explorations "of and in professional knowledge management practice" (Taylor, Ashford, & Jefferson, 2023). ACTA represents a natural advancement in methods for knowledge elicitation and representation, designed to capture complex judgments and attentional processes. Grounded in the framework of naturalistic decision making, the method investigates critical decisions and the use of information in dynamic environments, aiming to inform design and training.

ACTA is developed for training practitioners and systems designers to elicit and represent cognitive components of skilled task performance, and the means to transform those data into design recommendations. Systems designers are able to investigate and enquire more information and knowledge to improve the design of any system. The analysis may also expose implicit knowledge that is often difficult to verbalize as well as provide insights into non-observable tasks such as making decisions.

Applied Cognitive Task Analysis consist of four techniques, as shown in Figure 1. Generally, ACTA is split into three data collection stages, which are (1) task diagram, (2) knowledge audit, (3) simulation and one analytical stage which is (4) cognitive demand table. The techniques presented are intended to be complementary; each is designed to get at different aspects of cognitive skill. This set of three interview-based techniques served as a linear manuscript for probing each stage and provide a structure for targeted progressive deepening in the elicitation of critical elements of task execution by domain experts. The three data collection stages (task diagram, knowledge audit, simulation) may be repetitive and time-consuming processes.

Figure 1. ACTA techniques

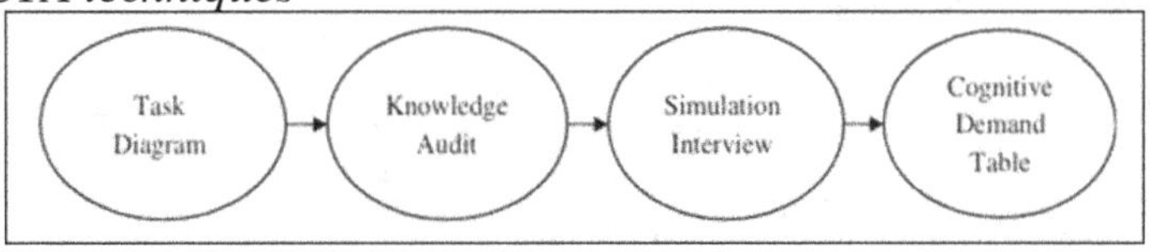

The four components of ACTA procedure are briefly introduced as followings (Militello & Hutton, 1998). Whilst each of the ACTA components is discussed in detail in the subsequent sections.

1. Task diagram interview providing the analyst with an in-depth overview of the task under analysis. During the interview, the analyst highlights those elements of the task that are cognitively challenging.
2. Knowledge audit interview highlighting those parts of the task under analysis where expertise is required. Once examples of expertise are highlighted, the subject matter expert is probed for specific examples within the context of the task.
3. Simulation interview, probing the cognitive processes used during a simulated incident by the subject matter expert.
4. Cognitive demands table integrating the data obtained from the task diagram, the knowledge audit and simulation interviews.

To facilitate interviewers and interviewees in completing every technique in the ACTA process, Stanik et al. (2021) introduced a guideline capable of overseeing the entire ACTA process. The guideline consists of a table that emphasizes the four ACTA sequential stages represented as the columns. Whereas, description of each stage's goal, the employed method, the number of participants and their expertise, and the preliminary results are discussed in the row. Table 1 below presented a sample of the guideline.

Table 1. Overview of the ACTA process

	Stage 1 Task diagram	Stage 2 Knowledge audit	Stage 3 Simulation interview	Stage 4 Cognitive demand table
Goal	*Document the task of interest*	*Document information about cognitive difficulties, as well as personal experiences for each task*	*Collect ideas and design sketches for the user interface of a combined medical device*	*Systematically refine the sketches*
Method	*Hierarchical task analysis*	*semi-structured interviews*	*Design scenarios*	*Design model*
Participants	*End-users*	*Experts*	*Experts*	*Experts*
Preliminary results	*One hierarchical task diagram with main tasks and corresponding subtasks*	*Details about the process, possible difficulties, and personal experiences*	*sketches of the device's user interface*	*Refined the model for the teams' sketches*

Note. Adapted from "Lessons Learned from Customizing and Applying ACTA to Design a Novel Device for Emergency Medical Care" by Stanik et al., 2021, *Proceedings of theIEEE International Conference on Requirements Engineering.*

The goal of ACTA is to identify and depict the critical cognitive components from subject matter experts (Militello & Hutton, 1998). The ACTA was used to gain insight into the different decisions and cognitive tasks, where tasks of interest are probably not standardized in execution. In-depth interviews with subject matter experts are used to gather most of the data. The quality of data produced by ACTA methods is partially influenced by the interviewer's skill. The purpose of these interviews is to learn more about the cognitive strategies used to complete the task, such as methods for assessing the scene and for identifying and interpreting key cues. It is also used to highlight differences between expert and novice performance, thus examining how novices can more quickly be brought to the performance level of experts. The ACTA toolkit recommended three to five experts in conducting this method (Militello et al., 1997). Interviewing a limited number of experts can lead to an increase in the knowledge of the interviewer, information can be verified among experts and a buffer can be provided for unsuccessful interviews (Mohd Yusoff, 2015).

The advantages of ACTA over the rest of the methods are its characteristics as a structured approach, the use of three different interviews to ensure the comprehensiveness of the method and probes, and the questions provided to the researchers to facilitate relevant data extraction. The usability of questionnaires in the ACTA method is also found to be clear in its output and useful in knowledge representation. ACTA will be the medium for design recommendations from which critical cognitive elements are extracted (Militello & Hutton, 1998).

3.2 Task Diagram Technique

This technique is used to identify and elicit a broad overview of the cognitive skills required in the designing process. Simply put, this technique outlines the major steps of the task process and identifies those that involve the most cognitive demand. The goal is to identify and examine the perceived critical elements. Interviews will be conducted with the experts involved in the designing process, allowing designers to differentiate tasks that require cognitive skills from non-cognitive skills.

The task diagram technique is considered as a preliminary interview that offers only a surface-level view of the cognitive elements of the task. The task diagram step is generally employed to perform an initial assessment of practice areas that may present the most cognitive challenges. The time and resources spent for the first technique is crucial so as to unpack or uncover the most difficult and relevant of those cognitive elements. It enables the interviewer to focus more in-depth interviews in the next two techniques which are the knowledge audit interviews and simulation interviews.

The task diagram interview provides the interviewer or the analyst with the task under analysis. During the interview, the analyst highlights those elements of the task that are cognitively challenging or difficult to be probed further with in-depth interviews in the next two stages of ACTA; knowledge audit technique and simulation technique.

By conducting interviews with subject matter experts involved in the design process, the study distinguished between cognitive and non-cognitive skills required. Upon the completion of the task diagram interview, a diagram is developed representing the component task steps involved and those task steps that require the most cognitive skill.

3.3 Knowledge Audit Technique

Knowledge audit interview highlighting those parts of the task under analysis where expertise is required. The knowledge audit reveals the key expertise of SMEs relevant to the tasks and explores their personal experiences. This technique employs a set of probes designed to describe types of domain knowledge or skills and elicit appropriate examples. The Knowledge Audit is organized around knowledge categories that have been found to characterize expertise. These include diagnosing and predicting, situation awareness, perceptual skills, developing and knowing when to apply tricks of the trade, improvising, metacognition, recognizing anomalies, and compensating for equipment limitations (Militello et al., 1997).

The objective is not merely to determine whether each component is present in the task, but to understand the nature of these skills, the specific instances in which they were needed, and the strategies employed. The purpose of the knowledge audit is to identify elements of the task of interest which requiring judgment and decision making, accordingly, identifying the knowledge bases that underpin this expertise.

This technique is used to explore the cognitive task difficulties identified in the task diagram technique, where the natures of the skills, specific events where they were required, strategies that have been used are found out. The knowledge audit identifies ways in which expertise is used in a domain and provides examples based on actual experience. The knowledge audit offers a general survey of the cognitive challenges associated with a domain, along with the related knowledge, cues and strategies (Taylor et al., 2023).

The knowledge audit has been developed as a means for capturing the most important aspects of expertise while streamlining the intensive data collection and analysis methods that typify studies of expertise. The technique is subsequently organized according to knowledge categories identified as characteristic of expertise. These include diagnosing and predicting, situation awareness, perceptual skills, developing and knowing when to apply tricks of the trade, improvising, metacognition, recognizing anomalies, and compensating for equipment limitations (Militello & Hutton, 1998).

3.4 Simulation Technique

The data finding in the simulation technique is independent of the findings of the knowledge audit technique. Identifying and examining information related to high-stakes, challenging decisions can offer a solid foundation for developing effective system design. Building on the first two stages, the simulation technique sought to gain a deeper insight into the complex issues. During simulation technique interview, subject matter experts probe the cognitive processes based on simulated incident. The simulation can be either high fidelity or low fidelity.

The experts involved within the context of specific scenario will be presented with existing scenarios that address the challenges of designing and conducting the said scenario. Militello & Hutton (1998) recommend that the interviewer retrieves a scenario that already exists for use in simulation technique interview which the scenario may need to adapt or modify in order to conform to practical constraints. The simulation can be in the form of a paper-and-pencil exercise, using maps or diagrams, and even using video or computer-supported simulations. The elicited scenarios are designed to help novices enhance situational awareness for specific tasks and clearly outline cognitive complexity within an organizational repository of the scenarios.

In a standard situation of the simulation interview, the experts were asked with a list all major events along the timeline of a specific incident, which may be identified as key judgments or decision points, such as the diagnosis, situation assessment, sensitivity to critical cues, and selection of courses of action within the environment. As these were listed, subsequent columns were used to record actions, probed by: *"what actions did you take at this point?"*; situation assessment: *"what did you think was going on at this stage?"*; critical cues *"what information led you to believe this?"* and finally, potential errors: *"what potential errors would be made here?"*.

3.5 Cognitive Demand Table

A cognitive demands table is intended to provide a format for practitioners to use in focusing the analysis on project goals, providing a framework for evidence informed implications. All data and findings obtained upon conducting interviews with experts from three different techniques discussed earlier (task diagram, knowledge audit and simulation interviews) will be integrated and consolidated into cognitive demand table. This analytical stage may also be supported by mediating objects (i.e. visual aids such as charts, visual scene renderings from a simulator, and model) to focus the discussion. Clearly, not every bit of information discussed in an interview will be relevant for the goals of a specific project.

Cognitive demands table will act as a mean of communication between experts that provide a common format to be used in focusing the analysis on a specific project's goals. The table allow for the display of more subtle aspects of expertise in the form of key cues and strategies focusing on the outcomes of analysis that are pertinent to problem solving and decision making.

In addition, the table help practitioners synthesizing and get a glimpse of common themes in the data, as well as conflicting information given by multiple experts.

The construction of the cognitive demands table involved systematically working through the findings of all interviews for common themes as well as conflicting information (Martini et al., 2022). Information relevant to the aims of the study was categorized. The cognitive demands table offers a helpful overview that analyses important features of expert cognition in the specific domain and provide a complete picture of the cognitive challenges especially for novices, which includes a listing of difficult cognitive elements, descriptions of why they are difficult, common errors associated with them, and the cues and strategies used in them. It presents the cognitive demand underpinning each cognitive element and the cues and strategies used to support proficient decision-making.

CONCLUSION

This chapter has discussed and presented the key steps to conduct Applied Cognitive Task Analysis (ACTA). ACTA is developed for practitioners and systems designers to elicit and represent cognitive components of skilled task performance, and the means to transform those data into design recommendations. ACTA is generally separated into data collection stage and analytical stage. Data collection stage consists of three techniques which are task diagram, knowledge audit, and simulation. Whereas the analytical stage consists of cognitive demand table.

The Task Diagram technique is a valuable tool for gaining a comprehensive understanding of the cognitive skills involved in the design process. This method is particularly effective in providing a broad overview of the intellectual abilities required for various tasks. During structured interview sessions with experts, the technique systematically outlines the key steps of the task process, offering insight into the sequence and structure of tasks. Through this detailed mapping, the technique identifies which steps involve the most significant cognitive demands, such as problem-solving, decision-making, and analytical thinking. By visually representing the task flow, the Task Diagram helps designers distinguish between tasks that require cognitive effort and non-cognitive skills. This clear differentiation allows for a focused analysis of the specific areas where expert knowledge is most critical, enabling designers to allocate resources and attention more effectively. Overall, the Task Diagram technique provides a holistic view of the cognitive landscape of a task, making it an essential tool for optimizing performance and efficiency in the design process.

The next technique, known as the Knowledge Audit, plays a crucial role in identifying the essential expertise of Subject Matter Experts (SMEs) that is directly applicable to the tasks at hand. This method delves into their personal experiences, uncovering the nuanced knowledge that influences their decision-making and judgment during task execution. Through the Knowledge Audit, specific elements of the task that demand higher-level cognitive processes, such as critical thinking and problem-solving are highlighted. The Knowledge Audit interview is structured around a series of carefully designed probes aimed at uncovering various types of domain-specific knowledge and skill sets. By eliciting relevant examples, the Knowledge Audit ensures that tacit knowledge, which may not be readily apparent, is brought to the surface and effectively documented for further analysis and application.

The third technique, known as the Simulation Technique, builds upon the foundational insights gathered from the previous two stages. This approach is designed to provide a more in-depth understanding of complex, real-world issues by immersing SMEs in a simulated environment that mirrors critical scenarios or incidents relevant to their domain. The primary objective of this technique is to explore and analyse the cognitive processes that experts engage in when faced with challenging situations. During a simulation technique interview, SMEs are presented with a simulated incident that replicates a critical event from their field of expertise. Through this process, they are asked to probe the decision-making strategies they employed during similar real-life events. This interactive approach allows for the exploration of not just the actions taken, but also the underlying reasoning, judgment, and cognitive pathways that guided those actions. In a typical simulation interview, experts are tasked with identifying and listing all major events along the timeline of the specific incident. These events are critical junctures, often referred to as key judgment or decision points, where crucial choices were made, or critical assessments were performed. By outlining these vital moments, the Simulation Technique helps to uncover how experts navigate complex situations, make decisions, and apply their expertise to resolve problems.

Finally, the data and findings gathered through interviews with experts using the three techniques—Task Diagram, Knowledge Audit, and Simulation Interviews—are synthesized and compiled into a comprehensive Cognitive Demands Table. This table serves as a crucial tool, effectively integrating insights from all stages of the analysis to provide a clear, structured overview of the cognitive processes involved in a particular domain or project. The Cognitive Demands Table not only consolidates the diverse pieces of information uncovered through each technique but also functions as a key communication tool among experts. By providing a standardized format, it enables a shared understanding and common language for discussing the cognitive requirements of a project. This helps to focus the analysis on the specific goals and challenges of the project, ensuring that all stakeholders are aligned in their approach. Moreover, the Cognitive Demands Table provides a detailed, high-level overview of the essential cognitive features and challenges within the domain. It highlights the critical points where expert judgment, decision-making, and problem-solving are most necessary, offering a thorough analysis of expert cognition. By outlining the key cognitive challenges, it acts as a guide for developing the skills needed to navigate complex tasks, helping novices to understand the areas where they will need to focus their learning and development. In summary, the Cognitive Demands Table not only bridges communication between experts but also serves as an invaluable resource for analysing and understanding the cognitive complexities of a specific project or domain, particularly for those new to the field.

REFERENCES

Martini, N., Choong, J. W., Dela Cruz, P. D., Lau, H., Lim, H., Liu, R., & Marshall, D. (2022). Assessing antibiotic prescribing in nurse practitioners: Applied cognitive task analysis. *International Journal of Nursing Studies Advances*, 4. Advance online publication. DOI: 10.1016/j.ijnsa.2022.100101

Militello, L. G., & Hutton, R. J. B. (1998). Applied cognitive task analysis (ACTA): A practitioner's toolkit for understanding cognitive task demands. *Ergonomics*, 41(11), 1618–1641.

Militello, L. G., Hutton, R. J. B., Pliske, R. M., Knight, B. J., Klein, G., & Randel, J. (1997). *Applied Cognitive Task Analysis (ACTA) methodology.*

Mohd Yusoff, N. (2015). *A Shared Cognitive User Task Framework for Elearning Storyboard.*

Stanik, C., Puhlfurs, T., Mahler, A., Sasu, P. B., Reip, W., & Maalej, W. (2021). Lessons Learned from Customizing and Applying ACTA to Design a Novel Device for Emergency Medical Care. *Proceedings of the IEEE International Conference on Requirements Engineering*, (September 2021), 302–311. DOI: 10.1109/RE51729.2021.00034

Taylor, J., Ashford, M., & Jefferson, M. (2023). High performance coach cognition in the wild: Using applied cognitive task analysis for practical insights–cognitive challenges and curriculum knowledge. *Frontiers in Psychology*, 14. Advance online publication. DOI: 10.3389/fpsyg.2023.1154168

Chapter 4
Detailed Methodologies for Gathering Data

ABSTRACT

This chapter explores the methodologies used in the Applied Cognitive Task Analysis (ACTA) process, detailing how data is gathered to understand cognitive tasks. It covers structured interviews, direct observation, and work sample analysis, providing a step-by-step guide on applying each technique. The chapter discusses the strengths and limitations of these methods and how to combine them for a comprehensive view of cognitive processes. It highlights the importance of aligning data collection strategies with research objectives, offering practical examples such as using structured interviews for verbal explanations and direct observation for real-time task performance. Work sample analysis is emphasized for its ability to provide concrete evidence of task execution and cognitive demands. This chapter equips researchers and practitioners with the tools needed to conduct effective cognitive task analysis, providing rich, actionable data for task design, training, and system optimization.

INTRODUCTION

This chapter explores the methodologies used in the Applied Cognitive Task Analysis (ACTA) process, detailing how data is gathered to understand cognitive tasks. It covers structured interviews, direct observation, and work sample analysis, providing a step-by-step guide on applying each technique. The chapter discusses the strengths and limitations of these methods and how to combine them for a comprehensive view of cognitive processes. It highlights the importance of aligning data collection strategies with research objectives, offering practical examples such as using structured interviews for verbal explanations and direct observation for real-time task performance. Work sample analysis is emphasized for its ability to provide concrete evidence of task execution and cognitive demands. This chap-

DOI: 10.4018/979-8-3693-7733-8.ch004

ter equips researchers and practitioners with the tools needed to conduct effective cognitive task analysis, providing rich, actionable data for task design, training, and system optimization.

This chapter provides a comprehensive and detailed methodology for each stage of the Applied Cognitive Task Analysis (ACTA) process. It thoroughly outlines the design, structure, and implementation phases involved in each technique used within ACTA. By breaking down the methodology step-by-step, the chapter offers a clear and systematic approach to understanding how each technique is applied.

For each technique employed in ACTA, whether it is the Task Diagram, Knowledge Audit, or Simulation Interview, this chapter delves into the specific processes and considerations that guide their use. It highlights the critical phases that are essential for capturing the cognitive skills, decision-making processes, and expert knowledge required for complex tasks.

By providing this level of detail, the chapter equips readers with the necessary tools to implement ACTA effectively, ensuring that they can replicate the process in their own research or projects. This methodology serves as a roadmap, guiding practitioners through the details of each phase and helping them to extract meaningful insights into expert cognition. Ultimately, the chapter offers a robust framework for applying ACTA, making it accessible and practical for both researchers and professionals seeking to explore cognitive demands in real-world settings.

4.1 Task Diagram Technique

Task Diagram technique is typically used to conduct an initial screening of areas of practice that might pose the most difficult cognitive challenges. Task diagram technique elicits broad overview of the tasks and identifies difficult cognitive elements, where only surface-level views of the cognitive elements of the task are identified.

The Task Diagram technique offered a structured approach to understanding the intricate cognitive processes and knowledge requirements involved. To some extent, the task diagram represents a simplified mental model of how subject matter experts (SME) perceive the task. Its function is to embed the cognitive task analysis within the task and to focus the rest of the interviews on the cognitively challenging parts of the task (Militello et al., 1997).

The Task Diagram phase entailed visually representing and analysing the component tasks and interconnections of the tasks involved and those tasks that require the most cognitive skill. By creating a task diagram, the experts are able to gather comprehensive data and gained insights into the critical cognitive elements and their interrelationships.

Among possible techniques to be used in developing task diagram involved direct methods of watching and talking with the SMEs. Militello & Hutton (1998) recommended limiting decomposition of task between three to six elements in order to obtain a broad overview of the task. The goal is to get the experts to walk through the task in his/her mind, verbalizing major steps and to ensure that time is not wasted delving into minute detail during the surface-level interview.

Using the standard Task Diagram approach to gain an understanding of the general overview, the SMEs are asked to decompose the *'task of interests'* into steps or subtasks with a question such as, *"Think about what you do when you (task of interest). Can you break this task down into less than six, but more than three steps?*". Further, the interviewer may employ a semi-structured interview guide to understand the impacts of the tasks in discussion. From there, the interviewer asked questions for clarification and continually redirected the expert's focus to a broad-level overview of the process rather than delving into the details of factors that influence into different situations.

Nevertheless, likelihood that incidents will arise in which the initial Task Diagram will not provide sufficient information for the interviewer to progress with the subsequent techniques of Applied Cognitive Task Analysis. In these situations, a second stage of decomposition is recommended, and this decision is left to the interviewer's discretion.

While in the process of decomposing cognitive tasks into steps or subtasks, the steps or subtasks may possibly documented via a flip chart, whiteboard or cognitive map (or any other medium) which show 3-6 circles relate to the task. The interviewer then enquires about which step or subtasks of the task poses the greatest cognitive challenge and why novices might struggle with it. The interviewer is encouraged to confirm with the expert that the task diagram accurately represents a comprehensive overview of the task. Together, the interviewer and the expert identify which aspect of the task is the most cognitively complex and requires the most thinking, judgment, and decision-making.

Upon completion of the task diagram technique, a comprehensive task diagram was constructed. Additionally, it highlighted the specific elements that pose higher cognitive challenges.

A team of authors and engineers (Gore, Banks, & Mcdowall, 2015) had created a set of tips, presented in Table 1, to enhance the task diagram and knowledge-elicitation phase of ACTA, acknowledging the significance of socio-cognitive competence and insight.

Table 1. Accessing expertise in the field: Top tips for getting rich data/developing socio-cognitive competence/insight from ACTA

Tasks	Detailed Perceptions of Cues and Strategies
Redo and refine the task diagrams	Retrace your steps and redo the task diagrams as needed – you may need several drafts to get the detail level right
Listen actively throughout	ACTA works better if the interviewer listens actively: listen, summarise and then record the information
Stay focused and be clear about your roles	Reign in the temptation to share anecdotes, this can distract from the task, and remain clear about how interviews and who is interviewer
Bear with frustration	The process might entail some frustration about taking too long, or not getting the right level of detail. If in doubt or getting too tired, leave the task for a while, and come back to it the next day
Ask what is difficult and ask about thinking	One of the key objectives for ACTA is to highlight what experts think, but might not have shared explicitly. Ask to clarify, ask for more detail, or ask questions again in a different way
Don't assume and choose your pairings wisely	You might think that things are obvious (as interviewer or interviewee) but chances are that they are not. It can work well to work in pairs or triads who don't usually work with each other, rather than pairing up with close colleagues. This will allow you to ask important questions which team members may not ask, assuming that the answers should be obvious
Remember that detail is good	Each component of your task diagram should be annotated with detail
Be aware of when you stop recording information	If there is a time in the interview when you talk, but no information is recorded, then ask yourself 'why'. Are you not asking the right questions? Have you gone 'off track'?
Use the crib sheets	Each component of your task diagram should be annotated with detail. ACTA works best with structure, so don't be shy to use the crib sheets
Check your thinking	Do talk each other through your diagrams and knowledge audits again, for instance clarify anything which is not clear, and make sure the examples are specific, rather than general.

Note. Reprinted from "Advancing ACTA: Developing Socio-Cognitive Competence/Insight" by Gore, J., Banks, A., & Mcdowall, A. 2015, *Proceedings of the 12th International Conference on Naturalistic Decision Making.*

4.2 Knowledge Audit Technique

Knowledge Audit technique is the second technique of ACTA which employs a set of probes designed to describe types of domain knowledge or skills and elicit appropriate examples. The goal is to ascertain the nature of these skills, specific events and strategies that have been used. This technique is used to explore the cognitive task difficulties and extract the expertise necessary to perform tasks as identified in the task diagram technique. Those tasks are then explored and probed in great detail. Knowledge audit table is used to present the data findings. By utilizing the probes and conducting a knowledge audit table, this phase aimed to gain a deeper understanding of the cognitive challenges and knowledge necessary for effective task performance.

This technique is designed to discern the nature of the skills, particular events, and strategies employed within the context of the tasks. By employing this method, the study sought to investigate the cognitive challenges associated with the tasks identified through the task diagram methodology.

The use of probes in eliciting and capturing detailed information about the cognitive processes and knowledge required to complete tasks. The use of probes allows for a systematic exploration of the knowledge and skills employed by experts in the field. By collecting data on the specific events, strategies, and knowledge utilized during task performance, a comprehensive understanding of the cognitive demands and requirements can be achieved (Militello & Hutton, 1998).

The data gathered through this technique were systematically organized and presented in a knowledge audit table. The knowledge table provides a structured and coherent format for organizing and analysing the information obtained from the various probes used during the knowledge audit interviews. By arranging the data in this manner, it becomes easier to identify emerging patterns, recurring themes, and key insights related to the cognitive aspects of the tasks under investigation.

The Knowledge Audit Table not only helps in classifying the expert knowledge but also enables a deeper analysis of the cognitive elements involved in specific tasks. The table facilitates the recognition of critical cognitive demands, allowing researchers or analysts to identify which tasks require higher levels of problem-solving, or decision-making. This structured approach is particularly helpful when dealing with complex or multi-faceted tasks, where cognitive load and expertise play a significant role.

To guide the conversation during interviews or discussions with experts, Berlin & Praetorius (2023) proposed a matrix table that provides a flexible and non-linear framework for exploring the relevant cognitive skills. It allows interviewees to move through different aspects of the task without being confined to a rigid structure, promoting a more dynamic exchange of ideas. The matrix serves as a visual aid that encourages experts to reflect on various dimensions of their cognitive processes, helping them articulate the underlying knowledge that supports their task performance.

A sample of such a matrix table is illustrated in Table 2, demonstrating how the data can be mapped out to enhance understanding and facilitate further analysis of the cognitive demands associated with the tasks. This approach ensures that the knowledge gathered is not only comprehensive but also accessible for future reference and application.

Table 2. Visual matrix of knowledge audit technique

[Task At Hand]		
Aspects of expertise	**Cues and strategies**	**Why difficult?**
Past and future		
Big Picture		
Noticing cues and patterns		
Job Smarts		
Opportunities/Improvising		
Self-monitoring		

Note. Reprinted from "Applied Cognitive Task Analysis (ACTA) of Marine Piloting in a Swedish Context" by Berlin, C., & Praetorius, G., 2023, *Human Factors in Transportation.*

4.3 Simulation Technique

The simulation interview technique, as described by Militello & Hutton (1998), provides a valuable means to explore the thought processes of subject matter experts (SMEs) within a specific scenario. This technique offers unique insights that complement the findings obtained through previous methods such as the knowledge audit. Using simulations or scenarios in interviews allows for in-depth probing into topics like situation assessment, the impact of situation assessment on decision-making, and potential errors that novices may make when faced with similar situations (Militello & Hutton, 1998). This contextual information is often challenging to gather using other interview techniques.

The simulation technique provides a unique method for understanding the cognitive processes of experts within the context of specific incidents, offering insights that are independent of the findings gathered from the knowledge audit. By presenting experts with a scenario that highlights particularly difficult and challenging aspects of the task, the interviewer is able to explore how experts react, make decisions, and apply their knowledge in the actual or presumed situations. This technique is especially valuable for gaining a deeper understanding of how experts navigate complex scenarios and overcome cognitive demands.

To facilitate this exploration, a matrix table can be used to guide interviewees or experts. This table helps structure the conversation in a way that allows experts to discuss relevant cognitive skills without being confined to a rigid, linear format. It also articulate experts thought processes and decision-making strategies. This non-linear approach allows for a more holistic understanding of the task, emphasizing the interplay between different cognitive elements. A sample of the matrix table is shown in Table 3 below.

Once the interview is complete, the data gathered from the simulation technique is systematically organized and presented in a simulation table. This table serves as a structured format for analyzing the information collected, offering a clear overview of the expert's responses, decision points, and cognitive challenges encountered during the simulated scenario. Ultimately, the simulation table provides a comprehensive summary of the expert's cognitive processes, offering insights that can be compared or integrated with findings from other techniques, such as the task diagram and knowledge audit, to form a well-rounded understanding of the task at hand.

Table 3. Visual matrix of simulation technique

[Task At Hand]				
Steps (events) where evaluation occurs	**What do you do (actions)?**	**What is your judgement of the situation?**	**What clues do you use?**	**Possible mistakes made**

Note. Reprinted from "Applied Cognitive Task Analysis (ACTA) of Marine Piloting in a Swedish Context" by Berlin, C., & Praetorius, G., 2023, *Human Factors in Transportation.*

By utilizing the simulation interview technique, this stage aimed to delve deeply into the perspectives of experts, offering a comprehensive understanding of their cognitive processes and decision-making strategies. The simulation method enabled a more immersive and dynamic exploration of real-world situations, allowing experts to reflect on their experiences and decisions in environments that closely replicate actual conditions. This technique not only highlighted the steps experts take during decision-making but also uncovered critical insights into the potential challenges and pitfalls they encounter. These valuable insights provided by the experts played a vital role in evaluating the effectiveness of the practice. By revealing how experts navigate complex scenarios, the interviewer was able to identify strengths and weaknesses of the current tasks.

Additionally, the insights gathered from these expert interviews played a crucial role in refining and enhancing the overall scenario. The findings helped pinpoint key areas for improvement, ensuring that the scenario is well-aligned with best practices while also addressing practical, real-world challenges. Ultimately, the use of the Simulation Interview technique provided a valuable, data-driven basis for improving the design, implementation, and effectiveness, making it more efficient, focused, and user-friendly.

4.4 Cognitive Demand Table Technique

The final technique is to integrate the data obtained from the data findings from the previous techniques: task diagram, knowledge audit and simulation, where the data will be presented in a cognitive demand table. The table provides a framework for evidence informed implications to enable the presentation of more tacit features of expertise in the form of key cues and strategies focusing on the outcomes of analysis that are pertinent to problem solving and decision making (Taylor, Ashford, & Jefferson, 2023).

The cognitive demand table technique, as supported by research conducted by Militello et al. (1997), entails the synthesis of data obtained from earlier techniques namely, the task diagram, knowledge audit, and simulation. The results are systematically presented using a structured format. The data presented in the cognitive demand table contains difficult cognitive elements, with "Why Difficult", "Common Errors" and "Cues and Strategies" used as the headings. It provides a format to analyze the types of information needed to design a new system.

The use of cognitive demand tables offers a systematic approach to analyse and categorize the diverse information required for the development of a new system or projects. These tables provide a means to capture and consolidate critical insights into the difficulties and common errors associated with cognitive tasks, as well as the cues and strategies that can support successful task performance.

By utilizing the cognitive demand table technique, it aims to harness the research-supported benefits of this structured approach. It enables the analysis of the multifaceted information obtained from the task diagram, knowledge audit, and simulation techniques, facilitating the identification of key cognitive challenges and the formulation of targeted strategies.

CONCLUSION

This chapter has discussed and presented a detail methodology for each stage of the Applied Cognitive Task Analysis (ACTA) process, which equips readers with the necessary tools to implement ACTA effectively.

By employing the standard Task Diagram approach to gain an understanding of the general overview, the SMEs are asked to decompose the *'task of interests'* into three to six steps or subtasks. This decomposition is intended to provide a broad overview of the task. The goal is to get the experts to walk through the task in his/her mind, verbalizing major steps and to ensure that time is not wasted delving into minute detail during the surface-level interview.

The incorporation of probes in the knowledge audit interview helps elicits and captures detailed information about the cognitive processes and knowledge necessary for completing tasks. The use of probes allows for a systematic exploration of the knowledge and skills utilized by experts in the field. The data gathered through this technique are then systematically organized and presented in a knowledge audit table. The knowledge audit table provides a structured and coherent format for organizing and analysing the information obtained from the various probes used during the knowledge audit interviews.

Complementing the findings obtained through previous techniques (task diagram and knowledge audit), the simulation interview provides a valuable approach to explore the thought processes of subject matter experts (SMEs) within a specific scenario. In this technique, the experts are presented with a scenario that addresses difficult and challenging elements of the task at hand, enabling the interviewer to gain a deeper understanding of the experts' cognitive processes within the context of an incident. The insights gathered from these expert interviews played a crucial role in refining and enhancing the overall scenario. These findings helped pinpoint key areas for enhancement, ensuring that the scenario is well-aligned with best practices while also addressing practical, real-world challenges.

Lastly, the final technique known as cognitive demand table consolidates the data gathered from the data findings from the three preceding methods (task diagram, knowledge audit and simulation). The use of cognitive demand tables offers a systematic approach to analyse and categorize the diverse information required for the development of a new system or projects. These tables provide a means to capture and consolidate critical insights into the difficulties and common errors associated with cognitive tasks, as well as the cues and strategies that can support successful task performance.

REFERENCES

Berlin, C., & Praetorius, G. (2023). Applied Cognitive Task Analysis (ACTA) of marine piloting in a Swedish Context. *Human Factors in Transportation*, 95, 709–718. DOI: 10.54941/ahfe1003856

Gore, J., Banks, A., & Mcdowall, A. (2015). Advancing ACTA: Developing Socio-Cognitive Competence/Insight. *Proceedings of the 12th International Conference on Naturalistic Decision Making*, 1–4. Retrieved from http://www2.mitre.org/public/ndm/papers/GoreBanksMcDowall41315.pdf

Militello, L. G., & Hutton, R. J. B. (1998). Applied cognitive task analysis (ACTA): A practitioner's toolkit for understanding cognitive task demands. *Ergonomics*, 41(11), 1618–1641.

Militello, L. G., Hutton, R. J. B., Pliske, R. M., Knight, B. J., Klein, G., & Randel, J. (1997). *Applied Cognitive Task Analysis (ACTA) methodology*.

Chapter 5
Data Analysis Techniques

ABSTRACT

This chapter explores the data analysis techniques used in the Applied Cognitive Task Analysis (ACTA) process, detailing how the cognitive data is being analysed. It covers structured interviews, direct observation, and work sample analysis, providing a step-by-step guide to analysed all data gathered from each technique in ACTA. The chapter discusses the strengths and limitations of these methods and how to combine them for a comprehensive view of cognitive processes. It highlights the importance of aligning data analysis strategies with research objectives, offering practical examples such as using structured interviews for verbal explanations and direct observation for real-time task performance. Work sample analysis is emphasized for its ability to provide concrete evidence of task execution and cognitive demands. This chapter equips researchers and practitioners with the tools needed to conduct effective cognitive task analysis, providing rich, actionable data for task design, training, and system optimization.

INTRODUCTION

Data analysis is an integral part of any research process, and in the context of Applied Cognitive Task Analysis (ACTA), it involves a meticulous and systematic approach to ensure the accuracy and relevance of the insights derived. According to Saunders, Lewis, & Thornhill (2009), the data analysis process consists of developing categories, identifying connections within and across these categories, and formulating hypotheses that lead to well-supported conclusions. In this chapter, the data collected through the various ACTA techniques—Task Diagram, Knowledge Audit, Simulation, and Cognitive Demand Analysis—is examined in detail. Each

DOI: 10.4018/979-8-3693-7733-8.ch005

of these techniques was applied to address specific cognitive challenges associated with task performance in educational settings.

The data analysis is designed to map out the cognitive demands of the tasks as identified by experts, providing a clear understanding of the complexities involved in system design, content creation, and assessment methods. This chapter highlights the structured methodology used to break down tasks into smaller, manageable components, allowing for a deeper analysis of cognitive skill requirements. The findings are then systematically organized using the Cognitive Demand Table and triangulated across various methods to validate the insights. By doing so, the analysis not only aligns with the objectives of ACTA but also contributes to refining the instructional design strategies and system functionality.

Through the application of triangulation methods, which involve cross-referencing data from multiple sources, the reliability of the findings is enhanced. This chapter does not merely present raw data; it synthesizes expert feedback and situational assessments to provide a robust framework that can inform practical applications in both educational and professional development contexts. Ultimately, this analysis supports the goal of improving instructional strategies and system design by addressing cognitive challenges more effectively, ensuring that learners can engage with the material in a meaningful and productive way.

5.1 Task Diagram Technique

The data analysis involved a thorough and organized examination of the recommendations and guidelines provided by experts, all of which were rooted in well-established academic theories and principles from previous research. This comprehensive review was carried out by leveraging the Task Diagram technique from the Applied Cognitive Task Analysis (ACTA). By applying this method, the analysis was able to systematically explore the cognitive demands of the tasks, ensuring that expert insights were thoroughly considered in light of existing scholarly frameworks. This approach not only validated the findings but also provided a deeper understanding of how expert knowledge aligns with theoretical foundations, ultimately enhancing the analysis and its applicability to real-world contexts.

A preliminary interview is conducted to divide and separate tasks into smaller steps or subtasks. Interview questions will be based on sample questions presented by Militello et al. (1997). By utilizing established sample questions, the interview can gather insightful information from the experts regarding the subtasks and their relationship to the overall tasks. The interview will start with a question such as, '*Think about what you do when you (task of interest)*'. The aim is to have the expert mentally outline the task and articulate the key steps involved and elicit a very broad overview of the task. Subject matter experts (SME) are requested to decompose the ‘tasks of interest’ into relevant subtasks between three to six subtasks to avoid wasting time on minor details during the preliminary interview that are better addressed by other interview techniques (Militello & Hutton, 1998).

After the steps of the task have been articulated, the SME is asked to identify which of the steps require cognitive skill, with a question such as, *‘Of the steps you have just identified which require difficult cognitive skills?’*. Cognitive skills encompass judgment, assessment, problem-solving, and critical thinking abilities. The resulting task diagram acts as a guide for future interviews by outlining the main steps involved in the task, their sequence, and identifying which steps demand the highest level of cognitive skill.

A sample of task diagram for a troubleshooting process (Nauert & Gillan, 2019) is shown in Figure 5.1 below. The SME indicated that the first three steps (*“notice problem”, “explore for cause of problem”, and “try fix”*) were cognitively demanding, highlighting the *“explore for cause of problem”* step as the one requiring the most cognitive skill. The steps illustrated in the task diagram aligns well with a general conceptualization consisting of fault symptom identification (*“notice problem”*), fault diagnosis (*“explore for cause of problem”*), and solution generation and verification (*“try fix”, “check fix success”, “monitor for same problem”*).

Figure 1. Task diagram of a troubleshooting process

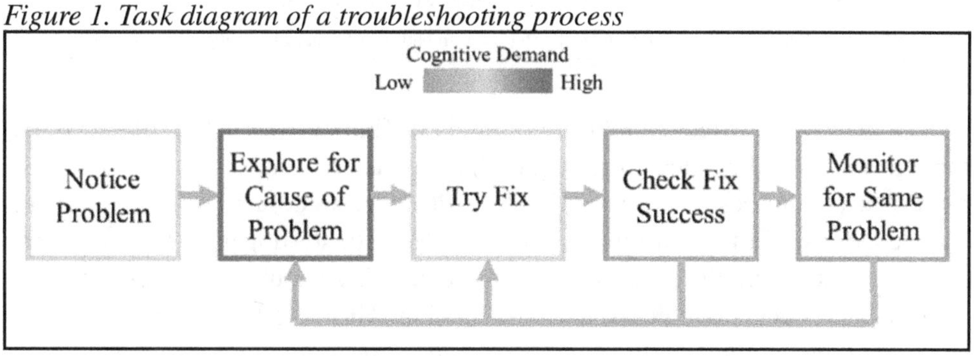

5.2 Knowledge Audit Technique

The knowledge audit technique highlights how expertise is applied within a domain and includes examples drawn from real experiences. Detail interviews need to be conducted for the knowledge audit to explore the cognitive task difficulties identified in task diagram technique. This process provides enough detail to retain the appropriate context of the incident.

The knowledge audit employs a set of probes designed to describe types of domain knowledge or skill and elicit appropriate examples. Based on those tasks identified from task diagram interview, experts must determine the cues and strategies they rely on. The expert needs to also identify the cause to the difficulties. To guide the interviews and ensure a thorough exploration of the issues, a list of probes was utilized. List of probes are a series of well-developed questions which are based on extensive research on expert thinking (Militello & Hutton, 1998). The list of probes is the starting point for conducting knowledge audit interview. The probe questions are shown in Table 1 below (Militello et al., 1997), which have been recognized as effective for eliciting valuable information in similar contexts. Utilizing these probes allows the interviews to methodically collect in-depth responses from the experts, offering valuable insights into their experiences and knowledge.

Table 1. Questions asked in knowledge table technique

Column title	Questions
Examples	Examples of a situation where you experience cognitive task difficulties
Cues/strategies	In this situation, what cues and strategies would you rely on?
Why difficult	In what way would this be difficult for an end-user? What makes it difficult to do?

The goal of knowledge audit interview is to find out the nature of the skills, specific events where they were required, strategies that have been used, and so forth. A set of probes as shown in Table 2, designed to describe types of domain knowledge or skill and elicit appropriate examples is the starting point for conducting this interview. Then, the interviewer asks for specifics about the example in terms of critical cues and strategies of decision making. This is followed by a discussion of potential errors that a novice, less-experienced person might have made in this situation.

Table 2. Knowledge audit probes

Basic probes	Questions
Past and future: experts can figure out how a situation developed, and they can think into the future to see where the situation is going. This can allow experts to head off problems before they develop.	*Is there a time when you walked into a middle of a situation and knew exactly how things got there and where they are headed?*
Big Picture: Novices may only see bits and pieces. Experts can quickly build an understanding of the whole situation – the Big Picture view. This allows expert to think about how different elements fit together and affect each other.	*Do you have an example that illustrates all of the different kinds of things you have to keep track of the Big Picture for this task?*
Noticing: Experts can detect cues and see meaningful pattern that less-experienced personnel may miss altogether.	*Have you had an experience where something just popped out at you that other people didn't notice?*
Job Smarts: Experts learn how to combine procedures and work the task in the most efficient way possible. They don't cut corners, but they don't waste time and resources either.	*Are there any special ways you use to accomplish more with less; that you have found especially useful?*
Opportunities/Improvising: Experts are comfortable improvising – seeing what will work in this particular situation; they are able to shift directions to take advantage of opportunities.	*Can you think of a time where you had to improvise in this task or notices opportunity to do something better?*
Self-Monitoring: Experts are aware of their performances; they check how they are doing and make adjustments. Experts notice when their performance is not what it should be (this could be due to stress, fatigue, high workload, etc) and are able to adjust so that the job gets done.	*Have there been times where you noticed that you need to change the way you were performing in order to get the job done?*
Anomalies: Novices don't know what is typical, so they have a hard time identifying what is atypical. Experts can quickly spot unusual events and detects deviations. They can notice when something that ought to happen, doesn't.	*Can you describe an instance when you noticed something unusual, a deviation from the norm or knew something was amiss?*
Equipment Difficulties: Equipment can sometimes mislead. Novices usually believe whatever the equipment tells them; they don't know when to be sceptical.	*Do you have an example of a situation where some piece of equipment was giving misleading information?*

The output of the Knowledge Audit is a table containing an inventory of task-specific expertise. A sample of knowledge audit table for fireground command (Militello & Hutton, 1998) is shown in Table 3. This table includes examples of situations in which experience has been called into play, cues and strategies used in dealing with these difficult situations, and an explanation why such situations present a challenge to less-experienced operators.

Table 3. Knowledge audit table of a fireground command

Aspects of expertise	Cues and strategies	Why difficult?
Past and future e.g. Explosion in office strip; search the office areas rather than source of explosion	Material safety data sheets (MSDS) tell you that explosion in area of dangerous chemicals and information about chemicals Start where most likely to find victims and own safety considerations	Novice would be trained to start at source and work out May not look at MSDS, to find potential source of explosion, and account for where people are most likely to be
Big picture Big picture includes source of hazard, potential location of victims, ingress/egress routes, other hazards	Senses, communication with others, building owners, MSDS, building pre-plans	Novice gets tunnel vision, focuses on one thing, e.g. victims
Noticing Breathing sounds of victims	Both you and partner stop, hold your breath, and listen Listen for crying, talking to themselves, victims knocking things over	Noise from own breathing in apparatus, fire noises Don't know what kinds of sounds to listen for

5.3 Simulation Technique

In order to gain valuable insights into the experiences and opinions of experts a simulation interview was conducted. In simulation technique, the data finding is independent of the findings from the knowledge audit technique. A simulation interview will be conducted towards the experts in order to discover their experiences and opinions of task design. They will also be allowed to interact and experience the simulation.

It is advised that the interviewer obtain a pre-existing scenario to use during the simulation interview. Often, simulations and scenarios exist for training purposes. It may be necessary to adapt or modify the scenario to conform to practical constraints such as time limitations. Creating a new simulation specifically for the interview is a complex undertaking that will likely require an initial cognitive task analysis to collect the essential information needed to present a challenging scenario. The simulation can be in the form of a paper-and-pencil exercise, using maps or other diagrams and it may also be possible to use video or computer-supported simulations.

Allowing the experts to directly engage with and interact within the simulation, offers a more refined and detailed understanding of its strengths, weaknesses, and areas for improvement. This hands-on experience enables the experts to provide a deeper evaluation, identifying potential issues or challenges that may not be immediately apparent through observation alone. Their insights, grounded in professional expertise, are essential in highlighting specific areas where the simulation can be enhanced. Additionally, their feedback offers valuable suggestions for refining the system to ensure it performs more effectively, addresses any existing gaps, and aligns with the intended objectives. This collaborative approach ensures that the simulation benefits from expert-driven enhancements, resulting in a more robust and well-rounded system.

After exposure to the simulation, the SME is asked to identify major events, including judgements and decisions, with a question such as, *'As you experience this simulation, imagine you are the (job you are investigating) in the incident. How you would think and act in this situation?'*. Feedback are recorded based on the question that are related to "Events", "Actions", "Situation Assessments", "Critical Cues" and "Potential Errors", as shown in Table 4 below (Militello et al., 1997).

Table 4. Questions asked in simulation technique

Column title	Questions
Event	The events that you identified as difficult and challenging
Actions	What actions would you take at this point?
Situation assessments	What is your assessment of the current situation?
Critical cues	What pieces of information led you to this situation assessment and these actions?
Potential errors	What errors would a user make?

The information gathered during the simulation interviews is systematically documented in the simulation table. Feedback provided by Subject Matter Experts (SMEs) is meticulously recorded and analyzed to uncover recurring themes, patterns, and insights. By using the same simulation across interviews with different SMEs, this approach enables a deeper understanding of scenarios where multiple courses of action are viable, as well as instances where different interpretations of the same situation may occur. This technique proves invaluable in highlighting the diverse perspectives and decision-making processes of SMEs. The results offer critical insights that can inform the development of more effective training programs and system design recommendations, ensuring a well-rounded and comprehensive approach to learning and problem-solving. A sample of simulation table is shown in Table 5.

Table 5. Simulation table of a fireground scenario

Events	Actions	Assessment	Critical cues	Potential errors
On-scene arrival	Account for people (names) Ask neighbours (but don't take their word for it, check it out yourself) Must knock on or knock down to make sure people aren't there	It's a cold night, need to find place for people who have been evacuated	Night time Cold -> 15° Dead space Add on floor Poor materials wood (punk board), metal girders (buckle and break under fire) Common attic in whole building	Not keeping track of people (could be looking for people who are not there)
Initial attack	Watch for signs of building collapse If signs of building collapse, evacuate and throw water on it from outside	Faulty construction, building may collapse	Signs of building collapse include: What walls are doing: cracking What floors are doing: groaning What metal girders are doing: clicking, popping Cable in old buildings hold walls together	Ventilating the attic, this draws the fire up and spreads it through the pipes and electrical system

5.4 Cognitive Demand Technique

Upon completing interviews with multiple Subject Matter Experts (SMEs), the final phase of the Applied Cognitive Task Analysis (ACTA) process culminated in the development of the Cognitive Demands Table. This table serves as a structured framework that translates the insights gained from these experts into evidence-based implications. It highlights essential, often implicit, aspects of expertise, presenting them as key cues, strategies, and decision-making processes. By organizing this critical information, the Cognitive Demands Table emphasizes the most relevant findings related to problem-solving and decision-making, allowing for a deeper understanding of expert cognitive processes and guiding future practice.

All the data obtained were integrated, consolidated and synthesized from the data findings of the previous techniques: task diagram, knowledge audit and simulation, to sort through and analyse the data. The data will be presented in a cognitive demand table intended to provide a format for the practitioner to use in focusing the analysis on project goals. By systematically organizing and categorizing the cognitive demands linked to various learning tasks and activities, the table offers a thorough overview of the cognitive requirements involved in the entire learning process. The cognitive demand table provides a comprehensive view of the cognitive task difficulties and expertise experienced by experts.

Cognitive demand table contains difficult cognitive elements, with "Why Difficult", "Common Errors" and "Cues and Strategies" used as the headings. It provides a format to analyze the types of information needed to design a new system. A sample of cognitive demand table for a development of e-learning storyboard (Mohd Yusoff & Salim, 2012) is shown in Table 6 below.

Table 6. Cognitive demand table for a development of e-learning storyboard

Difficult cognitive elements	Why difficult?	Common errors	Cues and strategies used
Storyboard templates *VO scripts* *Animation and graphics* *Storyboard review process*	Difficult to translate content materials to VO scripts. Difficult to handle the storyboard templates. Difficult to differentiate and prioritize graphic elements with relation to the VO scripts. Difficult to reach SMEs. Some SMEs are resident in other institutions. IDs are not SMEs. Thus, some of them do not understand the subject content of the storyboard. It is difficult to determine the number of entries for graphics and VO scripts in the screen areas, due to the style of storyboard templates.	Ignore the guidelines in the storyboard. Not well-informed of the procedure. Not enough skills in VO scripts, graphics and animation SMEs are not trained as instructional designers Some SMEs do not have skills in technologies which makes it difficult for them to create storyboard contents Some SMEs do not attend the training provided for storyboard designers which makes it difficult for them to comprehend the task of storyboarding development. Unable to identify and differentiate tasks of storyboard designers and subject content providers. Delay review process.	*Cues:* SMEs do not have knowledge in handling voice scripts. SMEs have problems with storyboard templates. The storyboard templates show the SCORM no., which makes it difficult to proceed with longer VO scripts. SMEs also have difficulties in handing graphics and its relation to VO scripts. SMEs tend to avoid doing amendments after the review process. SMEs do not understand their roles as storyboard designers, and left all tasks to the ID. SMEs assume ID will guide and refine the storyboard. The storyboard reviews need to be reviewed by many members of the team, which includes the instructional designers, SMEs, multimedia designers, and graphics designers. *Strategies:* Most of them will convert all materials from textbooks to the storyboard. SMEs tend to copy a lot of graphics to the storyboard. IDs need to communicate very often with the SMEs to get a clear view of the contents. ID need to discuss every SCO with SMEs

5.5 Triangulation Methods

Triangulation refers to the use of multiple methods or data sources to gain a well-rounded and in-depth understanding of a particular phenomenon. This approach enhances the reliability and richness of the research by drawing from different perspectives, techniques, or datasets. By comparing and cross-referencing information from these varied sources, triangulation acts as a robust strategy for verifying the validity of the data, ensuring that the findings are more accurate, credible, and comprehensive.

Triangulation method requires data to be collected and analysed separately and the results were compared during the interpretation of data. Abdalla, Oliveira, Azevedo, & Gonzalez (2018) stated in their report, triangulation enhance the validity of assessments when researchers use manifold methodologies, different ways of data collection, different methods of data analysis or even involves many participations, with many different points of view. The effect rests on the proposition that the weaknesses in each single data source, method, evaluator, theory, or data type will be counterbalanced by the strength of another (Reeves, 2020). By leveraging multiple perspectives and methods, triangulation provides a more comprehensive and balanced understanding of the phenomenon under study, ultimately increasing the reliability and credibility of the research findings. This approach ensures that any biases or limitations inherent in a single method are mitigated, leading to more robust and well-rounded conclusions.

Most research employing various data collection methods and participants, such as interviews and surveys in each phase of the research. As such, to gain deeper understanding and to apprehend the cognitive skills require, triangulation method is used to assist in enhancing the reliability and internal validity of the findings. Combining different and multiple methods in a research study can avoid fundamental biases that may arise from using a single method in a study. In addition, triangulation can improve research by providing a variety of data sets to accurately capture and describe different elements of a phenomenon.

CONCLUSION

In conclusion, Chapter 5 has demonstrated the critical importance of a structured and multi-layered approach to data analysis, particularly when applied to the techniques used in the Applied Cognitive Task Analysis (ACTA). Each technique—task diagram, knowledge audit, simulation, cognitive demand table, and triangulation—has been explored in detail, highlighting the value of systematically processing data to uncover the complexities involved in task performance. By adhering to established academic principles, as outlined by Saunders, Lewis, & Thornhill (2009), the data were carefully categorized, connections were identified, and hypotheses were formulated, leading to well-supported conclusions.

The task diagram technique provided a clear breakdown of tasks and cognitive demands, while the knowledge audit offered deeper insights into the experts' experiences and the cues and strategies they relied on to manage difficult tasks. The simulation technique allowed for real-time interaction with training systems, giving experts the opportunity to share valuable feedback, and the cognitive demand table consolidated the insights gained across techniques into actionable strategies. The use of triangulation further enhanced the reliability and validity of the data, ensuring that a wide range of perspectives and methods were integrated into the analysis.

Overall, this chapter has underscored the rigorous processes involved in data analysis, providing a framework for interpreting the findings in a way that supports the development of more effective systems and instructional designs. By systematically addressing cognitive challenges and task performance difficulties, the analysis methods detailed in this chapter contribute significantly to enhancing both learning and system functionality in professional development and microlearning environments.

REFERENCES

Abdalla, M. M., Oliveira, L. G. L., Azevedo, C. E. F., & Gonzalez, R. K. (2018). Quality in Qualitative Organizational Research: Types of triangulation as a methodological alternative. *Administração: Ensino e Pesquisa*, 19(1), 66–98. DOI: 10.13058/raep.2018.v19n1.578

Militello, L. G., & Hutton, R. J. B. (1998). Applied cognitive task analysis (ACTA): A practitioner's toolkit for understanding cognitive task demands. *Ergonomics*, 41(11), 1618–1641.

Militello, L. G., Hutton, R. J. B., Pliske, R. M., Knight, B. J., Klein, G., & Randel, J. (1997). *Applied Cognitive Task Analysis (ACTA) methodology.*

Nauert, E., & Gillan, D. (2019). The Cognitive Process of Wildland Fire Chainsaw Troubleshooting: Structure, Content, and Training. *Proceedings of the Human Factors and Ergonomics Society Annual Meeting*, 63(1), 1676–1680. DOI: 10.1177/1071181319631536

Nor'ain, M. Y., & Salim, S. S. (2012). Investigating cognitive task difficulties and expert skills in e-learning storyboards using a cognitive task analysis technique. *Computers & Education*, 58, 652–665. DOI: 10.1016/j.compedu.2011.09.009

Reeves, T. C. (2020). Design research from a technology perspective. In *Educational Design Research* (pp. 64–78). DOI: 10.4324/9780203088364-13

Saunders, M., Lewis, P., & Thornhill, A. (2009). Research methods for business students. *Business* (Fitfh, Vol. 5th). Pearson Education. DOI: 10.1017/CBO9781107415324.004

Section 3
ACTA in Educational Settings

Chapter 6
Introduction of ACTA in Educational Settings

ABSTRACT

This chapter explores the application of Applied Cognitive Task Analysis (ACTA) within educational settings, focusing on two distinct case studies that demonstrate its versatility and effectiveness. The first case study examines Electronic Training for Professional Development, highlighting how ACTA can enhance the design and delivery of digital training programs, ensuring they meet the evolving needs of professionals in various fields. The second case study delves into Developing Pedagogical Practices of Microlearning Educational Experience in Higher Education. It showcases how ACTA can inform the development of innovative microlearning strategies, leading to more personalized, engaging, and effective learning experiences for students in higher education. By analyzing these case studies, the chapter underscores the potential of ACTA to revolutionize educational practices, providing educators and trainers with robust tools to optimize learning outcomes and foster continuous improvement in educational environments.

INTRODUCTION

As the educational landscape continues to evolve, the need for more effective and targeted instructional strategies has become paramount. In this context, Applied Cognitive Task Analysis (ACTA) provides a structured method to identify and address the cognitive demands placed on learners and educators in various educational environments. By breaking down complex tasks and analysing the cognitive skills

DOI: 10.4018/979-8-3693-7733-8.ch006

required to perform them, ACTA offers valuable insights into how educational programs can be designed to better support learning and professional development.

In this chapter, we explore the introduction of ACTA into two critical educational settings: electronic training systems for professional development and microlearning pedagogical practices in higher education.

Section 6.1 focuses on the application of ACTA in the design and implementation of an electronic training system aimed at enhancing professional development. As professionals increasingly turn to online platforms for continuous learning, it is crucial to understand the cognitive challenges they face when interacting with these systems. ACTA helps identify these challenges and provides strategies to design training systems that are more intuitive, effective, and aligned with the learners' professional needs.

Section 6.2 shifts the focus to the use of ACTA in developing microlearning pedagogical practices within the context of higher education. Microlearning, which delivers content in concise, manageable units, has gained significant popularity for its flexibility and adaptability to modern learners. ACTA plays a vital role in ensuring that these microlearning experiences are designed to meet the cognitive demands of students, promoting engagement, retention, and the practical application of knowledge.

This chapter introduces the key ways in which ACTA can be utilized to optimize educational strategies in both professional development and higher education settings. By applying ACTA to these contexts, educators and instructional designers can create more effective, learner-centred environments that foster meaningful learning outcomes.

6.1 Electronic Training System for Professional Development

In order to boost the economy of any country, improving the quality of human resources that universities produce is vital. Thus, organizing and administering professional development activities among the educators should become a major agenda. However, in the rapid growth of information and communication technology, the deliverables of electronic training system were obstructed by problems that are related to the training needs and design, which are influenced by culture, religious and technology resistance. This radical change of technology in training has enable educators to participate in online learning process without having to attend a conventional facility (Ismail, Zaharudin, Hashim, & Ariffin, 2020). Electronic training system provides opportunities for continuous learning for educators

(Shahzad, Hassan, Aremu, Hussain, & Lodhi, 2021) as the system is available over the web at any time and anywhere the users has access to internet.

The design and development of an electronic training system prototype known as TaSPoD (e-Training System for Professional Development) is significant in establishing the professional development of higher learning institutions, where TaSPoD assist the concern users: institution, experts and trainees to integrate the electronic training into the professional development plan. Moreover, educators comprehend the use of such technologies to collaboratively communicate with peers in a learning community as most adults entwined their daily activities with computerized innovation (Tawafak, Malik, & Alfarsi, 2021).

The introduction section emphasizes the importance of higher learning institutions in shaping economic and social growth by producing high-quality employees. With the integration of technology into daily activities, there is a need for continuous professional development for educators. The TaSPoD system is proposed to meet this need by providing personalized learning experiences. The objectives of the system are categorized into stages: Analysis, Design, Development & Evaluation. These include identifying training needs, analysing existing systems, designing the TaSPoD system, developing prototypes, and evaluating cognitive elements through expert feedback.

In response to the digital age and technological advancements, the study adopts a Design-Based Research (DBR) approach. DBR is chosen for its systematic study of design, development, and evaluation processes to create effective educational products. The goal is to integrate technology into education through quality professional development programs.

Table 1 presents the research design of this study, offering a detailed overview of the study's objectives, research questions, data collection methods, instruments, and expected result. The table acts as a guide for understanding the structure and execution of the research. Each section specifies the goals and inquiries, the methods used to gather relevant data, the tools for data collection, the target populations or samples, and the projected results. By outlining these elements, Table 1 provides a well-organized framework that supports the systematic design and development of TaSPoD, intended for use by educators in higher learning institutions.

Table 1. Research design

Research Objective	Research Question	Data Collection Technique	Instrument	Expected Result
1. Analysis i. To identify training needs for designing electronic training system ii. To analyse problems with regard to the design of the existing electronic training system	1. Analysis i. What are the training needs for end-users (trainees and trainers) in electronic training system? ii. What are the issues in the design of the existing electronic training system?	Interview Content analysis	Semi-structured Interview Protocol Criteria for Evaluating the Quality of Online Courses (Clayton R. Wright)	End-users requirements for designing an electronic training system. Design refinement of the existing electronic training system.
2. Design i. To design TaSPoD system according to the design refinement ii. To validate the design of TaSPoD system according to experts' and end-users' recommendation.	2. Design: i. What is the design of TaSPoD system according to the design refinement? ii. What is the most suitable e-training design recommended by the experts and end-users to be implemented in the TaSPoD system?	Storyboard narration ACTA Technique (3-5 participants)	Narration i. Task Diagram ii. Knowledge Audit iii. Simulation Interview iv. Cognitive Demand Table	Screen design TaSPoD Design Recommendation
3. Develop & Evaluation i. To develop TaSPoD system prototype based on design recommendation ii. To evaluate cognitive element of task of TaSPoD according to experts' and end-user's recommendation	3. Develop & Evaluation i. What is the TaSPoD prototype evaluation according to the experts and end-users?	System narration ACTA Technique (3-5 participants)	Narration i. Task Diagram ii. Knowledge Audit iii. Simulation Interview iv. Cognitive Demand Table	TaSPoD prototype Best attributes recommendation for TaSPoD system
4. Innovation To produce TaSPoD system prototype based on 'empirical-based design principle'				

DBR approach integrates the identifiable of design principles with development of solution for TaSPoD, with the purpose to explore the needs of e-training environments, designing, developing, and evaluating the TaSPoD. First phase of analysis articulating the problem arises in training by analysing of the users' training needs. Design phase focused on designing effective training modules and designing the framework and architecture for the TaSPoD. Next, the instructional and system designers investigate and enquire more information and knowledge to improve the TaSPoD design using an applied cognitive task analysis (ACTA) approach. Upon completion of the validation of cognitive tasks, TaSPoD prototype will be developed, followed by evaluation stage where experts investigate and refine to determine its effectiveness.

DBR involves continuous collaboration between researchers and practitioners to address complex problems in real educational contexts. It aims to generate and test design principles iteratively to refine learning environments. The TaSPoD design framework follows six iterative phases: Analysis, Design, Validate, Develop, Evaluate, and Innovate. Each phase involves specific tasks like needs analysis, theoretical embedding, empirical testing, and iterative refinement based on user feedback and cognitive task analysis.

The research is significant for higher learning institutions as it addresses barriers to professional development through innovative e-training solutions. It contributes to the field by establishing a framework for integrating technology into education effectively. The section details the comprehensive analysis, design, and evaluation phases undertaken for the TaSPoD prototype, aimed at addressing training needs and improving electronic training environments.

Analysis Phase: Unveiling the Training Needs

The analysis phase involved interviews with various participant groups including managers, instructional designers, system developers, trainees, and trainers. The interviews focused on identifying training needs and problems in electronic training environments, as well as issues with existing electronic training systems. It is important to emphasize on the design and development of course structure, page design and content engagement as they are important to build a sustainable electronic training system (Saintika, Astiti, Kusuma, & Muhammad, 2021). Key findings of training needs and problems that were discussed include lack of innovation, pedagogical skills and feedback mechanisms. Directed content analysis was employed to evaluate an existing e-training system based on specific criteria. This method aimed to assess the system's current effectiveness in delivering professional development activities within a university context.

Design and Development Phase: Designing the TaSPoD and Constructing the Prototype

Following the analysis phase, the TaSPoD prototype was developed to address identified training needs and system design issues. Applied Cognitive Task Analysis (ACTA) techniques were utilized to refine the prototype, where ACTA consists of four techniques: Task Diagrams, Knowledge Audit Tables, Simulation Interviews and Cognitive Demand Tables. Post-development, the TaSPoD prototype underwent evaluation to assess its effectiveness and usability. Similar ACTA techniques were employed to evaluate cognitive demands and user experiences.

Overall, the analysis, design, and evaluation phases of the TaSPoD prototype provided a structured approach to addressing training needs and enhancing electronic training systems. Key findings highlighted the importance of innovative teaching methods, effective feedback mechanisms, and user-friendly interfaces. The application of ACTA techniques throughout the design and evaluation phases ensured a thorough understanding of cognitive demands and user experiences, leading to recommendations for improving the prototype's functionality and usability.

Evaluation Phase: Decoding Expert Perspectives

Among key design issues identified in the existing electronic training system are trainees lack adequate support in terms of general information needed to complete the e-training sessions. This includes a lack of clear descriptions of courses, goals, learning objectives, timelines, guidelines for online discussions, and group work activities. While technologies used align with some goals and objectives, there is a lack of variety in technology use to achieve these goals effectively. The system minimally supports instructional strategies and lacks variety in learning activities. There is also insufficient interactivity between trainees and trainers, which could be enhanced through activities like online discussions, conferencing, and collaborative assignments. While learning resources are generally appropriate, there is minimal use of multimedia components to cater to different learning styles and interests. Consistency of content across the organization is also essential for successful implementation as it may increase user engagement (Rebelo & Isaías, 2020). The course content alone does not fully meet the learning needs of participants. There is also a need to ensure instructional materials are inclusive and free from bias related to age, culture, ethnicity, gender, etc. The interface and organization of the system are not very effective. Presentations and illustrations are not sufficiently engaging. Improvements could include consistent navigational icons, better use of colours and images, and a clearer structure throughout the course. While evaluation activities are aligned with goals and objectives, there is a lack of participation in online activities, evaluative exercises, and feedback from instructors. Enhancing evaluation processes could involve increasing the frequency and depth of evaluations, providing clear criteria for assessment, and ensuring feedback mechanisms are robust.

Overall, addressing these design issues could significantly enhance the effectiveness and engagement of the electronic training system, making it more aligned with the needs of both trainers and trainees in professional development contexts. By integrating insights from stakeholders and employing rigorous analysis methods, the TaSPoD prototype aims to offer a robust platform for professional development training in electronic environments, catering to the diverse needs of educators and learners alike.

The research introduces TaSPoD aimed at enhancing educators' skills and knowledge through workshops and training. It follows a framework comprising needs analysis, design, development, enhancement, and innovation stages. TaSPoD addresses cultural and technological barriers in education, especially in times like the COVID-19 pandemic, emphasizing a structured approach to professional development. It seeks to bridge cultural gaps in higher education by incorporating diverse instructional methods and learning styles.

Eventually, TaSPoD is a platform to bridge the culture and religious gap, as well as technological resistance gap, that enable educators regardless of gender, to maximize their potential through professional development activities.

6.2. Review of Electronic Training System for Professional Development (TaSPoD)

This research aimed to establish the professional development and emphasizes its importance in the higher learning institutions by systematically analysing the training needs, analysing existing systems, designing effective TaSPoD interface screens, developing TaSPoD prototype, and evaluating the cognitive elements. Through the use of the Applied Cognitive Task Analysis (ACTA) technique, this study offered an in-depth exploration of the tasks and cognitive processes associated with teaching and learning.

Through comprehensive analysis, the study identified the training needs, as well as issues with the design of existing electronic training systems. The ACTA method played a crucial role in identifying these needs and understanding the cognitive demands placed on educators. By deconstructing complex tasks into their core components, ACTA provided a detailed perspective on the teaching and learning processes, emphasizing the cognitive aspects embedded within the TaSPoD prototype.

By consulting experts in instructional design, system developers, trainers and trainees the research constructed a robust electronic training system to be used as professional development activities for educators in higher learning institutions.

ACTA played a critical role in the design and development phases, ensuring that the design of the electronic training system was grounded in a deep understanding of cognitive tasks and learning processes for enhancing electronic training systems.

The evaluation phase utilized ACTA to gather expert feedback on the TaSPoD prototype's usability, and practicality. By simulating the TaSPoD user interface designs as the scenarios and analysing cognitive demands, ACTA provided a thorough assessment of how well the TaSPoD could be integrated into actual professional development settings.

To conclude, this research offers important insights and practical recommendations for educators and institutions. The application of ACTA across the design, development, and evaluation stages ensured a thorough, cognitively informed approach to the electronic training system. This approach facilitates the improvement of educators' skills and knowledge through workshops and training, enabling them to meet the evolving demands of 21st-century learners and ultimately enhancing personalized learning experiences for educators in higher education.

6.3 Developing Pedagogical Practices of Microlearning Educational Experience in Higher Education

The rapid evolution of information and communication technologies (ICT) has fundamentally transformed the creation, access, dissemination, and sharing of knowledge and skills. In today's educational environment, there is a strong emphasis on the efficiency of knowledge delivery and transfer. However, as we progress further into the digital age, it is essential to rethink how learners acquire and generate their own knowledge and essential skills for the 21st century (Collins & Haverson, 2018). This shift calls for the development of learning environments that support lifelong learning, with a particular focus on informal learning settings where self-directed activities play a critical role. The rise of intelligent devices and advancements in web technologies have profoundly changed how learning and interaction occur (Haleem, 2022). Technology has become a central element in modern education, providing educators with the opportunity to integrate diverse and innovative teaching methods. The Internet and the World Wide Web now enable a student-centred learning approach, where educators facilitate rather than merely transmit information. Among the emerging trends is microlearning, which involves short, targeted learning activities delivered through visual, auditory, and video formats (Sankaranarayanan et al., 2023). This approach aligns well with contemporary learners' needs for flexible, on-the-go learning solutions and represents a significant shift towards more dynamic and accessible educational practices.

As the world shifts toward a more global perspective, educators need to be flexible, innovative, and adept at meeting the evolving educational needs of their students (Sharif & Cho, 2015). A notable shift in educational paradigms is the adoption of microlearning (Zhao et al., 2023), which involves delivering small, concentrated learning units designed to support efficient knowledge acquisition. Microlearning has been recognized for decades as a valuable educational approach (Gabrielli et al., 2006; Hug, 2005; Major & Calandrino, 2018, cited in Moore, 2020). According to Major and Calandrino (2018), microlearning in an educational context combines specific learning objectives with small, practical learning units. Bleich (2018) and Callisen (2019) characterize microlearning as fragmented content paired with adaptable technologies. Aitchanov et al. (2012) and Giurgiu (2017) emphasize that microlearning is effective for deconstructing complex information into more digestible parts for learners.

Although opinions on the exact purpose of microlearning vary, researchers generally agree on its fundamental components: focusing on a single topic, breaking it into smaller segments, fostering personal growth, providing on-demand access, and utilizing web or mobile platforms for real-time delivery (Hug, 2005; Park & Kim, 2018; Trowbridge et al., 2017; Moore, 2020). This approach not only enhances learning efficiency but also caters to the needs of modern learners, who benefit from flexible, on-the-go educational solutions.

Microlearning, also known as bite-sized learning, involves delivering brief and targeted learning materials and activities that collectively form mini-courses. This method employs small, well-organized segments of content, allowing learners to access information at any time, from any location, and in their preferred media format (Mohammed et al., 2018). It is designed to align with the human brain's natural attention span and to mitigate cognitive overload. The increasing demand for microlearning is driven by two key factors, as identified by So, Lee, and Roh (2020). The first factor is the shift from "just-in-case" to "just-in-time" learning, which reflects the growing need to acquire knowledge efficiently in response to an expanding range of information. Microlearning offers the advantage of flexible, on-the-go access via computers and mobile devices, thus alleviating temporal and spatial constraints for learners. The second factor pertains to the distinct characteristics of 21st-century learners, particularly Generation Z. To create effective modern learning strategies, it is essential to understand their evolving needs and preferences. Generation Z is noted for its social and mobile nature, making the integration of microlearning a key component for successful educational experiences. Research by Choudhary and Pandita (2023) and Van Wyk (2023) supports the view that microlearning is particularly effective for Generation Z, enhancing their ability to recall and retain information, engage with content, and stay motivated. Despite its benefits, Van Wyk (2023) highlights that educators and academic support staff are often not fully equipped to meet the needs of this demographic. Consequently, educational institutions must prioritize providing adequate training and support to effectively implement microlearning and address the specific requirements of Generation Z learners. By doing so, they can better cater to the demands of modern education and improve learning outcomes for this generation.

The transition to microlearning is primarily driven by the need to combat the decreasing ability of individuals to maintain focus and resist distractions (Diaz Redondo et al., 2020). Studies reveal a noticeable decline in student attention spans, reflective thinking, concentration, and comprehension skills (Hari, 2022; Kim & Maloney, 2020; Lang, 2020; Mark, 2023). Strom et al. (2023) found that many young people struggle with managing their attention and focus. Microlearning, which emphasizes brief, focused learning sessions, helps students remain engaged and attentive, leading to more profound learning experiences and better retention of knowledge (Shail, 2019). Furthermore, traditional lecture-based teaching has been found to be less effective in maintaining student engagement and enhancing performance (Kozanitis & Nenciovici, 2023; Ettarh et al., 2018). As a result, educators are increasingly in need of innovative teaching strategies that keep students motivated and actively involved in their learning. Microlearning, characterized by its small, targeted content and practical, interactive activities, has been shown to significantly boost student engagement, satisfaction, and academic performance (Kossen & Ooi, 2021). The growing recognition of these benefits underscores the urgent need to adapt educational pedagogy to incorporate microlearning. This shift aims to better address the evolving demands and learning preferences of today's generation, ultimately fostering a more effective and engaging educational environment.

Research has demonstrated that microlearning can significantly boost students' confidence, interest, self-efficacy, learning abilities, and overall academic performance (Fitzgerald & Tisdell, 2019). As a supplementary educational approach, microlearning enhances the learning experience and supports the achievement of personal learning objectives. Despite its advantages, current implementations of microlearning face several practical and pedagogical challenges. These challenges include the absence of clear guidelines for effectively conducting microlearning activities and the need for a more cohesive integration of pedagogical principles with technological tools (Wang et al., 2021; Alqurashi, 2017).

Although microlearning holds considerable promise, its widespread adoption is hindered by the lack of well-defined fundamental elements and characteristics (Shabadurai et al., 2022). To overcome these barriers, it is essential to develop comprehensive frameworks and guidelines that clarify how to best implement microlearning strategies. This development should also focus on integrating these strategies seamlessly with existing pedagogical approaches and technology. By addressing these issues, educators can better harness the potential of microlearning, making it a more effective and widely adopted educational tool.

In the ever-evolving landscape of higher education, the quest for innovative and effective pedagogical practices is unceasing. This research study embarks on a journey to unlock the potential of microlearning, seeking to amplify students' learning experiences in higher education. Microlearning, renowned for its ability to facilitate focused and rapid learning, holds immense promise for educational contexts. However, its application in higher education often encounters challenges stemming from a lack of robust pedagogical foundations.

Recognizing this gap, this study is structured into four distinct phases, each meticulously designed to address specific research questions and objectives. The initial phase delves into an in-depth analysis of current teaching and learning practices in higher education. This involves identifying and scrutinising academic challenges faced by students, setting the stage for a targeted intervention. The subsequent design phase focuses on determining fundamental principles for microlearning approaches. These principles aim to foster the effective creation and delivery of microlearning experiences within the dynamic setting of a classroom.

Moving forward to the development phase, this study centres on crafting pedagogical practices that harness the potential of microlearning in higher education, aiming to transform students' learning experiences. Finally, the evaluation phase rigorously assesses the usability, practicality, and feasibility of the microlearning framework through the lens of experts in the field.

At the heart of this research is the application of Applied Cognitive Task Analysis (ACTA) throughout the entire research process – from analysis to design, development, and evaluation. The culmination of this endeavour is the creation of the AMPLIFY framework, an acronym representing Advancing Microlearning Practices for Enhanced Learning Experiences. This framework encapsulates eight key elements that promise to reshape the landscape of microlearning in higher education.

The AMPLIFY framework addresses essential components such as clear and concise learning objectives, personalised learner-centric activities, robust feedback and assessment mechanisms, engaging content delivery, seamless incorporation of microlearning delivery methods, infusion of interactive learning experiences, integration of social interaction, and alignment with the curriculum.

This study not only offers educators practical guidance on implementing tailored microlearning strategies but also provides a systematic approach to bridge the existing gap in the literature. As we present the AMPLIFY framework, we aim to contribute valuable recommendations for educators and institutions, paving the way for further exploration and advancement in the field of microlearning and its profound impact on higher education.

Table 2 illustrates the research design of this study, providing a comprehensive overview of the research objectives, questions, data collection techniques, instruments, sampling methods, and expected results. This table serves as a roadmap for understanding how the study was structured and conducted. Each section outlines specific goals and inquiries, the methodologies employed to gather relevant data, the tools used for data collection, the target populations or samples, and the anticipated outcomes. By detailing these components, Table 2 offers a clear and organized framework that underpins the study's systematic approach to developing AMPLIFY microlearning framework in higher education.

Table 2. Research design

Research Objectives	Research Questions	Data Collection Technique	Instrument	Sampling	Expected Result
1. Analysis 1.1 To explore current practices in teaching and learning within higher education. 1.2 To identify the academic challenges faced by higher education students	1. Analysis 1.1 What are the current practices of teaching and learning in higher education? 1.2 What are the academic problems faced by higher education students?	Applied Cognitive Task Analysis (ACTA) Technique	Task Diagram	Higher education students	This research investigated the current practices in teaching and learning within higher education, while also identifying and analysing the academic challenges faced by higher education students.
2. Design Examine the fundamental elements in the design of microlearning in higher education for creating and delivering effective microlearning experiences in a classroom setting according to experts.	2. Design What are the fundamental elements in the design of microlearning in higher education that contribute to the effective creation and delivery of microlearning experiences in a classroom setting according to experts?	Applied Cognitive Task Analysis (ACTA) Technique	Task Diagram	Experts of higher learning education from the following areas of expertise: - Microlearning - Curriculum and Instruction - Instructional Technology	This research constructed a framework for the crucial design principles for microlearning approach to be incorporated in instructional framework in higher education learning. This outcome involved the design of fundamental elements for microlearning pedagogical practices.
3. Develop To develop effective pedagogical practices for microlearning in higher education to enhance students' learning experiences according to experts.	3. Develop How can pedagogical practices of microlearning in higher education be best developed to enhance students' learning experiences according to experts?	Applied Cognitive Task Analysis (ACTA) Technique	Knowledge Audit	Experts of higher learning education from the following areas of expertise: -Microlearning - Curriculum and Instruction - Instructional Technology	This research developed a set of microlearning pedagogical practices named AMPLIFY (Advancing Microlearning Practices for Enhanced Learning Experiences) that enhance students' learning experiences.

continued on following page

Table 2. Continued

Research Objectives	Research Questions	Data Collection Technique	Instrument	Sampling	Expected Result
4. Evaluation Evaluate the usability, practicality and feasibility of the microlearning framework's design principles in enhancing students' learning experience in higher education based on expert perspectives.	3. Evaluation How do experts perceive the usability, practicality, and feasibility of the microlearning framework's design principles for integration into higher education?	Applied Cognitive Task Analysis (ACTA) Technique	Simulation Interview Cognitive Demand Table	Experts of higher learning education from the following areas of expertise: -Microlearning - Curriculum and Instruction - Instructional Technology	This research obtained reviews and feedback from higher learning experts on the developed AMPLIFY instructional framework to assess its usability, practicality, and feasibility.

Analysis Phase: Unveiling the Educational Complexities

This research aimed to explore the current practices in teaching and learning within higher education and to identify the academic challenges encountered by students. By employing the Applied Cognitive Task Analysis (ACTA) technique, the study utilized task diagrams to capture detailed insights from higher education students regarding their learning experiences and the difficulties they face. The investigation into current teaching and learning practices revealed a complex landscape of methodologies and approaches utilized in higher education.

The use of the ACTA technique provided a nuanced understanding of how these practices are implemented and perceived by students. The research also aimed to identify and analyse the academic challenges faced by students in higher education. The ACTA technique allowed for a detailed examination of these challenges by capturing students' cognitive processes and difficulties through task diagrams. Overall, this research contributes to a deeper understanding of the current state of teaching and learning in higher education and highlights areas for improvement that can enhance the academic experience and success of students.

Design Phase: Constructing Microlearning Foundations

The focus of this research was to examine the fundamental elements involved in the design of microlearning for higher education, with the goal of creating and delivering effective microlearning experiences in classroom settings. The research specifically aimed to identify these core design principles by consulting experts in the fields of microlearning, curriculum and instruction, and instructional technology.

To achieve this, the study utilized the Applied Cognitive Task Analysis (ACTA) technique along with task diagrams. This approach facilitated a comprehensive analysis of how microlearning can be effectively structured and implemented within higher education contexts. The outcome of the research was the development of a framework detailing the crucial design principles necessary for integrating microlearning into instructional practices. This framework provides a structured guide for educators seeking to incorporate microlearning strategies into their teaching methodologies, ensuring that the design and delivery of microlearning experiences are both effective and aligned with educational goals.

Development Phase: Crafting the AMPLIFY Framework

The objective of this research was to develop effective pedagogical practices for microlearning in higher education to enhance students' learning experiences. To achieve this, the study focused on identifying the best strategies and practices for implementing microlearning effectively, as determined by experts in microlearning, curriculum and instruction, and instructional technology.

Using the Applied Cognitive Task Analysis (ACTA) technique namely Knowledge Audit, the research aimed to explore and refine pedagogical approaches that optimize microlearning. This methodological approach provided a thorough understanding of how microlearning can be leveraged to improve educational outcomes.

The outcome of this research was the development of a set of pedagogical practices under the framework named AMPLIFY (Advancing Microlearning Practices for Enhanced Learning Experiences). The AMPLIFY framework outlines a comprehensive approach to integrating microlearning into higher education, focusing on strategies that enhance student engagement, retention, and overall learning experiences. By aligning these practices with contemporary educational needs, the AMPLIFY framework provides a structured approach to effectively utilizing microlearning in classroom settings, aiming to create more impactful and responsive learning environments.

Evaluation Phase: Decoding Expert Perspectives

The goal of this research was to evaluate the usability, practicality, and feasibility of the microlearning framework's design principles in enhancing students' learning experiences in higher education, based on expert perspectives. Specifically, it sought to understand how experts perceive these aspects of the AMPLIFY (Advancing Microlearning Practices for Enhanced Learning Experiences) instructional framework.

To gather comprehensive insights, the study employed the Applied Cognitive Task Analysis (ACTA) technique alongside simulation interviews and the Cognitive Demand Table. This multi-faceted approach allowed for a detailed examination of the framework's effectiveness from various expert viewpoints.

Experts from the fields of microlearning, curriculum and instruction, and instructional technology were consulted. Their feedback was instrumental in assessing the AMPLIFY framework's potential for integration into higher education. Through these evaluations, the research aimed to refine the framework to ensure it meets the practical needs of educators and effectively enhances student learning experiences.

The findings from this evaluation phase provided critical reviews and feedback from higher education experts, highlighting the strengths and areas for improvement within the AMPLIFY framework. This iterative process of review and refinement aimed to ensure that the framework is both theoretically sound and practically viable for widespread adoption in higher education settings.

Review of Developing Pedagogical Practices of Microlearning Educational Experience in Higher Education

This research aimed to advance the implementation of microlearning in higher education by systematically exploring current practices, designing effective frameworks, developing pedagogical strategies, and evaluating their impact. Using the Applied Cognitive Task Analysis (ACTA) technique, the study provided a detailed examination of the tasks and cognitive processes involved in teaching and learning.

Through comprehensive analysis, the study identified key academic challenges and opportunities for enhancing learning experiences. ACTA was instrumental in mapping out these challenges and understanding the cognitive demands placed on students and educators. By breaking down complex tasks into their fundamental components, ACTA allowed for a nuanced view of the learning and teaching processes, highlighting areas where microlearning could be most effectively implemented.

By consulting experts in microlearning, curriculum, and instructional technology, the research constructed a robust framework for integrating microlearning into higher education. The development phase resulted in the AMPLIFY (Advancing Microlearning Practices for Enhanced Learning Experiences) framework, designed to improve student engagement and learning outcomes. ACTA played a critical role in the design phase, ensuring that the framework was grounded in a deep understanding of cognitive tasks and learning processes.

The evaluation phase utilized ACTA to gather expert feedback on the framework's usability, practicality, and feasibility. By simulating real-world scenarios and analysing cognitive demands, ACTA provided a thorough assessment of how well the AMPLIFY framework could be integrated into actual educational settings. This feedback loop was essential in refining the framework and ensuring its relevance and applicability.

Overall, this research provides valuable insights and actionable guidelines for educators and institutions. The use of ACTA throughout the analysis, design, development, and evaluation phases ensured a comprehensive and cognitively informed approach to microlearning. This supports the effective integration of microlearning to meet the evolving needs of 21st-century learners, ultimately enhancing the educational experience and outcomes for students in higher education.

SUMMARY

In this chapter, we have explored the introduction and application of Applied Cognitive Task Analysis (ACTA) in two distinct yet interconnected educational settings: electronic training systems for professional development and microlearning pedagogical practices in higher education. The use of ACTA as a tool for understanding and addressing the cognitive demands of learners has proven to be essential for optimizing educational practices across these varied contexts.

In Section 6.1, we discussed how ACTA helps enhance electronic training systems by identifying the cognitive challenges faced by professionals engaging in online learning platforms. By using ACTA to map out task difficulties and required cognitive skills, instructional designers can develop more intuitive and effective training systems that support the learner's journey in a structured and meaningful way. This not only improves professional development outcomes but also contributes to more efficient use of digital learning environments.

Section 6.2 highlighted the role of ACTA in shaping microlearning pedagogical practices within higher education. ACTA's focus on breaking down complex learning tasks into manageable components allows educators to design microlearning experiences that align with the cognitive needs of students. By doing so, educators ensure that students can engage more effectively with content, leading to better retention, understanding, and application of knowledge.

In conclusion, ACTA offers a powerful framework for enhancing both professional development and higher education through thoughtful and informed instructional design. By integrating ACTA into educational settings, instructors and designers are better equipped to create learner-centred environments that cater to the specific cognitive demands of their audiences, resulting in more engaging, effective, and impactful educational experiences.

Linking to Subsequent Chapters

Building on the foundational understanding of ACTA's application in electronic training systems and microlearning pedagogical practices introduced in this chapter, the subsequent chapters will delve deeper into the specific methodologies of ACTA and their practical applications in these educational contexts. These methodologies provide the tools necessary to break down complex cognitive tasks, address learner challenges, and create effective educational interventions.

- **Chapter 7** explores how Task Diagrams are used to map out the cognitive steps and decision points involved in electronic training and microlearning tasks, helping designers visualize and address task complexity.
- **Chapter 8** focuses on Knowledge Audits, detailing how expert insights are gathered to identify the critical knowledge and skills required for effective learning and task performance.

- **Chapter 9** examines the use of Simulations to recreate realistic scenarios for training and learning, providing a dynamic platform for understanding and addressing cognitive challenges.
- **Chapter 10** concludes this analysis with a discussion of Cognitive Demand Tables, showcasing how they are employed to evaluate and optimize cognitive workload in educational tasks.

Together, these chapters provide a cohesive and detailed exploration of ACTA methodologies, demonstrating their transformative role in enhancing educational practices. By bridging theory with actionable strategies, they offer valuable insights into designing more effective and learner-centred educational interventions.

REFERENCES

Aitchanov, B., Nussipbekov, A., and Zhaparov, M. (2012). Microlearning of web fundamentals based on mobile learning. *IJCSI International Journal of Computer Science Issues, 9*, (6:3), 148150.

Alqurashi, E. (2017). Microlearning: A pedagogical approach for technology integration. *The Turkish Online Journal of Educational Technology*, 16, 942–947.

Bleich, C. (2018). [*Types of Microlearning for Your Employees.* EdgePoint Learning. https://www.edgepointlearning.com/blog/types-of-microlearning/]. *Top (Madrid)*, 11.

Callisen, L. (2019). *Why Micro Learning is the Future of Training in the Workplace*. eLearning Industry. https://elearningindustry.com/micro-learning-future-of-trainingworkplace

Choudhary, H., & Pandita, D. (2023). Maximizing learning outcomes in the digital age: The role of microlearning for Gen Z. *Development and Learning in Organizations*, Vol. ahead-of-print No. ahead-of-print. DOI: 10.1108/DLO-02-2023-0038

Collins, A., & Halverson, R. (2018). *Rethinking education in the age of technology: The digital revolution and schooling in America*. Teachers College Press.

Diaz Redondo, R. P., Rodríguez, M., Escobar, J., & Vilas, A. F. (2021). Integrating micro-learning content in traditional e-learning platforms. *Multimedia Tools and Applications*, 80, 3121–3151. DOI: 10.1007/s11042-020-09523-z

Ettarh, R., Al-hussaini, H., Hassan, S., & Kilarkaje, N. (2018). Declining attention span in medical students: Fact or fiction? *The FASEB Journal*, 32(S1). Advance online publication. DOI: 10.1096/fasebj.2018.32.1_supplement.506.6

Fitzgerald, J. H., & Tisdell, C. C. (2019, June). The impact of educational microcontent on the student learning experience. In *Proceedings of the 2019 International Conference on Mathematics, Science and Technology Teaching and Learning* (pp. 17-22). Association for Computing Machinery. DOI: 10.1145/3348400.3348412

Giurgiu, L. (2017). Microlearning an evolving trend. *Science Bulletin*, 22(1), 18–23. DOI: 10.151/bsaft-2017-003

Haleem, A., Javaid, M., Qadri, M. A., & Suman, R. (2022). Understanding the role of digital technologies in education: A review. *Sustainable Operations and Computers*, 3, 275–285. DOI: 10.1016/j.susoc.2022.05.004

Hari, J. (2022). *Stolen Focus: Why You Can't Pay Attention – and How to Think Deeply Again*. Crown.

Hug, T. (2005). Microlearning: A new pedagogical challenge. In Hug, T., Lindner, M., & Bruck, P. A. (Eds.), *Microlearning: Emerging concepts, practices and technologies after e-learning* (pp. 7–12). Innsbruck University Press.

Ismail, A., Zaharudin, R., Hashim, N., & Ariffin, J. (2020). The Impact of e-NPQEL on the Continuance Intention of Using e-Training among Aspired School Leaders in Malaysia. *International Journal of Interactive Mobile Technologies*, 14(19), 109–123. DOI: 10.3991/IJIM.V14I19.15965

Kim, J., & Maloney, E. J. (2020). *Learning Innovation and the Future of Higher Education*. Johns Hopkins University Press.

Kossen, C., & Ooi, C.-Y. (2021). Trialling micro-learning design to increase engagement in online courses. *Asian Association of Open Universities Journal*, 16(3), 299–310. DOI: 10.1108/AAOUJ-09-2021-0107

Kozanitis, A., & Nenciovici, L. (2023). Effect of active learning versus traditional lecturing on the learning achievement of college students in humanities and social sciences: A meta-analysis. *Higher Education*, 86(6), 1377–1394. DOI: 10.1007/s10734-022-00977-8

Lang, J. (2020). *Distracted: Why Students Can't Focus and What You Can Do About It*. Basic Books.

Major, A., & Calandrino, T. (2018). Beyond chunking: Micro-learning secrets for effective online design. *FDLA Journal, 3*(13). https://nsuworks.nova.edu/fdla-journal/vol3/iss1/13

Mark, G. (2023). *Attention Span: A Groundbreaking Way to Restore Balance, Happiness, and Productivity*. Harper Collins.

Mohammed, G. S., Wakil, K., & Nawroly, S. S. (2018). The effectiveness of microlearning to improve students' learning ability. *International Journal of Educational Research Review*, 3(3), 32–38. DOI: 10.24331/ijere.415824

Moore, M. L. (2020). *Qualitative Exploration of Instructional Designers' Use of Microlearning for Formal Workplace Training* [Doctoral dissertation, Capella University]. ProQuest Dissertations Publishing

Park, Y., & Kim, Y. (2018). A design and development of micro-learning content in e-learning system. *International Journal on Advanced Science, Engineering and Information Technology*, 8(1). Advance online publication. DOI: 10.18517/ijaseit.8.1.2698

Rebelo, S., & Isaías, P. (2020). Gamification As An Engagement Tool In E-Learning Websites. *Journal of Information Technology Education*, 19, 834–854. DOI: 10.28945/4653

Saintika, Y., Astiti, S., Kusuma, D. J. A., & Muhammad, A. W. (2021). Analysis of E-learning readiness level of public and private universities in central Java, Indonesia. *Register: Jurnal Ilmiah Teknologi Sistem Informasi*, 7(1), 15–30. DOI: 10.26594/register.v7i1.2042

Sankaranarayanan, R., Leung, J., Abramenka-Lachheb, V., Seo, G., & Lachheb, A. (2023). Microlearning in diverse contexts: A bibliometric analysis. *TechTrends*, 67, 260–276. DOI: 10.1007/s11528-022-00794-x

Shabadurai, Y., Chua, F. F., & Lim, T. Y. (2022). Investigating the employees' perspectives and experiences of microlearning content design for online training. *International Journal of Information and Education Technology (IJIET)*, 12(8), 786–793. DOI: 10.18178/ijiet.2022.12.8.1685

Shahzad, A., Hassan, R., Aremu, A. Y., Hussain, A., & Lodhi, R. N. (2021). Effects of COVID-19 in E-learning on higher education institution students: The group comparison between male and female. *Quality & Quantity*, 55(3), 805–826. DOI: 10.1007/s11135-020-01028-z

Shail, M. (2019). Using micro-learning on mobile applications to increase knowledge retention and work performance: A review of literature. *Cureus*, 11(8), e5307. DOI: 10.7759/cureus.5307

Sharif, A., & Cho, S. (2015). 21st-century instructional designers: Bridging the perceptual gaps between identity, practice, impact and professional development. *RUSC. Universities and Knowledge Society Journal*, 12(3), 72–85. DOI: 10.7238/rusc.v12i3.2176

So, H. J., Lee, H., & Roh, S. Z. (2020) Examining the design of microlearning for Korean adult learners. *Computer-Based Learning in Context, 2*(1), 40-53. DOI: 10.35542/osf.io/dmu4c

Strom, P. S., Strom, R. D., Sindel-Arrington, T., & Rude, R. V. (2023). Student attention and distraction in community college. *Journal of Educational and Developmental Psychology*, 13(1), 41–49. DOI: 10.5539/jedp.v13n1p41

Tawafak, R. M., Malik, S. I., & Alfarsi, G. (2021). Impact of technologies during the COVID-19 pandemic for improving behavioral intention to use e-learning. *International Journal of Information and Communication Technology Education*, 17(3), 137–150. DOI: 10.4018/IJICTE.20210701.oa9

Trowbridge, S., Waterbury, C., & Sudbury, L. (2017). Learning in bursts: Microlearning with social media. *EDUCAUSE Review*, 4.

Van Wyk, B. (2023). New Kids on the Block?: Exploring Technological Preferences of a New Generation. *Journal of Higher Education Theory and Practice*, 23(3), 13–19. DOI: 10.33423/jhetp.v23i3.5837

Wang, T., Towey, D., Ng, R. Y., & Gill, A. S. (2021). Towards post-pandemic transformative teaching and learning: Case studies of microlearning implementations in two post-secondary educational institutions. *SN Computer Science*, 2(4), 271–279. DOI: 10.1007/s42979-021-00663-z

Chapter 7
Task Diagram

ABSTRACT

This chapter discusses the use of task diagrams as a key tool in Applied Cognitive Task Analysis (ACTA) within educational settings, divided into two sections. The first, "Task Diagrams for Electronic Training for Professional Development," explains how task diagrams break down complex tasks into manageable components, optimizing the design of electronic training programs. These diagrams visually represent cognitive processes, helping to address challenges and enhance professional development. The second section, "Task Diagrams for Designing Microlearning Pedagogical Practices," shows how task diagrams aid in developing microlearning strategies in higher education, offering insights into structuring content, personalizing learning, and aligning pedagogy with educational goals. Together, these sections demonstrate how task diagrams streamline the design process and improve educational outcomes.

INTRODUCTION

Task diagrams are a fundamental method in Applied Cognitive Task Analysis (ACTA), used to break down complex tasks into their cognitive components and facilitate a deeper understanding of how individuals perform and acquire skills. By visually mapping out the sequence and structure of tasks, task diagrams allow instructional designers and educators to identify key cognitive processes, challenges, and decision points that learners encounter. This method is particularly valuable in fields that require precise training and development of expertise, as it offers insight into how learners interact with and process information.

In this chapter, we will explore two specific applications of task diagrams within the context of professional development and microlearning design.

Section 7.1 examines the use of task diagrams in electronic training for professional development. Task diagrams are used here to capture the essential cognitive processes involved in professional tasks and translate them into structured learning

DOI: 10.4018/979-8-3693-7733-8.ch007

experiences. By breaking down these tasks, trainers can tailor electronic training programs to better support learners' cognitive needs, ensuring that they develop the necessary skills and expertise more effectively.

Section 7.2 focuses on using task diagrams to design microlearning pedagogical practices. Microlearning, which emphasizes delivering content in short, focused bursts, benefits from the detailed cognitive breakdown provided by task diagrams. This approach helps educators design microlearning modules that align with the specific cognitive steps required for learning, promoting better engagement, retention, and practical application of knowledge.

Throughout this chapter, task diagrams will be demonstrated as a powerful tool for enhancing instructional design through the lens of Applied Cognitive Task Analysis. By visually representing the steps and cognitive demands of learning tasks, educators and instructional designers can create more targeted, effective, and learner-centred educational experiences.

7.1 Task Diagrams for Electronic Training System for Professional Development

In the design and development of the Electronic Training System for Professional Development (TaSPoD), Applied Cognitive Task Analysis (ACTA) was employed in two different phases of the research design framework which are the design phase and evaluation phase. The subsequent sections discussed how task diagram technique was applied in these two different phases.

7.1.1 Task Diagrams for System Design

Task Diagram technique as the first phase of the ACTA session is to elicit the broad overview of the TaSPoD system design into component task steps that require the most cognitive skill. TaSPoD system was designed with four main components: (i) system design and functionality, (ii) course content design, (iii) voice over, and (iv) downloadable and pop-up documents.

This phase is dedicated to addressing Research Question 2.1: How is the design of TaSPoD system according to the design refinement? and Research Question 2.2: What is the most suitable electronic training design recommended by the experts and end-users to be implemented in the TaSPoD system? The task diagram technique serves as a powerful tool in unravelling the system complexities, providing a detailed map of the cognitive processes involved in the design of the system.

Objective:

The primary objective of utilising the task diagram technique is to conduct a comprehensive investigation of the tasks difficulties in electronic training system through the exploration of experts' knowledge. Through task diagram, critical cognitive elements of particular tasks of the electronic training system were extracted.

Methodology:

The interface screen design for TaSPoD were developed based on a design guidelines framework adapted from Mohd Yusoff and Salim (2012). The guidelines include multimedia component, medium communication, well-structured storyboard, embedded training, automated problem-solving. Task diagrams were then created through interviews and discussions with experts. Task diagrams which are visual representations outlining the cognitive steps derived from user interface screen designs. The study captures the evolving nature of cognitive processes, offering a valuable dataset for analysis.

Data Collection:

The data collection process emphasizes capturing the complexities of cognitive activities while interacting with the system through its interface screen designs. Interviews with instructional designers, trainers and trainees were conducted to gain insights into decision-making, problem-solving approaches, and method for information processing.

Insights and Findings:

Experts identified two system components that require the most cognitive skills which are system design and functionality, and course content design. Upon completing the task diagram interview, the task components in the e-training design were classified into two categories; tasks that required most cognitive skill and tasks that required less cognitive skill. The two distinct tasks; most cognitive skill and less cognitive skill are represented with a task diagram. The task diagram for TaSPoD design is shown in Figure 1 below.

Figure 1. Task diagram for TaSPoD design

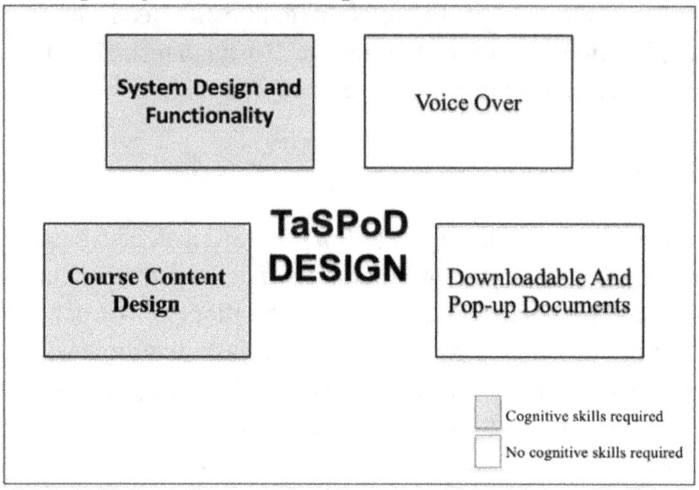

Note: Cognitive task skills differentiated from the non-cognitive task skills

7.1.2 Task Diagrams for System Prototype

First phase of the ACTA session is to elicit the broad overview of the TaSPoD prototype into component task steps that require the most cognitive skill. TaSPoD prototype was constructed with five major components: (i) system functionality, (ii) course content design, (iii) communication medium, (iv) downloadable and pop-up documents and (v) assessment design. This phase is dedicated to addressing Research Question 3: What is the TaSPoD prototype evaluation according to the experts and end-users?

Objective:

The main aim of using the task diagram technique is to thoroughly examine task challenges in the electronic training system by tapping into experts' knowledge. The task diagram helped identify critical cognitive components of specific tasks within the electronic training system.

Methodology:

The TaSPoD prototype was developed based on the recommendation provided by experts from the ACTA process conducted in design phase. Task diagrams were then created through interviews and discussions with experts. Task diagrams which are visual representations outlining the cognitive steps derived from prototype system navigation. The study captures the evolving nature of cognitive processes, offering a valuable dataset for analysis.

Data Collection:

The data collection process emphasizes capturing the complexities of cognitive activities while interacting with the prototype system. Interviews with system analyst, instructional designers, trainers and trainees were conducted to gain insights into decision-making, problem-solving approaches, and method for information processing.

Insights and Findings:

Experts identified three system components that require cognitive skills which are system functionality, course content design and assessment design. They believed that these system components require heavy cognitive skills for the HLEs either as trainer or trainee in operating and utilizing the online training system. It is important to draw learners' attention and ignite trainees' excitement in completing the course (Mohd Mokhtar, Adnan, Mohd Shazali, & Ahmad, 2020). Thus, satisfy one of the electronic training benefits for independent learning. On the other hand, all experts agreed that communication medium and downloadable documents do not require as many cognitive skills as the three identified components. Figure 2 shows the task diagram of TaSPoD prototype.

Figure 2. Task diagram for TaSPoD prototype

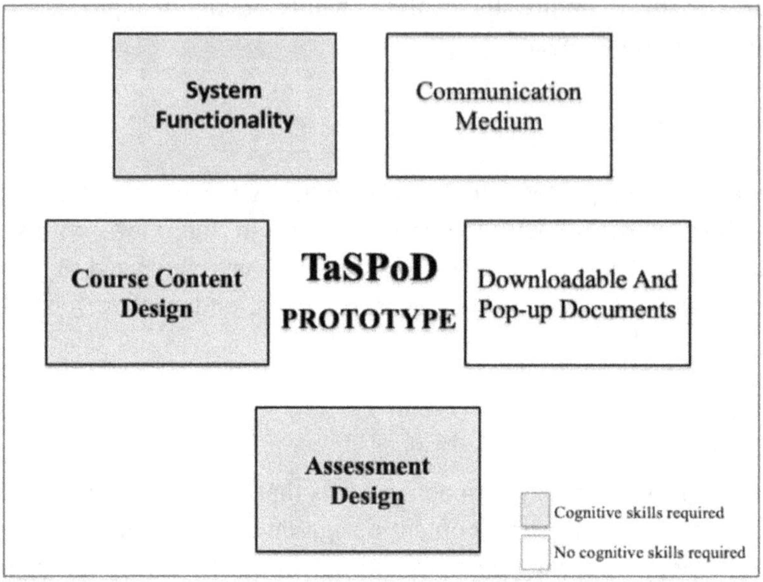

Note: Cognitive task skills differentiated from the non-cognitive task skills

7.1.3 Review of Task Diagram for Electronic Training for Professional Development

In this section, we explored the use of task diagrams to support the design and development of the TaSPoD system and its prototype within the context of electronic training for professional development. By applying the Applied Cognitive Task Analysis (ACTA) method, the key task components that require high cognitive skills were identified, enabling a structured understanding of the system's complexity and the cognitive demands placed on both trainers and trainees.

For the system design, the task diagram highlighted the cognitive load in system functionality and course content design, areas deemed critical for successful system operation. In contrast, components like voice over and downloadable documents were categorized as requiring fewer cognitive skills, allowing designers to focus efforts where the cognitive load is heaviest.

Similarly, the system prototype task diagram revealed that system functionality, course content design, and assessment design demand the most cognitive skills from users, particularly in creating an engaging and effective training system. Experts concluded that components such as the communication medium and downloadable documents were less cognitively demanding, further reinforcing the need to concentrate on the areas requiring higher levels of cognitive processing.

Through these task diagrams, this section emphasizes how ACTA provides a methodical approach for identifying and visualizing the most cognitively intensive components in system design. This process ultimately aids instructional designers and system developers in refining their focus to enhance both the training experience and overall system usability.

7.2 Task Diagrams for Designing Microlearning Pedagogical Practices

7.2.1 Application of Task Diagram Technique in the Analysis Phase

In the Analysis Phase, the study employs the Applied Cognitive Task Analysis (ACTA) technique, specifically leveraging the task diagram methodology, to navigate the intricate landscape of contemporary teaching and learning practices in higher education. This phase is dedicated to addressing Research Question 1.1: What are the current practices of teaching and learning in higher education? and Research Question 1.2: What are the academic problems faced by higher education students? The task diagram technique serves as a powerful tool in unravelling the educational complexities, providing a detailed map of the cognitive processes involved in academic tasks.

Objective:

The primary objective of utilising the task diagram technique is to conduct a comprehensive exploration of the educational environment. This goes beyond a mere snapshot; it represents an expedition to decode the nuanced challenges faced by higher education students. Through ACTA, the study seeks to elucidate the underlying intricacies that shape the educational journey, aiming to identify and address the challenges that require innovative solutions.

Methodology:

The methodology involves the creation of task diagrams, graphical representations that delineate the cognitive steps and interactions inherent in each chosen educational task. Through interviews and interactions with higher education students, the study captures the dynamic nature of cognitive processes, providing a rich dataset for analysis.

Data Collection:

Data collection focuses on capturing the intricacies of cognitive activities during teaching and learning. Interviews with higher education students are conducted to gain insights into decision-making, problem-solving strategies, and information processing. The goal is to unearth both explicit and implicit cognitive processes that influence the teaching and learning experience.

Data Analysis:

The collected data is subjected to rigorous analysis using the task diagram technique. Patterns, relationships, and dependencies within the cognitive processes are identified and visually represented. This analysis not only unveils the complexities but also aids in understanding the cognitive load associated with different aspects of the educational tasks.

Insights and Findings:

The application of the task diagram technique yields valuable insights into the cognitive dimensions of teaching and learning in higher education. It provides a detailed understanding of how students and educators navigate complex academic scenarios, revealing points of friction and areas where cognitive resources are heavily taxed. For instance, Asma and Dallel (2020) highlights the importance of recognizing and addressing cognitive overload in academic learning processes, where students often struggle to manage their mental resources effectively due to poorly designed instructional methods that do not account for their cognitive limitations. These findings form the basis for informed decision-making and intervention strategies in subsequent phases of research. Interventions informed by such insights can help streamline cognitive tasks, improving both teaching efficiency and student learning outcomes.

I. Students' Perspectives on Teaching and Learning Practices:
   a. Student Preferences and Learning Environment

The insights drawn from in-depth interviews with students illuminate nuanced perspectives on teaching and learning practices. Students articulate a distinct preference for a diverse learning environment, emphasizing the significance they attribute to integrating a broad spectrum of resources, including online platforms, media resources, and social networks. Recent research by Yadav (2024) supports this, showing that students' learning success is increasingly tied to the ability to access and integrate digital resources and multimedia content. This emphasis on varied learning tools suggests an awareness among students that successful learning is intricately tied to the incorporation of diverse resources. Moreover, the recognition that lecturers employ a range of teaching activities, such as individual and group work, oral presentations, and research, signals an alignment with diverse learning styles. Han (2021) found that varied teaching methods enhance student engagement, particularly when instructors adapt their approach to cater to different learning preferences. The revelation that lecturers provide tasks stimulating students' interest resonates with a sense of student engagement, indicating that tailored tasks aligning with individual interests significantly contribute to motivation. Additionally, the acknowledgment that lecturers encourage critical thinking activities reflects an emphasis on higher-order cognitive skills, crucial for both academic and professional success. El Yazidi (2023) found that fostering critical thinking through collaborative learning significantly enhances students' critical thinking skills, which are essential for academic success and beyond. The use of diverse teaching aids, including visual aids, texts, and online materials, coupled with a broad range of resources, underscores a comprehensive approach to enriching the learning environment. Overall, the insights from interviews affirm that lecturers foster an environment promoting active engagement and involvement, highlighting the effectiveness of the lecturer in cultivating a participatory classroom setting.

b. Cognitive Activities in Student Learning

In addition, the interview unveils a nuanced perspective on students' perceptions of their teaching and learning experiences, emphasizing various cognitive activities. Through a qualitative lens, students express engagement in mental activities that span the spectrum of cognitive processes. Notably, they articulate a moderate emphasis on the memorization of facts, suggesting a foundational element in their learning experiences. Fengyu (2023) found that while rote memorization aids in vocabulary retention, it is less effective for practical vocabulary application, with contextual learning and multimodal approaches leading to deeper engagement and better vocabulary use. This suggests that a balanced approach, incorporating both memorization and immersive, context-based methods, can better support learning. The acknowledgment of analysing basic elements of ideas, experiences, or theories indicates a commitment to exploring and understanding content at a deeper level. Additionally, the synthesis and organization of information into more complex interpretations and relationships reflect a cognitive process aimed at fostering a holistic understanding. Amanda et al. (2023) emphasize that critical thinking and problem-solving skills are enhanced when students are encouraged to synthesize information and interconnect ideas, facilitating a more comprehensive grasp of complex concepts. Students also report a propensity for applying theories or concepts to practical problems or in new situations, showcasing a practical and applicable dimension to their learning experiences. While the interview suggests a slightly lower emphasis on making value judgments, the qualitative interpretation underscores the richness of cognitive engagement within the learning environment. Overall, these findings reveal a balanced integration of various mental activities in students' teaching and learning experiences, encompassing both foundational memorization and higher-order cognitive processes like synthesis and application. This balance not only cultivates a diverse set of cognitive skills but also positions students for success in their future academic and professional pursuits.

c. Perceptions of Academic Learning Quality

Furthermore, the interview provides valuable insights into students' perceptions of academic learning quality and their overall learning experience at the institution. The data reveals a predominantly positive picture, with an overwhelming majority of students expressing satisfaction and providing positive evaluations for these aspects. This positivity is particularly evident in the evaluation of academic learning quality, where a substantial proportion of students considered it to be of high quality. Malik et al. (2023) emphasized that enhancing academic satisfaction can positively impact the student experience, and increasing satisfaction was recommended to support overall student well-being. This nuanced perspective suggests not only satisfaction but a significant appreciation for the calibre of academic learning experiences. Similarly, the majority of students expressed positive sentiments about the overall learning experience, indicating contentment with their holistic learning journey at the institution. However, the data also highlights a smaller subset of students who described their academic learning quality and overall learning experience less favourably. This underscores an area for potential improvement, signalling opportunities for the institution to enhance teaching methods or resources. Adarkwah (2021) emphasizes that effective assessment feedback—when specific, timely, and constructive—plays a crucial role in improving teaching and learning outcomes, suggesting that continuous refinement of feedback mechanisms can help address diverse student needs. In essence, while the majority of students express general satisfaction with their academic learning quality and overall experience, the nuanced data emphasizes the importance of continuous improvement to cater to the diverse needs of all students, ensuring a holistic and enriching educational environment.

Figure 3 illustrates the Task Diagram of Students' Perspectives on Teaching and Learning, highlighting key aspects such as a diverse learning environment, lecturers' varied teaching activities, student engagement, the development of critical thinking and cognitive skills, the use of diverse teaching aids and resources, the balanced integration of cognitive activities, and the overall positive academic learning quality and experience. This visual representation encapsulates the multifaceted elements that shape students' perceptions and experiences in higher education settings.

Figure 3. Task diagram of students' perceptions of teaching and learning

In addition, the insights gathered from students' perspectives shed light on their academic concerns and challenges. These qualitative findings hold significant implications for the thoughtful design and implementation of microlearning practices, designed to enrich the learning experiences of higher education students. By addressing specific cognitive challenges and tailoring content to individual learning needs, microlearning can provide a more adaptable and responsive educational framework that aligns with the diverse preferences and demands of today's learners.

II. Academic Concerns and Challenges:
   a. Preference for Engaging Learning Experiences

Students consistently express a strong desire for engaging and interactive learning experiences, recognizing that traditional, lecture-based methods often fail to hold their attention or promote deep understanding. They emphasize the importance of activities that are not only educational but also entertaining, allowing for more active participation and involvement in their learning process. This preference for dynamic and participatory learning aligns with current educational trends that advocate for more student-centred approaches, which prioritize engagement, collaboration, and experiential learning.

This shift towards interactive learning is supported by recent research, which indicates that students are more likely to retain information and develop critical thinking skills when they are actively involved in their learning through interactive activities that promote engagement and problem-solving (Blyznyuk & Kachak, 2024). As Hodges (2020) explains, active learning methods—such as discussions, problem-solving exercises, and group work—can improve students' performance and engagement, but their effectiveness depends on factors such as cognitive engagement, activity design, and group dynamics, compared to passive, lecture-based instruction. Together, these findings highlight the need for carefully designed interactive learning environments that actively engage students and promote deeper cognitive processing to achieve optimal learning outcomes.

One approach that directly responds to students' demand for engaging learning is microlearning, a strategy that delivers content in small, bite-sized pieces that are easier to digest and retain. Microlearning is characterized by brief, focused sessions that incorporate multimedia, quizzes, and interactive elements, fostering greater student involvement and retention of information. As research by Ghafar et al. (2023) highlights, microlearning not only caters to shorter attention spans but also supports better cognitive processing by allowing learners to engage with content at their own pace.

Furthermore, with the rise of digital platforms and e-learning technologies, educators have more tools at their disposal to create engaging and interactive learning environments. Game-based learning, simulations, and augmented reality (AR) tools are being increasingly incorporated into curricula, offering immersive and interactive experiences that further enhance student engagement. According to a study by Kaur et al. (2019), the use of AR in classrooms has been shown to increase motivation and promote deeper learning, providing students with hands-on, interactive opportunities to explore complex concepts.

In summary, students' preference for engaging and participatory learning experiences reflects the broader trend towards active, student-centred educational practices. By incorporating strategies such as microlearning, interactive activities, and emerging technologies, educators can create learning environments that not only resonate with students' preferences but also lead to improved academic outcomes and a more enriched learning experience overall.

b. Challenges with Workload and Information Retention

Students frequently express concerns about the overwhelming volume of assignments and the associated difficulty of retaining information without regular, structured revision. These challenges point to a common issue in modern education: the cognitive overload that results from attempting to process and retain large amounts of information within limited timeframes. Without proper instructional guidance, students may struggle to engage effectively with online course content, leading to challenges in completing tasks and maintaining performance (Al Mamun & Lawrie, 2023).

Recent research by Skulmowski and Xu (2022) suggests that design factors in learning environments can contribute to cognitive overload, supporting Sweller's Cognitive Load Theory (CLT), which highlights the limitations of working memory. When overwhelmed by excessive information, learners struggle to process and retain material effectively. To address this, implementing microlearning—characterized by short, focused learning modules—can help reduce cognitive load and enhance learning outcomes.

Microlearning provides a solution to the problem of cognitive overload by breaking down complex information into smaller, more manageable segments. These bite-sized learning units allow students to engage with content in shorter bursts, making it easier to digest and retain information. Research by Nikkhoo et al. (2023) found that microlearning strategies were particularly effective in improving information retention among students, as the segmented format encourages regular revision and reinforces knowledge over time. By offering focused, high-impact content, microlearning helps students navigate large volumes of material without becoming overwhelmed.

Additionally, microlearning supports spaced repetition, a technique that has been shown to enhance long-term memory retention by reintroducing material at strategic intervals, as highlighted by Shail (2019). Micro-lessons effectively combat the Ebbinghaus forgetting curve, helping learners retain information for longer periods through repeated exposure to key concepts. By revisiting material at regular intervals, spaced repetition ensures that information remains fresh in the learner's mind, mitigating the natural tendency to forget newly learned concepts. This aligns with students' needs for structured revision opportunities, as regular engagement with bite-sized content allows for more effective reinforcement of learning.

Microlearning also integrates seamlessly into digital learning platforms, providing students with on-demand access to content that can be reviewed and practiced at their own pace. This flexibility allows students to fit learning into their schedules more easily, reducing the burden of managing multiple assignments and revision sessions. For example, a study by Al-Zahrani (2024) found that a mobile-based microlearning approach in a flipped learning environment led to significantly better outcomes compared to traditional learning methods. Students reported improved accessibility, engagement, and knowledge retention, with an overall more positive learning experience. Additionally, the group using mobile-based microlearning scored higher on their final exams, demonstrating the effectiveness of this method in enhancing the learning process.

Moreover, the incorporation of gamified elements—such as quizzes and interactive exercises—into microlearning can further enhance student engagement and motivation. According to a study by Göschlberger and Bruck (2017), the use of gamification in microlearning has been shown to enhance student engagement and understanding of course material, making the learning process more interactive and motivating. By combining the benefits of microlearning with gamified elements, educators can create a more dynamic and effective learning environment that not only motivates students but also promotes deeper understanding and long-term retention.

In conclusion, the challenges students face with assignment volumes and information retention underscore the need for innovative learning approaches that provide structured revision opportunities. Microlearning, with its concise and targeted content delivery, addresses these concerns by facilitating more manageable workloads and promoting regular engagement with material, ultimately improving retention and reducing cognitive overload. As education continues to evolve, the integration of microlearning into traditional curricula offers a promising avenue for enhancing student learning experiences and outcomes.

c. Challenges with Academic Overload and Attention Span

Students frequently report experiences of academic overload, feeling overwhelmed by the extensive demands of pursuing a degree, and struggling with short attention spans. This overwhelming sense of burden often stems from the need to balance multiple academic tasks—such as assignments, exams, and required readings—alongside personal responsibilities. Research by Wilson et al. (2021) highlights the importance of time-management skills in improving academic performance, where a significant impact on academic success was observed. Time-management interventions, such as workshops, were found to reduce subject failures and foster a more positive attitude towards time management for both male and female students. These findings underscore the crucial role that effective time management plays in alleviating the burden of academic overload, helping students navigate the competing demands of coursework and personal responsibilities.

These challenges are becoming increasingly prevalent in modern education, as students face greater volumes of information and tighter deadlines than ever before. Increased workload is a significant factor contributing to student burnout, which in turn affects academic performance, as found in a study on management students (Pham Thi & Duong, 2024). This burnout often arises from managing both the volume of work and associated stressors like English anxiety, ultimately impacting student well-being and learning outcomes. Compounding this is the decline in attention spans, which recent research attributes to various factors, including classroom dynamics, teaching methods, and student behaviour (Cicekci & Sadik,

2019). The study highlights that educators often perceive attention problems as disciplinary issues, while students suggest that more engaging teaching methods and positive interactions with educators could help mitigate these challenges. For students, managing the ability to concentrate on long-form content, such as dense textbooks and extended lectures, presents an additional challenge.

In response to these issues, microlearning emerges as a highly effective strategy to mitigate academic overload and better align with students' attention spans. Microlearning refers to the practice of delivering educational content in short, focused segments that allow students to engage with the material in bite-sized portions, rather than being overwhelmed by extensive content all at once (Fitria, 2022). By breaking down large volumes of information into smaller, digestible units, microlearning offers a solution that promotes better time management and accommodates the realities of modern academic life.

Flexibility is one of the most critical benefits microlearning offers to students facing academic overload. With traditional long-form learning methods, students are often required to dedicate large, uninterrupted blocks of time to study. However, microlearning allows students to learn in short bursts, which can be easily integrated into their busy schedules. A study by Senadheera et al. (2024) demonstrated that students who engaged in microlearning showed significantly higher academic performance compared to those who followed traditional macro-learning methods. The study's meta-analysis revealed that microlearning contributed to improved post-test scores by reducing cognitive load and promoting a flexible, self-directed learning environment. This flexibility allows students to engage with the content during moments of downtime, such as between classes or during commutes, reducing the pressure to allocate extended periods of time to study.

Additionally, microlearning is highly adaptable to students' unique learning preferences, offering flexibility that traditional long-form content often lacks. In a study by Fidan (2023), pre-service teachers (PTs) participating in a microlearning-supported flipped classroom (m-FC) model showed improved intrinsic motivation, emotional engagement, and learning performance compared to those using traditional flipped classrooms or teacher-centred methods. The microlearning format, which presents content in bite-sized, manageable chunks, not only facilitated greater engagement with pre-class activities but also enhanced PTs' willingness to participate in the learning process. These findings highlight how microlearning can effectively address challenges such as disengagement and amotivation, making it an adaptable and engaging approach for students with varying learning needs.

Moreover, microlearning supports the concept of just-in-time learning, which allows students to access content exactly when they need it (Nikkhoo et al., 2023). This approach is particularly helpful for those who experience academic overload, as it provides the opportunity for on-demand learning that can be integrated into real-world applications. A study by Kossen and Ooi (2021) supports this, showing that segmenting course materials into bite-sized modules significantly reduced cognitive overload and increased student engagement, satisfaction, and performance. This approach of delivering content in smaller, essential units aligns with just-in-time learning, offering students the flexibility to access and apply knowledge as needed, particularly in situations where they face high academic demands.

In summary, students' struggles with academic overload and short attention spans point to the need for more adaptable and flexible learning solutions. Microlearning addresses these challenges by offering shorter, targeted learning units that align with students' schedules and cognitive capacities. By allowing students to engage with educational materials in a more structured, manageable way, microlearning provides a promising approach to enhancing academic success and reducing stress.

Figure 4 illustrates the Task Diagram of Students' Academic Needs and Problems, outlining key areas such as their preference for engaging learning, challenges with assignments and retention, and issues related to academic overload and time management. The diagram highlights the need for interactive and participatory learning, the burden of assignments, the difficulty in information retention, and the importance of microlearning elements for manageable workloads and structured revision. It also addresses the struggles with short attention spans, the feeling of being overwhelmed, and the emergence of microlearning as a valuable strategy for flexibility, adaptability, and shorter learning units.

Figure 4. Task diagram of students' academic needs and problems

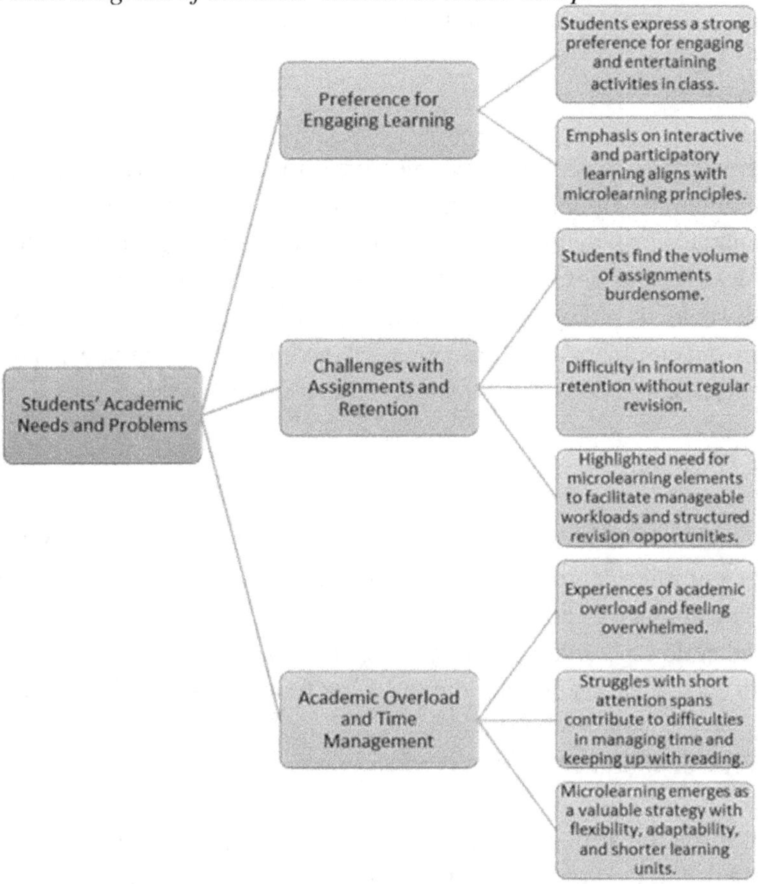

7.2.2. Application of Task Diagram Technique in the Design Phase

As the study transitions into the Design Phase, the spotlight is redirected towards decoding the foundational elements crucial for the effective creation and delivery of microlearning experiences within a classroom setting. Applied Cognitive Task Analysis (ACTA) remains the cornerstone of this phase, and once again, the study leverages the power of task diagrams. This phase is dedicated to addressing Research Question 2: What are the fundamental elements in the design of microlearning in higher education that contribute to the effective creation and delivery of microlearning experiences in a classroom setting according to experts? Supported by the insights of experts in microlearning, curriculum and instruction, and instructional technology, this phase undertakes a decoding process, aiming to extract and decipher the collective wisdom of these professionals. The objective is to construct a comprehensive framework for microlearning pedagogical practices.

Objective:

In the Design Phase, the primary objective is to construct a robust framework for the implementation of microlearning experiences in higher education. The utilisation of ACTA, namely task diagram, is directed towards capturing the intricate cognitive processes involved in designing and delivering microlearning content effectively. This phase seeks to amalgamate the expertise of professionals in microlearning, curriculum development, and instructional technology to create a pedagogical foundation that aligns with the principles of cognitive task analysis.

Task Selection:

Tasks chosen for analysis in this phase are those directly related to the creation and delivery of microlearning experiences. These may include content design, assessment development, and the integration of technology into microlearning modules. The selection is guided by the goal of understanding the cognitive intricacies involved in each aspect, ensuring the resulting framework is holistic and effective.

Expert Collaboration:

Experts in microlearning, curriculum and instruction, and instructional technology play a pivotal role in providing insights during the task diagram creation process. Through interviews their expertise is harnessed to ensure that the task diagrams accurately represent the cognitive processes inherent in designing and delivering microlearning content.

Methodology:

The task diagram technique is applied to represent the cognitive processes associated with each selected task. Through a collaborative approach involving experts and practitioners, the study captures the diverse perspectives and best practices in microlearning design. The task diagrams serve as visual maps, illustrating the flow of cognitive activities involved in creating and delivering effective microlearning experiences.

Data Integration:

Insights from both the Analysis Phase and expert contributions are integrated into the task diagrams. This holistic approach ensures that the framework not only addresses identified challenges and complexities but also incorporates the collective wisdom of professionals in the field. The resulting task diagrams become a synthesis of empirical observations and expert knowledge.

Iterative Design:

The design phase is iterative, allowing for refinement based on continuous feedback from experts. Task diagram is revisited and revised to reflect evolving insights and to ensure that the framework is responsive to the dynamic nature of microlearning practices.

Framework Construction:

The final output of the Design Phase is a comprehensive framework for microlearning pedagogical practices. This framework is constructed based on the insights derived from the task diagram interviews, expert contributions, and iterative design processes. It serves as a guide for educators, instructional designers, and curriculum developers seeking to integrate effective microlearning strategies into their teaching practices.

Implications for Implementation:

The constructed framework holds direct implications for the implementation of microlearning in educational settings. It provides a structured approach to designing and delivering microlearning experiences, taking into account the cognitive demands associated with each task. The insights gained from the task diagrams and expert collaboration guide the effective integration of microlearning into the broader educational landscape.

Future Development:

Concluding the Design Phase, the study outlines potential avenues for future development. This may include further refinement of the microlearning framework, exploration of additional cognitive analysis methods, and the application of the framework in diverse educational contexts. The insights gained from this phase contribute to the ongoing evolution of microlearning practices and the integration of cognitive task analysis in educational design.

Insights and Findings:

This research question investigated the essential elements in the design of effective microlearning experiences in higher education. The study identified eight fundamental themes that are pivotal for crafting dynamic and adaptive microlearning environments. Firstly, clear learning objectives and their alignment with learners' specific needs and aspirations were recognized as foundational. This practice provides a structured path for learners and fosters a deeper connection with the content. Recent research by Kossen and Ooi (2021) supports this, noting that microlearning, when designed with clear objectives and practical value, led to significant increases in student engagement and satisfaction.

Personalization of microlearning content and activities emerged as a second key element, resonating with contemporary educational theories that stress learner-centred approaches. Personalization not only enhances engagement but also empowers learners to take ownership of their educational journey. Research by Makhambetova et al. (2020) highlights that personalized learning experiences significantly improve student outcomes by catering to individual learning preferences and paces, making the learning process more meaningful and effective.

Effective feedback and assessment strategies, the third element, were deemed essential for guiding students, enabling them to track progress and make informed adjustments. Diverse assessment methods and alignment with learning objectives reinforced the effectiveness of microlearning. Studies by Bellhäuser et al. (2023) demonstrates that frequent, formative feedback encourages self-regulated learning by improving goal setting, planning, time management, and self-efficacy.

The fourth theme emphasized microlearning delivery methods that cater to the "just-in-time" learning paradigm. The flexibility of mobile microlearning, which allows learners to access content at their point of need, aligns with the principles of just-in-time learning and has been shown to increase knowledge and confidence in practical applications (Lee et al, 2021). The integration of multimedia elements further engaged learners and met their preferences for flexible and accessible learning, as supported by Mayer's (2014) multimedia learning theory, which suggests that using multiple modes of content (e.g., videos, infographics) improves comprehension and retention.

Structured and engaging content delivery, the fifth element, aligned with cognitive load theory, enhancing comprehension by breaking down content into manageable units. Logical progression and interactivity enriched learners' understanding and promoted active participation. This is supported by Liu (2024), who found that chunking information into small, structured learning units through segmentation techniques reduces cognitive load, improves comprehension, retention, and overall learning outcomes.

The integration of interactive learning experiences, the sixth element, emphasized learner engagement and participation. This element enriched understanding, critical thinking, and the application of knowledge. Research by Yin et al. (2020) supports this, demonstrating that a chatbot-based microlearning system significantly enhanced students' intrinsic motivation and engagement compared to traditional learning methods. The study showed that students in the chatbot learning group exhibited higher levels of motivation and autonomy, suggesting that interactive microlearning environments, such as chatbot-based systems, can foster deeper engagement and a greater sense of control over learning. This, in turn, promotes the development of critical thinking and the application of knowledge in real-world scenarios.

The seventh element, social interaction, aligned with social constructivist theories, emphasizing collaborative learning and peer interaction. Collaborative activities fostered diverse perspectives and valuable discussions, enhancing learning outcomes and social skills. A study by Gherghel et al. (2023) found that peer interaction in learning environments fosters higher levels of engagement, which in turn promotes self-directed study and enhances learning outcomes.

Lastly, the seamless integration of microlearning into the curriculum, the eighth element, supported educational alignment and coherence, promoting consistent and interconnected learning experiences. Recent research by Taylor and Hung (2022) supports this by demonstrating that the integration of microlearning into academic and professional training environments enhances learner performance and supports a more interconnected and consistent learning experience. The study highlights the growing trend of using microlearning to promote coherence across educational and training settings, emphasizing its adaptability and alignment with current instructional design practices.

These fundamental elements in microlearning design offer actionable insights for educators to create dynamic, learner-centred, and effective learning environments that meet the diverse needs and preferences of higher education students. By aligning content, delivery methods, and feedback mechanisms with these principles, educators can foster more engaging, personalized, and efficient learning experiences that promote student success.

Figure 5 illustrates the Task Diagram of AMPLIFY, outlining the components of the AMPLIFY framework aimed at advancing microlearning practices for enhanced learning experiences. This diagram highlights various elements such as focused learning goals, personalized learning experiences, effective feedback and assessment strategies, interactive learning methods, social interaction, and the seamless integration of these practices within the curriculum. These components collectively contribute to creating a dynamic and effective microlearning environment that enhances student engagement and learning outcomes.

Figure 5. Task diagram of AMPLIFY

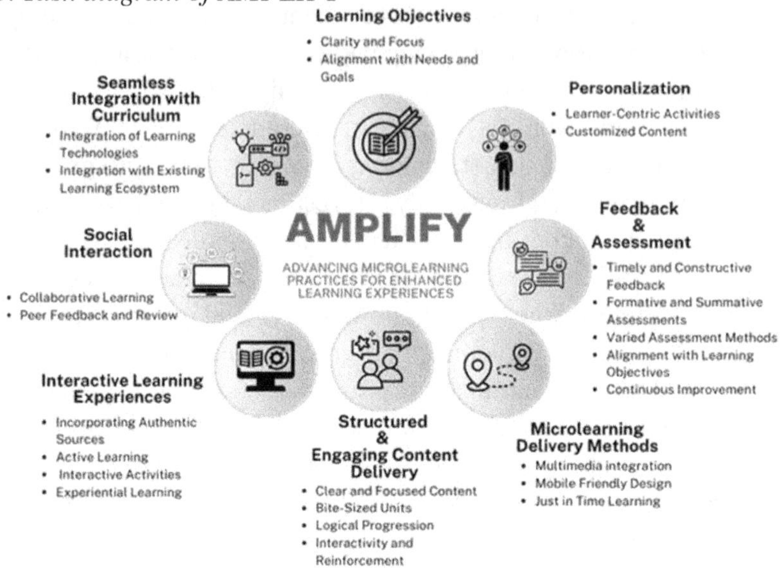

7.3 Review of Task Diagram of Designing Microlearning Pedagogical Practices

This study's exploration into higher education teaching and learning practices emphasizes the innovative use of the Applied Cognitive Task Analysis (ACTA) technique across the Analysis and Design Phases. Through these phases, the study has provided a foundational understanding and framework for effective microlearning practices.

In the Analysis Phase, the study employed the ACTA technique with a focus on task diagrams to dissect the current state of teaching and learning in higher education. By addressing Research Questions 1.1 and 1.2, this phase successfully mapped the cognitive processes involved in academic tasks and identified the significant academic challenges faced by students. This detailed analysis not only highlighted the complexities within the educational landscape but also underscored the need for innovative solutions to enhance learning experiences.

Moving into the Design Phase, the study continued to harness the power of ACTA, with a particular emphasis on decoding the fundamental elements necessary for the creation and delivery of effective microlearning experiences. Addressing Research Question 2, this phase involved gathering and synthesizing insights from experts in microlearning, curriculum and instruction, and instructional technology.

The resulting task diagrams provided a comprehensive framework, capturing the essential components for designing impactful microlearning pedagogical practices.

Together, these phases demonstrate the potential of the ACTA technique in uncovering and structuring the cognitive dimensions of educational practices. By systematically analysing current practices and designing a robust framework for microlearning, the study lays a strong foundation for the development and implementation of innovative teaching methodologies in higher education.

CONCLUSION

In this chapter, we have seen how task diagrams, as a core method of Applied Cognitive Task Analysis (ACTA), play a pivotal role in both professional development and the design of microlearning pedagogical practices. By providing a structured visualization of cognitive processes, task diagrams help educators and instructional designers dissect complex tasks into actionable learning components, ensuring more effective and targeted training.

The use of task diagrams in electronic training system for professional development highlights their utility in creating detailed learning experiences that align with professional requirements. This approach ensures that learners are equipped with the necessary skills and cognitive tools to succeed in real-world professional settings. Similarly, task diagrams in the microlearning context allow for the design of focused, manageable learning experiences, enhancing engagement, retention, and overall learning outcomes.

The chapter illustrates that task diagrams not only facilitate deeper understanding of cognitive processes but also support the development of learner-centred and effective instructional designs. As learning environments evolve, particularly in digital and professional contexts, the continued application of task diagrams will remain crucial in addressing the growing complexity of educational needs.

Looking ahead, further exploration of task diagrams in diverse educational environments, particularly through the lens of emerging technologies and digital tools, will enhance their value as a critical method for cognitive task analysis. As such, educators and instructional designers are encouraged to leverage the insights provided by task diagrams to create more adaptive, interactive, and personalized learning experiences.

REFERENCES

Adarkwah, M. A. (2021). The power of assessment feedback in teaching and learning: A narrative review and synthesis of the literature. *SN Social Sciences*, 1, 75. DOI: 10.1007/s43545-021-00086-w

Al Mamun, M. A., & Lawrie, G. (2023). Student-content interactions: Exploring behavioural engagement with self-regulated inquiry-based online learning modules. *Smart Learning Environments*, 10(1), 1. DOI: 10.1186/s40561-022-00221-x

Al-Zahrani, A. M. (2024). Enhancing postgraduate students' learning outcomes through flipped mobile-based microlearning. *Research in Learning Technology*, 32. Advance online publication. DOI: 10.25304/rlt.v32.3110

Amanda, F. F., Sumitro, S., Lestari, S. R., & Ibrohim, I. (2023). Enhancing critical thinking and problem-solving skills by complexity science-problem based learning model. *Multidisciplinary Journal of Educational Research*. Advance online publication. DOI: 10.17583/remie.9409

Asma, H., & Dallel, S. (2020). Cognitive load theory and its relation to instructional design: Perspectives of some Algerian university teachers of English. *Arab World English Journal*, 11(4), 110–127. DOI: 10.24093/awej/vol11no4.8

Bellhäuser, H., Dignath, C., & Theobald, M. (2023). Daily automated feedback enhances self-regulated learning: A longitudinal randomized field experiment. *Frontiers in Psychology*, 14, 1125873. DOI: 10.3389/fpsyg.2023.1125873

Blyznyuk, T., & Kachak, T. (2024). Benefits of interactive learning for students' critical thinking skills improvement. *Journal of Vasyl Stefanyk Precarpathian National University*, 11(1), 94–102. DOI: 10.15330/jpnu.11.1.94-102

Cicekci, M. A., & Sadik, F. (2019). Teachers' and students' opinions about students' attention problems during the lesson. *Journal of Education and Learning*, 8(6), 15–27. DOI: 10.5539/jel.v8n6p15

El Yazidi, R. (2023, March-April). Strategies for promoting critical thinking in the classroom. International. *Journal of English Literature and Social Sciences*, 8(2). Advance online publication. DOI: 10.22161/ijels

Fengyu, Z. (2023). The impact of vocabulary learning methods on students' vocabulary application skills. *English Language Teaching and Linguistics Studies*, 5(4), 206. DOI: 10.22158/eltls.v5n4p206

Fidan, M. (2023). The effects of microlearning-supported flipped classroom on pre-service teachers' learning performance, motivation and engagement. *Education and Information Technologies*, 28, 12687–12714. DOI: 10.1007/s10639-023-11639-2

Fitria, T. N. (2022). Microlearning in teaching and learning process: A review. *Cendekia: Jurnal Ilmu Sosial Bahasa dan Pendidikan, 2(*4), 114-135. DOI: 10.55606/cendikia.v2i4.473

Ghafar, Z. N., Abdullkarim, S. T., & Mahmud, I. (2023). Microlearning as a learning tool for teaching and learning in acquiring language: Applications, advantages, and influences on the language. *Canadian Journal of Educational and Social Studies*, 3(2), 45–62. DOI: 10.53103/cjess.v3i2.127

Gherghel, C., Yasuda, S., & Kita, Y. (2023). Interaction during online classes fosters engagement with learning and self-directed study both in the first and second years of the COVID-19 pandemic. *Computers & Education*, 200, 104795. DOI: 10.1016/j.compedu.2023.104795

Göschlberger, B., & Bruck, P. A. (2017, December). Gamification in mobile and workplace integrated microlearning. In *Proceedings of the 19th International Conference*. DOI: 10.1145/3151759.3151795

Han, F. (2021). The relations between teaching strategies, students' engagement in learning, and teachers' self-concept. *Sustainability*, 13(9), 5020. DOI: 10.3390/su13095020

Hodges, L. C. (2020). Student engagement in active learning classes. In Mintzes, J. J., & Walter, E. M. (Eds.), *Active learning in college science* (pp. 35–46). Springer., DOI: 10.1007/978-3-030-33600-4_3

Kaur, D. P., Mantri, A., & Horan, B. (2019). Enhancing student motivation with use of augmented reality for interactive learning in engineering education. *Procedia Computer Science*, 172(3), 881–885. DOI: 10.1016/j.procs.2020.05.127

Kossen, C., & Ooi, C.-Y. (2021). Trialling micro-learning design to increase engagement in online courses. *Asian Association of Open Universities Journal*, 16(3), 299–310. DOI: 10.1108/AAOUJ-09-2021-0107

Lee, Y. M., Jahnke, I., & Austin, L. (2021). Mobile microlearning design and effects on learning efficacy and learner experience. *Educational Technology Research and Development*, 69, 885–915. DOI: 10.1007/s11423-020-09931-w

Liu, D. (2024). The effects of segmentation on cognitive load, vocabulary learning and retention, and reading comprehension in a multimedia learning environment. *BMC Psychology*, 12, 4. DOI: 10.1186/s40359-023-01489-5

Makhambetova, A., Zhiyenbayeva, N., & Ergesheva, E. (2020). Personalized learning strategy as a tool to improve academic performance and motivation of students. *International Journal of Web-Based Learning and Teaching Technologies*, 16(6), 1–17. DOI: 10.4018/IJWLTT.286743

Malik, A., Iqbal, M., & Sultan, A. (2023). Impact of academic satisfaction on university students' academic achievement. *Journal of Education and Social Studies*, 4(3), 806–812. DOI: 10.52223/jess.2023.4341

Mayer, R. E. (2014). Cognitive theory of multimedia learning. In Mayer, R. E. (Ed.), *The Cambridge handbook of multimedia learning* (2nd ed., pp. 43–71). Cambridge University Press. DOI: 10.1017/CBO9781139547369.005

Mohd Yusoff, , NSalim, , S. S. (2012). Investigating cognitive task difficulties and expert skills in e-learning storyboards using a cognitive task analysis technique. *Computers & Education*, 58, 652–665. DOI: 10.1016/j.compedu.2011.09.009

Mokhtar, S. M., Nur, A. A. A., Shazali, N. M., & Ahmad, N. (2020). Why Are Students Involved In E-Learning? A Reasoning Study At Univerisity Kebangsaan Malaysia (UKM). *International Journal of Education and Pedagogy*, 2(1), 152–159. http://myjms.moe.gov.my/index.php/ijeap

Nikkhoo, I., Ahmadi, Z., Akbari, M., Imannezhad, S., Anvari Ardekani, S., & Lashgari, H. (2023). Microlearning for today's students: A rapid review of essentials and considerations. *Medical Education Bulletin*, 4(1), 673–685. DOI: 10.22034/meb.2022.355659.1066

Pham Thi, T. D., & Duong, N. T. (2024). Investigating learning burnout and academic performance among management students: A longitudinal study in English courses. *BMC Psychology*, 12(1), 219. DOI: 10.1186/s40359-024-01725-6

Senadheera, V. V., Muthukumarana, C. K., Ediriweera, D., & Rupasinghe, T. (2024). Impact of microlearning on academic performance of students in higher education: A systematic review and meta-analysis. *Journal of Medical and Technology Research*, 9(1). Advance online publication. DOI: 10.4038/jmtr.v9i1.2

Shail, M. S. (2019). Using micro-learning on mobile applications to increase knowledge retention and work performance: A review of literature. *Cureus*, 11(8), e5307. DOI: 10.7759/cureus.5307

Skulmowski, A., & Xu, K. M. (2022). Understanding cognitive load in digital and online learning: A new perspective on extraneous cognitive load. *Educational Psychology Review*, 34, 171–196. DOI: 10.1007/s10648-021-09624-7

Taylor, A., & Hung, W. (2022). The effects of microlearning: A scoping review. *Educational Technology Research and Development*, 70, 363–395. DOI: 10.1007/s11423-022-10084-1

Wilson, R., Joiner, K., & Abbasi, A. (2021). Improving students' performance with time management skills. *Journal of University Teaching & Learning Practice*, 18(4). https://ro.uow.edu.au/jutlp/vol18/iss4/16

Yadav, N. (2024). The impact of digital learning on education. I*nternational Journal of Multidisciplinary Research in Arts. Science and Technology*, 2(1), 2584–0231. DOI: 10.61778/ijmrast.v2i1.34

Yin, J., Goh, T.-T., Yang, B., & Xiaobin, Y. (2020). Conversation technology with micro-learning: The impact of chatbot-based learning on students' learning motivation and performance. *Journal of Educational Computing Research*, 59(1), 073563312095206. DOI: 10.1177/0735633120952067

Chapter 8
Knowledge Audit

ABSTRACT

This chapter explores conducting a knowledge audit within Applied Cognitive Task Analysis (ACTA) in educational settings. It is divided into two sections. The first, "Knowledge Audit for Electronic Training for Professional Development," focuses on assessing existing knowledge and identifying gaps in electronic training programs, ensuring content aligns with professional development needs. The second section, "Knowledge Audit for Developing Microlearning Pedagogical Practices," discusses how knowledge audits can help design microlearning strategies in higher education, tailoring experiences to learners' needs and improving engagement. The chapter emphasizes the role of knowledge audits in refining educational practices and optimizing content delivery for both training and microlearning.

INTRODUCTION

In the realm of Applied Cognitive Task Analysis (ACTA), the knowledge audit is the second critical step. It plays a pivotal role in identifying and evaluating the specific knowledge required to perform complex tasks effectively. A knowledge audit serves to uncover the critical knowledge that experts use to perform tasks efficiently, as well as pinpoint areas where knowledge gaps exist. This process is essential for optimizing both the design and delivery of educational programs, whether in professional development or microlearning environments.

This chapter explores how knowledge audits are applied in two significant contexts: electronic training for professional development and the development of microlearning pedagogical practices.

Section 8.1 focuses on conducting a knowledge audit for electronic training for professional development. In this context, the audit is used to assess the essential knowledge and cognitive skills professionals need to master in digital training systems. It evaluates the flow and accessibility of knowledge within the training

DOI: 10.4018/979-8-3693-7733-8.ch008

process, ensuring that the learning content is aligned with the professional demands and addresses the knowledge gaps that may hinder optimal performance.

Section 8.2 examines the role of knowledge audits in shaping microlearning pedagogical practices. Microlearning, with its focus on short, targeted learning modules, requires careful auditing to ensure that the content delivered is relevant and applicable to learners' immediate needs. By identifying key areas where learners may struggle or need reinforcement, the knowledge audit informs the development of microlearning modules that promote more focused and efficient learning experiences.

Through both of these applications, the knowledge audit as a method in ACTA provides invaluable insights into how knowledge is structured, distributed, and utilized within educational systems. The chapter underscores the importance of conducting knowledge audits to refine instructional strategies, close knowledge gaps, and support the cognitive needs of learners in various professional and educational contexts.

8.1 Knowledge Audit for Electronic Training for Professional Development

8.1.1 Knowledge Audit Table for System Design

The knowledge audit table technique is used to explore the cognitive task difficulties identified in the task diagram technique. Experts were asked about cues and strategies they rely on when they face difficulties in navigating TaSPoD. Using the task diagram described in previous phase, the cognitive skills identified by the experts are further explored.

The knowledge audit table shown in Table 1 below presents the detailed aspects of system design and functionality, and course content design. Experts probed further that user might also face difficulties in screen view design and navigation system. Among the main task difficulties in system design and functionality involved inexperience users with online learning application. It will be challenging for users that are not familiar with similar online learning application to perform some of the system functionalities, such as downloading or uploading external contents, adjusting browsers to meet some specific requirement, as well as being active in forum. The screen views need to be simple, straight-forward, user-friendly and attractive for most users to use the system. Experts also gauge that users will not be in favor to use the online system that is not utilizing variety types of multimedia. Having a front page with video narration or animation will draw user's attention.

In the system navigation, experts analysed that higher learner educators (HLE) will face difficulty to navigate the system if the mouse rollover function is used. TaSPoD did not provide a clear guide using the system while it allows non-linear navigation throughout the system. HLEs will face a similar difficulty with mouse rollover in course content design as the content will not be displayed completely. Thus, this will not contribute to active interactions between users and the system. Uniformly, HLEs will also encounter difficulty to communicate with TaSPoD, with only forum as the medium.

Table 1. Knowledge audit table

TaSPoD components	Cues & Strategies	Why difficult
System design and Functionality	***Cues:*** HLEs who are not familiar with similar online applications or Learning Management System Platform would find TasPoD challenging to follow. Navigation and placement of buttons should be more self-explanatory. Provide option to go to previous screen. The downloading and uploading of the documents via PDF and word processing application to be customised within the TaSPoD system. ***Strategies:*** A simple, straight-forward, user-friendly and attractive document should be designed. Identifies type or activities and tasks that will be performed by users.	Difficult to use due to lack of experience (especially download / upload external content). Difficult for users who don't know how to adjust their browsers. Difficult to be active in the forum, assignment, etc. along with their real work commitment.
Screen View Design	***Cues:*** The fonts should be easy to read on various screen resolutions. The standard text size should be used for visitors. Home page can be made more interesting with animations or pictures which can draw the attention of users, as well as including a short video narration about TaSPoD, what is it and how to use it, before users login or registers for TaSPoD ***Strategies:*** The screen views need to be designed using graphical user interface that contains simple and appropriate objects, tabs and icons, and video upload tools.	Difficult to focus on the learning process. Difficult to acquire users' engagement to the system.
System Navigation	***Cues:*** TaSPoD did not give the clear guide on how to use the system for training. HLEs need to navigate non-linear throughout TaSPoD ***Strategies:*** TasPoD should automate learning process in sequential manner to give clear navigation, guidance and engagement in learning to users. Avoid using complex navigation button and menu. Every page should have help/hint buttons to assists users.	Difficult to navigate as it has mouse rollover option. Difficult to motivate to complete the task.

8.1.2 Knowledge Audit Table for System Prototype

The knowledge audit table technique is used to explore the cognitive task difficulties identified in the task diagram technique. Experts were asked about cues and strategies they rely on when they face difficulties in navigating TaSPoD. Using the task diagram described in previous phase, the cognitive skills identified by the experts are further explored.

The knowledge audit table shown in Table 2 below presents the detailed aspects of system functionality, course content design and assessment design. The task difficulties in system functionality include exploring variety of exciting and constructive activities to facilitate the class. In the course content design, HLEs will face difficulty to organize or arrange their learning when the system does not show course completion status within the ongoing lessons. While in assessment design, expert believed it will be difficult to motivate HLEs in completing the assessment when there is inadequate information about the assessment and evaluation method. Expert also convinced that obtaining reward upon completing a course should be based on certain condition.

Table 2. Knowledge audit table

TaSPoD components	Cues & Strategies	Why difficult
System functionality	***Cues:*** Web functions in the TaSPoD prototype requires HLEs to activate 'thought process' to continue the learning process. HLEs must be able to identify downloadable links. TaSPoD must be able to check if hypertext link is access during the session. ***Strategies:*** Enable easier process for downloading and uploading of the documents. TaSPoD enable a log to examine or inspect that hypertext link is being assessed.	HLEs are required to search more fun ways and to try different constructive things to facilitate the class.
Course content	***Cues:*** Interactive contents with audio would be preferred. Possibility of incompletion of course as the system does not shows course completion status. Possible of information / content overload. ***Strategies:*** Trainers tend to use only text and images, and upload videos to deliver the course content. Course content should be folded inside the 'accordion' or other Html structure.	The system does not show course completion status within the ongoing lessons which will cause difficulty for HLEs to organize or arrange their learning.

continued on following page

Table 2. Continued

TaSPoD components	Cues & Strategies	Why difficult
Assessment module	***Cues:*** Missing instruction about assessments to be taken upon completion or during each lesson or level. One assessment strategy will not fit all. HLEs will not be encouraged to complete the course to obtain badges. ***Strategies:*** Instructions screen should be provided about minimum score (for certification) to achieve to move to next lesson/level. There can be different assessments strategies (approaches) for different level of users.	Difficult to provide motivation with missing instructions. The assessments should come with clear instructions to evaluate. This will be justified for trainee as well as trainer. Obtaining reward (badge) from completion course be on certain condition.

8.1.3 Review of Knowledge Audit for Electronic Training for Professional Development

In this section, we explored the use of knowledge audit tables to address the cognitive task difficulties encountered in the TaSPoD system design and its prototype. Through expert feedback and analysis, the knowledge audit process helped identify critical areas where higher learner educators (HLEs) may struggle, such as navigating the system, interacting with content, and completing tasks efficiently.

For the system design, the knowledge audit table emphasized challenges related to user experience, particularly for those unfamiliar with online learning platforms. The audit highlighted the need for more intuitive navigation, better screen design, and the integration of multimedia elements to engage users and improve functionality. Experts pointed out the importance of simplifying interactions with the system, such as downloading, uploading, and navigating through non-linear content. The findings underscore the need for user-friendly design and the role of visual and interactive cues in improving overall usability.

For the system prototype, the knowledge audit further revealed challenges in areas such as course content design and assessment modules. The audit table identified the importance of providing clear guidance to learners, including visible course completion statuses and comprehensive assessment instructions. Moreover, HLEs are likely to benefit from interactive, multimedia-rich content, and a more structured approach to assessments, where learners are clearly informed about progress and evaluation criteria.

Overall, the knowledge audit technique has proven invaluable in uncovering key cognitive challenges and offering strategic insights for improving both the TaSPoD system and its prototype. By addressing these task difficulties and implementing the strategies outlined in the knowledge audit tables, educational and training systems can be optimized to meet the needs of learners, thereby enhancing both engagement and performance in professional development settings.

8.2 Knowledge Audit for Developing Microlearning Pedagogical Practices

8.2.1 Application of Knowledge Audit Technique in the Development Phase

In the third phase of this study, the Knowledge Audit Technique, the second methodology of Applied Cognitive Task Analysis (ACTA), was employed to delve deeper into the cognitive challenges identified in the Task Diagram Technique. This phase is dedicated to addressing Research Question 3: How can pedagogical practices of microlearning in higher education be best developed to enhance students' learning experiences according to experts? The Knowledge Audit Technique builds upon the insights obtained from the Task Diagram Technique and contributes to the refinement of the AMPLIFY Instructional Framework.

Objective

The primary objective of employing the Knowledge Audit Technique is to conduct a thorough examination of the cognitive skills and challenges identified by experts in the pedagogical practices of microlearning. By building upon the Task Diagram from the previous phase, the Knowledge Audit Table aims to provide a more nuanced understanding of the specific elements crucial for the successful implementation of microlearning. The ultimate goal is to enhance and refine the AMPLIFY framework based on expert insights.

Methodology

The Knowledge Audit Table Technique involves in-depth interviews with experts who possess significant experience in the field of microlearning, curriculum development, and instructional technology. During these interviews, experts share valuable insights into the cues and strategies they rely on when faced with difficulties in implementing pedagogical practices of microlearning to enhance students' learning experiences.

Data Collection

Through the Knowledge Audit Technique interviews, experts contribute insights into the cues and strategies they utilise when navigating challenges in microlearning implementation. Cues, in this context, are indicators or prompts guiding the design and implementation of cognitive tasks related to microlearning. Strategies are deliberate approaches employed to integrate cognitive tasks effectively.

Knowledge Audit Table

The Knowledge Audit Table serves as a comprehensive repository of critical cognitive elements derived from expert insights. It encompasses the cues and strategies recognized by experts as essential cognitive elements within the pedagogical practices of microlearning in higher education, aiming to enhance students' learning experiences. This table is a valuable resource, offering educators insights into effective approaches and techniques for overcoming cognitive challenges in microlearning.

Role in Framework Development

The insights derived from the Knowledge Audit Table Technique play a pivotal role in shaping and refining the AMPLIFY framework. By analysing and documenting the specific details of the identified cognitive elements, this technique aids in understanding their importance and impact on microlearning implementation. The Knowledge Audit Table serves as a bridge between expert knowledge and practical framework development, ensuring alignment with the identified cognitive requirements for successful microlearning experiences.

Implications for Design

The findings from the Knowledge Audit Technique interviews have direct implications for the design and structure of the AMPLIFY framework. Educators can leverage the insights provided in the Knowledge Audit Table to inform their practices, adopting effective approaches and techniques for enhancing the microlearning experience in higher education.

Conclusion of Development Phase

Concluding the Development Phase, the integration of the Knowledge Audit Technique enhances the overall depth and richness of the study. The refined AMPLIFY framework, informed by expert insights and cognitive elements, stands as a testament to the value of applying cognitive task analysis in the development of innovative educational practices.

Insights and Findings

The findings from the knowledge audit focused on creating effective pedagogical strategies for microlearning, emphasizing the detailed design and application of these practices to enhance students' learning experiences in higher education. The study explored how microlearning can be utilized to enhance learning by identifying cues, strategies, and challenges for each element, based on expert interviews. These insights align with educational theories, underscoring the transformative potential of microlearning in pedagogical practices. The findings highlight the critical components and their consistency with established educational principles and theories.

i. Key Element 1: Establishing Clear and Concise Learning Objectives Aligned with Students' Needs and Educational Goals

The first key element in creating impactful microlearning experiences in higher education is setting clear and concise learning objectives that align with students' needs and goals. This approach is supported by instructional design theory, particularly Biggs' (2014) work, which underscores the importance of clear objectives in guiding effective learning. Strategies such as clearly articulating learning outcomes and using specific, measurable language echo Kennedy et al.'s (2007) emphasis on precise instructional cues to facilitate learning. This element also ties into the cognitive presence aspect of the Community of Inquiry framework, providing a cognitive foundation that guides learners' understanding of the subject matter. When learning objectives align with students' needs and goals, it enhances motivation and engagement, leading to better knowledge retention (McNeill & Fitch, 2023).

Additionally, clear learning objectives align with Merrill's First Principles of Instruction, which advocate for the activation of prior experience and problem-centred learning. Defining objectives helps learners understand the purpose of their learning and how it applies to real-world problems (McNeill & Fitch, 2023).

The findings suggest essential strategies, such as incorporating input from stakeholders, including students, subject matter experts, and industry professionals, to ensure relevance and alignment with real-world needs. Another strategy is to regularly review and update learning objectives. These approaches are crucial, aligning with McNeill and Fitch's (2023) research, which shows that improved knowledge retention is linked to learning content with a clear purpose and specific real-world applications.

The challenges identified, such as subject complexity and varying learner needs, are consistent with contemporary literature on learner-centred education. Gardner (2006) and Vygotsky (1978) highlighted the importance of considering individual learner differences and tailoring educational objectives accordingly. Additionally, the difficulty in aligning objectives with evolving needs and goals reflects Senge's (2006) concept of the 'learning organization,' which emphasizes adaptability and continuous improvement in response to changing landscapes. Just as a learning organization adjusts its strategies based on new insights and circumstances, educational institutions must recognize and adapt to the shifting educational environment. Senge's notion stresses the importance of fostering a culture of adaptability in education, ensuring that institutions can thrive amid change by aligning objectives with the dynamic educational landscape.

ii. Key Element 2: Integrating Structured and Engaging Content Delivery in Microlearning

The second key element for developing effective microlearning experiences in higher education focuses on integrating structured and engaging content delivery. This aligns with existing research on multimedia learning and cognitive load theory. Identified elements such as clear and focused content, bite-sized units, logical progression, interactivity, and reinforcement resonate with Mayer's Cognitive Theory of Multimedia Learning (2014), which highlights managing cognitive load by presenting information coherently and incorporating interactivity to enhance understanding and retention. This is also consistent with Sweller's Cognitive Load Theory (2011), which aims to minimize extraneous cognitive load through organized content presentation. Breaking content into manageable units helps reduce intrinsic cognitive load, while logical progression aids schema acquisition. Interactivity and reinforcement optimize learning experiences by engaging learners and supporting memory retention. These principles collectively enhance the design and delivery of effective microlearning experiences. Studies by Kossen and Ooi (2021), Susilana (2022), and Fidan (2023) have emphasized microlearning's efficacy in reducing cognitive load and improving learning outcomes through bite-sized content delivery.

Additionally, structured and engaging content delivery supports cognitive presence within the Community of Inquiry framework, providing a structure for learners to explore, discuss, and understand content. According to Fiock (2020), instructors can enhance learners' cognitive presence through the four phases of the Practical Inquiry Model (PIM): triggering event, exploration, integration, and resolution. This allows students to engage in critical thinking, brainstorm, discuss problems, and reflect on the learning process. This component also aligns with Merrill's First Principles of Instruction, particularly the demonstration of skills principle. Breaking down content into manageable units with a logical progression enables learners to observe step-by-step demonstrations of skills or concepts, facilitating learning.

Experts noted challenges in this component that align with contemporary educational literature. Balancing depth and brevity in content, ensuring comprehensive coverage of essential knowledge and skills while maintaining focus, requires careful planning and organization. Designing engaging interactive elements involves intricate activity design and suitable technological integration. Creating effective reinforcement activities and assessments that cater to individual learner needs is also crucial. This approach aligns with Yin and Chai's (2020) perspective, which emphasizes understanding learners as individuals with unique characteristics and needs as the foundation of effective and responsive teaching.

iii. Key Element 3: Integrating Interactive Microlearning Experiences

The third essential component for creating impactful microlearning experiences in higher education centres on integrating interactive elements. This approach aligns with contemporary educational paradigms that emphasize active and experiential learning to foster deeper engagement and understanding. Strategies include using authentic sources, active learning techniques, interactive activities, and experiential learning opportunities. These methods align with constructivist theories, such as Dewey's experiential learning theory (1938), which advocates learning through direct experiences, and Piaget's theory of cognitive development (1971), which emphasizes hands-on engagement for knowledge assimilation. Interactive learning experiences also align with the social presence aspect of the Community of Inquiry (CoI) framework, fostering collaboration and communication among learners to create a sense of community. This is supported by Singh et al. (2022), who highlight the importance of significant dialogues among students. Additionally, interactive activities contribute to cognitive presence by encouraging active participation in knowledge construction. This component also aligns with Merrill's First Principles of Instruction (2002), particularly the principles of applying skills and integrating them into real-world activities, allowing learners to apply their knowledge in practical contexts. A study by Cardino and Ortega-Dela Cruz (2020) found that active learning approaches, including cooperative, deductive, and inductive methods, significantly impact academic performance. Integrating microlearning strategies within these approaches can enhance engagement, knowledge retention, and overall educational outcomes.

However, challenges in this component reflect practical difficulties discussed in the literature. Experts emphasize the complexities of resource constraints, which involve the time-consuming process of developing or sourcing authentic and relevant materials, often requiring access to specific resources, expertise, or technologies. Additionally, the successful implementation of interactive activities and experiential learning depends on the availability and accessibility of suitable technologies and platforms (Haleem et al., 2022). Curriculum alignment is another key hurdle, highlighting the difficulty of ensuring interactive learning experiences harmonize with learning objectives and the broader curriculum. This challenge demands meticulous planning and coordination across various topics, modules, or subjects. Shaltry (2020) noted that misalignment in the curriculum can lead to gaps in students' learning, disrupting their educational progression. Designing assessments that effectively measure learner progress and provide timely feedback also presents challenges, especially when traditional methods need adaptation. Moges (2018) emphasized the importance of prioritizing student learning assessment to enhance educational outcomes. Lastly, experts underscore the necessity of instructor competence in designing interactive learning experiences, requiring pedagogical skills, familiarity with technology tools, and the ability to create a supportive learning environment that fosters active participation. Valzado et al. (2023) affirmed that strong pedagogical skills enhance classroom instruction, promoting a more effective teaching and learning process.

iv. Key Element 4: Incorporating Personalization Through Learner-Centric Microlearning Activities and Customized Content

The fourth essential component for impactful microlearning experiences in higher education revolves around the personalization of learner-centric activities and customized content. This approach has gained attention for its potential to enhance engagement, knowledge retention, and overall learning outcomes.

A significant strategy in this component is learner profiling, which aligns with constructivist principles by considering the diverse characteristics of individual learners. Constructivism, as highlighted by Piaget (1971), emphasizes that learners construct their own understanding through meaningful interactions with the learning environment. Learner profiling involves gathering information on a learner's preferences, strengths, weaknesses, and prior experiences to personalize and tailor educational experiences, making the material more meaningful and relevant. Vygotsky (1978) also stressed the importance of scaffolding and adapting instruction to individual learners' zones of proximal development. Gacusan et al. (2023) emphasized tailoring education to individual learners and leveraging their preferred learning styles to maximize learning outcomes. Additionally, Latif, Subramaniam, and Khatab (2020) noted that initiatives designed to meet students' needs could boost their active participation.

The design of adaptive microlearning paths is another key strategy, supported by Vygotsky's Zone of Proximal Development (1978) and Sweller's cognitive load theory (2011). These theories highlight the importance of providing appropriately challenging tasks and content. Adaptive paths adjust based on learners' progress, maintaining an optimal level of challenge to promote effective learning. Romanenko et al. (2023) argued that microlearning offers a high degree of individualization and personalization, aligning closely with the unique characteristics and preferences of learners. This approach also aligns with Merrill's First Principles of Instruction (2013), which emphasize activating prior knowledge and skills and applying new knowledge in a problem-centred manner.

Another strategy involves offering varied content formats, as supported by Mayer's Multimodal Learning Theory. By providing content in text, audio, video, and interactive modules, this approach acknowledges different learning preferences and cognitive processing styles. This aligns with Merrill's principles of demonstration and integration, helping learners integrate new information into their existing cognitive structures. Ordu (2021) highlighted that incorporating visual aids such as pictures, video clips, and objects helps students form a tangible mental representation of the subject matter, reinforcing learning and understanding.

Empowering learners through choice and flexibility aligns with self-determination theory (Deci & Ryan, 1985), which suggests that intrinsic motivation is enhanced when learners perceive autonomy, competence, and relatedness. By allowing learners to select activities or topics of interest and providing options for pacing and sequencing, this strategy taps into learners' intrinsic motivations, leading to higher engagement and ownership of learning.

Incorporating feedback and reflection is grounded in the theories of metacognition (Flavell, 1979) and self-regulated learning (Zimmerman & Schunk, 2011). These theories propose that learners become more effective when they are aware of their cognitive processes and can regulate their learning strategies. Providing timely and meaningful feedback supports learners' self-monitoring and reflection, fostering a deeper understanding of their learning processes. Miller and Staley (2021) found that feedback and reflection practices enhanced students' capacity for self-reflection and improvement.

However, implementing personalized microlearning strategies presents challenges. Developing personalized content and activities requires substantial time, effort, and resources. Scalability is another issue, especially in large learning environments, necessitating advanced technology and innovative instructional design methods. Gathering and analysing learner data is crucial for effective personalization, and designing personalized learning experiences requires pedagogical expertise and a deep understanding of instructional design principles and cognitive processes. Finally, successful implementation often depends on a robust and adaptive technological infrastructure, requiring investment in technology and supportive platforms.

v. Key Element 5: Incorporating Feedback and Assessment for Effective Microlearning Experiences

The fifth key element for creating impactful microlearning experiences in higher education involves integrating feedback and assessment. This element includes timely and constructive feedback, formative and summative assessments, varied assessment methods, alignment with learning objectives, and a focus on continuous improvement.

Effective microlearning revolves around clear communication of expectations and assessment criteria, aligning with the principles of constructive alignment (Biggs, 2014). Transparent communication about assessment criteria enhances students' understanding of expectations, reduces anxiety, and fosters a conducive learning environment (Jönsson & Prins, 2019). Timely feedback is crucial, as emphasized by Hattie and Timperley's (2007) feedback model, which highlights the importance of providing actionable feedback to enhance learning. Constructive feedback, characterized by specificity and guidance for improvement, aligns with formative assessment principles (Black & Wiliam, 2009), promoting continuous learning and development. Cevallos Menéndez et al. (2019) found that formative assessment enhances academic performance and fosters collaboration among students, highlighting the importance of feedback in microlearning.

Experts also emphasize integrating both formative and summative assessments to meet diverse learning needs. A study by Ismail et al. (2022) demonstrated that both types of assessments are effective, with formative assessment being particularly effective in enhancing academic motivation, reducing test anxiety, and fostering self-regulation skills. Incorporating microlearning techniques within formative assessment can amplify these benefits. Diverse assessment methods, such as quizzes, projects, presentations, group work, and discussions, accommodate different learning styles, supported by Gardner's theory of multiple intelligences (2006), which suggests that students possess varied strengths that can be tapped through diverse assessment approaches.

Continuous improvement aligns with the principles of self-regulated learning (Zimmerman & Schunk, 2011), where students actively engage in their learning process. Zarestky et al. (2022) suggested that encouraging students to reflect on feedback and identify areas for growth aligns with metacognitive strategies that foster lifelong learning and self-improvement. This approach develops students' metacognitive awareness and nurtures self-regulated learning behaviors, problem-solving, and critical thinking skills.

The integration of feedback and assessment gains depth and significance when viewed through the lenses of the Community of Inquiry (CoI) framework and Merrill's First Principles of Instruction. The CoI framework (Garrison et al., 2000) highlights the role of cognitive presence in fostering meaningful learning experiences. Clear communication and alignment with learning objectives engage learners by providing context and purpose for learning, aligning with Merrill's principle of activation. Constructive feedback and varied assessment methods align with Merrill's principle of demonstration, where feedback serves as a demonstration of correct performance and diverse assessment methods exemplify knowledge application across contexts. Formative assessments and timely feedback resonate with the application principle, enabling learners to apply their knowledge and receive iterative feedback. Continuous improvement harmonizes with the integration principle, encouraging students to reflect on feedback and integrate new knowledge and skills into their cognitive frameworks. These theories reveal how feedback and assessment support collaborative engagement, reflective learning, and seamless integration for creating impactful microlearning experiences in higher education.

Several challenges arise in integrating feedback and assessment for impactful microlearning experiences. Firstly, the time and effort required from educators to provide timely, detailed feedback, especially for large student cohorts or complex assessments, align with Ropohl's (2019) concerns about balancing feedback quality and quantity. Secondly, tailoring feedback and assessments to diverse learning styles, abilities, and needs of students resonates with Gardner's (2006) research on learning styles, highlighting the necessity of equitable and personalized approaches. Thirdly, balancing deep understanding and covering a breadth of knowledge, as underlined by Bloom's taxonomy (1956), emphasizes designing microlearning assessments that encompass higher-order thinking skills and comprehensive feedback. Lastly, engaging students with feedback, facilitated by the CoI framework's emphasis on social presence (Garrison et al., 2000), is essential for meaningful learning experiences. Addressing these challenges requires a nuanced understanding of educational theories and best practices, shaping strategies for enhancing microlearning experiences.

vi. Key Element 6: Incorporating Effective Microlearning Delivery Methods

The sixth key element for creating impactful microlearning experiences in higher education emphasizes the incorporation of microlearning delivery methods through multimedia integration, mobile-friendly design, and just-in-time learning.

Integrating these delivery methods can significantly enhance educational effectiveness. Strategies like multimedia integration, mobile-friendly design, and just-in-time learning can be enriched through the lenses of the Community of Inquiry (CoI) framework and Merrill's First Principles of Instruction. Multimedia elements align with CoI's social presence, fostering engagement and interaction (Garrison et al., 2000), and support Mayer's Cognitive Theory of Multimedia Learning (2014) by engaging multiple sensory modalities. Ensuring mobile-friendly design for microlearning content resonates with the principles of ubiquitous learning, as advocated by Sharples et al. (2007), allowing learners to access materials flexibly across devices. This strategy also aligns with Merrill's activation principle (2013), which emphasizes engaging learners in real-world, authentic tasks that prepare them for subsequent learning experiences. Just-in-time learning, viewed through CoI's cognitive presence, enables learners to apply knowledge in real-world contexts (Garrison et al., 2000), aligning with Merrill's application principle (2013).

These strategies can be operationalized by creating multimedia content that supports CoI's community building, optimizing design for mobile access, and facilitating just-in-time resources that align with Merrill's activation principle. This comprehensive approach, informed by educational theories and research, holds promise for fostering learner engagement, flexibility, and practical application within microlearning contexts.

However, the effective incorporation of microlearning delivery methods presents several challenges for educators and instructional designers. Creating high-quality multimedia content requires expertise in instructional design, multimedia production, and pedagogical considerations. Ghimire (2020) highlights the need for professional development programs for instructors to cultivate the necessary knowledge and skills for effective teaching. Meeting the principles of effective multimedia learning (Mayer, 2013) demands careful consideration of cognitive load, visual and auditory coherence, and instructional flow.

Technological requirements, such as implementing multimedia integration and mobile-friendly design, may entail complex technological needs and compatibility challenges. Adapting microlearning content to suit different devices, platforms, and learning contexts can be challenging and time-consuming. Ensuring that just-in-time resources are relevant, up-to-date, and easily accessible requires continuous management and maintenance. Navigating these challenges necessitates a comprehensive approach that balances pedagogical expertise, technological proficiency, and a commitment to learner-centeredness. Addressing these hurdles effectively will enable educators to harness the full potential of microlearning delivery methods while ensuring impactful and engaging learning experiences.

vii. Key Element 7: Integrating Social Interaction Through Collaborative Learning and Peer Feedback

The seventh essential element for creating impactful microlearning experiences in higher education involves integrating social interaction through collaborative learning and peer feedback. Strategies such as group projects, peer assessments, online discussions, reflective practices, facilitator guidance, and establishing norms and expectations have shown potential to foster a vibrant and effective microlearning environment that extends beyond traditional teaching methods.

Research supports the positive impact of collaborative learning on social interaction and shared learning experiences. Vygotsky's Social Development Theory emphasizes that learning is enhanced through peer-to-peer interactions, where students collaboratively construct knowledge through discussions and activities. Microlearning complements this by breaking down complex concepts into manageable units, facilitating focused group discussions and interactions, which enhance engagement and knowledge construction. Fidan's 2023 study showed that microlearning content creates an engaging environment that fosters critical thinking, student interaction, collaborative learning, and comprehension of new concepts.

Incorporating peer feedback and review aligns with social constructivism principles, where learning is an active process shaped by social interactions. Bandura's Social Learning Theory suggests that observing and interacting with peers can lead to acquiring new skills and behaviours. Studies such as Kumar et al. (2023) demonstrate that peer assessment enhances understanding of the subject matter and encourages self-regulated learning, critical thinking, and problem-solving skills. Microlearning can facilitate peer feedback by allowing students to review smaller portions of work, provide targeted feedback, and gradually refine their evaluative skills.

The integration of online discussion forums is supported by the Community of Inquiry (CoI) framework (Garrison et al., 2000), which highlights the importance of cognitive, social, and teaching presence in online learning environments. Al-Areibi et al. (2022) noted that asynchronous and synchronous interactions offer students valuable learning opportunities, boosting motivation and success. Trespalacios and Uribe-Florez (2020) found that online discussions help build a supportive online community, enabling meaningful dialogues and relationship-building despite physical separation. Microlearning supports these discussions by presenting concise prompts or questions that encourage focused exchanges, aligning with the asynchronous nature of the platform. Reflective activities for students to analyse and discuss their learning experiences align with the cognitive and social presence elements of the CoI framework, encouraging critical thinking, self-assessment, and metacognitive reflection.

Another strategy is providing clear guidelines and facilitator support to ensure effective group interactions and a positive social learning environment, aligned with the principles of the CoI framework. The CoI framework emphasizes the importance of social, cognitive, and teaching presence in online and blended learning environments. Facilitator guidance involves providing instructors with tools and techniques to manage group interactions, guide discussions, and foster a positive learning community. These aspects align with the teaching presence component of the CoI framework, focusing on the design, organization, and facilitation of online and blended learning experiences. Armah et al. (2023) highlighted the significance of teacher presence as a main predictor of online course satisfaction, emphasizing the vital role of facilitator support and clear guidelines in successful microlearning experiences.

Within the CoI framework, social presence includes creating a sense of community and interaction among learners. Setting clear expectations and fostering a culture of respect, active participation, and constructive feedback aligns with the social presence aspect of the CoI framework. This contributes to creating an environment where learners feel comfortable engaging with peers and instructors, supporting meaningful learning interactions. The CoI framework emphasizes building a learning community where learners can engage in collaborative discussions, share ideas, and provide feedback. Establishing norms and expectations that promote respectful and productive interactions is key to fostering a positive social learning environment, a central aspect of the CoI framework. Mutezo & Maré (2022) demonstrated the importance of social presence for successful collaborative learning. Microlearning, when embedded within the learning ecosystem, supports learners with bite-sized content that reinforces their sense of belonging, fosters ongoing engagement, and facilitates seamless interactions within the learning community.

While integrating social interaction through collaborative microlearning and peer feedback enhances learning experiences, these strategies also present challenges. Managing group dynamics to ensure equitable participation and contributions demands clear communication and conflict resolution strategies. Ensuring assessment validity and reliability in group assessments requires transparent rubrics and recognition of individual contributions. Stanley et al. (2020) noted that well-written rubrics facilitate consistent marking across a student cohort, contributing to student progress and success. Selecting suitable technology platforms for effective interaction and feedback necessitates thorough research and user-friendly interfaces. Haleem et al. (2022) concurred that effective use of technology enhances educational experiences, student interest, and learning outcomes. Cultivating a supportive environment as advocated by the Community of Inquiry involves creating a safe and inclusive learning space where students feel comfortable providing and receiving feedback and engaging in collaborative learning. Microlearning activities can provide focused content that complements collaborative interactions, enhancing overall learning experiences.

viii. Key Element 8: Seamless integration with curriculum through integration of learning technologies and integration with learning ecosystem

The eight essential element focuses on effectively integrating microlearning into higher education by aligning it seamlessly with the curriculum through the use of learning technologies and integration within the learning ecosystem. This integration aims to align with educational objectives, promote coherence and continuity, and optimize existing resources.

Initially, integrating microlearning technologies with the curriculum necessitates aligning them deliberately with educational goals and desired outcomes. This alignment ensures that these tools complement the curriculum and contribute to achieving educational objectives. Hamdoun (2021) stresses the importance of aligning instructional methods with learning goals to enhance student achievement. Secondly, seamlessly integrating learning technologies within the curriculum enhances coherence and continuity (Panke et al., 2017), ensuring that microlearning experiences integrate smoothly into the educational framework and avoid disruptions. Thirdly, incorporating microlearning tools and technologies should enhance existing educational resources synergistically, where technology enhances the effectiveness of traditional materials. Díaz Redondo et al. (2021) found that integrating microlearning into traditional e-learning platforms not only improves knowledge transmission but also enhances student engagement, highlighting its potential as a valuable addition to blended learning frameworks.

Several strategic approaches are advocated for achieving seamless integration of microlearning into the curriculum. Curriculum Mapping is foundational, aligning learning technologies with specific objectives and topics to ensure a cohesive learning experience (Reniers et al., 2022). This approach identifies strengths and weaknesses in teaching particular outcomes, promising curriculum refinement for greater educational efficiency. Additionally, ongoing training and professional development are crucial for educators to adeptly integrate learning technologies (Germuth, 2018), ensuring effective teaching practices that enhance student learning. Conducting pilot projects and evaluations is also essential to refine integration strategies based on feedback (Stufflebeam, 2003), systematically gathering data to improve educational experiences continuously.

Assessing scalability and sustainability is crucial for effective integration, aligning with the principles of the Technology Acceptance Model (Davis, 1989) to ensure technologies are perceived as useful and user-friendly by educators and learners alike. This approach facilitates successful implementation and long-term maintenance of chosen technologies. Lastly, aligning microlearning and technologies within the broader learning ecosystem promotes a comprehensive, interconnected learning experience (Wang, 2023), preventing fragmentation and ensuring educational coherence.

Addressing integration challenges involves establishing robust technological infrastructure to support seamless integration of learning tools, providing comprehensive training and support for educators, and addressing resistance to change among stakeholders. Managing costs and resources, adapting to evolving technologies, and maintaining pedagogical alignment are essential for meaningful integration and positive learning outcomes.

Table 3 illustrates the knowledge audit, providing a detailed analysis of the critical cognitive components within the AMPLIFY framework, cues and strategies for effective microlearning, and the inherent challenges faced in implementing these strategies.

Table 3. Knowledge audit table

No.	AMPLIFY's Critical Cognitive Components	Cues and Strategies	Why Difficult?
i.	The establishment of clear and concise learning objectives and alignment with needs and goals	**Cues:** 1. Confusion or lack of understanding regarding learning objectives.	
		2. Mismatch between stated objectives and learners' needs.	
		3. Inconsistent alignment of learning objectives with overarching goals.	1. Subject Complexity: Some subject matters may be inherently complex, requiring careful consideration to define clear and focused learning objectives that adequately cover the required knowledge and skills.
		4. Varied levels of engagement or achievement related to learning objectives.	2. Varying Learner Needs: Designing learning objectives that align with the diverse needs and goals of learners can be challenging, as different learners may have different backgrounds, prior knowledge, and skill levels.
		Strategies: 1. Conducting a thorough analysis of learner needs, including their prior knowledge, skills, and aspirations.	3. Curriculum Alignment: Ensuring that learning objectives align with the broader curriculum or program goals requires careful coordination and consideration of the overall learning outcomes.
		2. Aligning the learning objectives with the overall goals and objectives of the course, program, or curriculum.	4. Emerging Needs and Changing Goals: Educational contexts are dynamic, and needs and goals may evolve over time. Designing learning objectives that can adapt to these changing needs and goals requires ongoing reflection and adjustment.
		3. Incorporating input from stakeholders, such as students, subject matter experts, and industry professionals, to ensure relevance and alignment with real-world needs.	

continued on following page

Table 3. Continued

No.	AMPLIFY's Critical Cognitive Components	Cues and Strategies	Why Difficult?
		4. Regularly reviewing and updating learning objectives to ensure they remain aligned with evolving needs and goals.	
ii.	The integration of structured and engaging content delivery through clear and focused content, bite-sized units, logical progression, and interactivity and reinforcement	**Cues:** 1. Clear and Focused Content: Lack of clarity in content organization and delivery.	
		2. Bite-Sized Units: Disengagement during lesson.	
		3. Logical Progression: Difficulty in following the content.	1. Balancing depth and brevity: Ensuring that the content is comprehensive enough to cover the necessary knowledge and skills while keeping it concise and focused can be challenging.
		4. Interactivity and Reinforcement: Limited engagement during class activities.	2. Maintaining coherence: Creating a seamless flow and logical progression of content requires careful planning and organization.
		Strategies: 1. Clear and Focused Content: • Conduct a thorough content analysis and mapping to determine the most essential and relevant information to be included.	3. Designing effective interactive elements: Designing interactive activities and incorporating appropriate technology tools to enhance engagement and learning effectiveness can be complex.
		2. Bite-Sized Units: • Breaking down the content into smaller, manageable chunks or modules.	4. Providing meaningful reinforcement: Designing reinforcement activities and assessments that effectively reinforce learning and address individual learner needs can be demanding.

continued on following page

Table 3. Continued

No.	AMPLIFY's Critical Cognitive Components	Cues and Strategies	Why Difficult?
		3. Logical Progression: • Plan and organize the learning materials in a progressive manner, building upon previously covered concepts and skills.	
		4. Interactivity and Reinforcement: • Incorporate interactive components such as quizzes, simulations, discussions, and multimedia to enhance learner engagement and participation. • Design activities, exercises, and assessments that allow learners to apply their knowledge and skills, receive feedback, and reinforce their understanding.	
iii.	The integration of interactive learning experiences through authentic sources, active learning, interactive activities, and experiential learning	**Cues:** 1. Authentic Sources: Limited engagement in activities that involve authentic sources.	
		2. Active Learning: Inactive participation in learning tasks.	
		3. Interactive Activities: Low involvement in class activities.	1. Resource Constraints: Developing or sourcing authentic and relevant materials can be time-consuming and may require access to specific resources, expertise, or technologies.
		4. Experiential Learning: Lack of opportunities for learners to directly apply theoretical knowledge in real or simulated environments.	2. Technology and Infrastructure: Implementing interactive activities and experiential learning may depend on the availability and accessibility of suitable technologies, platforms, and infrastructure to support such experiences.

continued on following page

Table 3. Continued

No.	AMPLIFY's Critical Cognitive Components	Cues and Strategies	Why Difficult?
		Strategies: 1. Authentic Sources: Identifying and curating authentic sources, ensuring their alignment with learning objectives, and integrating them into the learning materials and activities to enhance relevance and authenticity.	3. Curriculum Alignment: Ensuring that the interactive learning experiences align with the learning objectives and curriculum can be complex, as it requires careful planning and coordination across different topics, modules, or subjects.
		2. Active Learning: Designing and facilitating activities that promote learner engagement, critical thinking, problem-solving, and collaboration. It may include group discussions, hands-on experiments, role plays, or project-based learning.	4. Assessment and Feedback: Designing effective assessments that measure learner progress and provide timely feedback can be demanding, particularly in the context of interactive and experiential learning, where traditional assessment methods may need to be adapted.
		3. Interactive Activities: Designing and implementing interactive elements such as quizzes, simulations, games, and multimedia presentations to enhance learner engagement, motivation, and knowledge retention.	5. Instructor Competence: Facilitating interactive learning experiences requires instructors to possess the necessary pedagogical skills, knowledge of technology tools, and the ability to create a supportive learning environment that fosters active participation and engagement.
		4. Experiential Learning: Designing opportunities for learners to actively apply their knowledge and skills in real or simulated environments, providing hands-on experiences, reflection activities, and feedback to facilitate deeper learning.	
iv.	The incorporation of personalization through learner-centric activities and customized content.	**Cues:** 1. Lack of learner engagement in activities tailored to their preferences.	
		2. Limited participation in personalized learning paths.	

continued on following page

Table 3. Continued

No.	AMPLIFY's Critical Cognitive Components	Cues and Strategies	Why Difficult?
		3. Low utilization of varied content formats by learners.	1. Resource Intensiveness: Developing personalized content and activities for each learner requires significant time, effort, and resources.
		4. Limited choices or flexibility in learner selections.	2. Scalability: Ensuring personalization at scale, especially in large learning environments, can be challenging, as it may require sophisticated technology and instructional design approaches.
		5. Absence of mechanisms for feedback and reflection.	3. Data Collection and Analysis: Gathering and analysing learner data to inform personalization strategies
		Strategies: 1. Developing learner profiles by gathering information about preferences, interests, and prior knowledge.	4. Pedagogical Expertise: Designing effective personalized learning experiences requires expertise in instructional design and understanding of learners' cognitive processes and learning needs.
		2. Designing adaptive learning paths that adjust based on learners' progress and provide personalized recommendations and content.	5. Technological Infrastructure: Implementing personalized learning experiences often relies on robust and adaptive learning platforms, which may require investment in technology and infrastructure.
		3. Offering content in different formats (text, audio, video, interactive modules) to accommodate diverse learning styles.	
		4. Allowing learners to have choices in selecting activities or topics of interest and providing options for pacing and sequencing.	
		5. Providing timely and meaningful feedback, encouraging self-reflection, and enabling learners to track their progress and set personal goals.	

continued on following page

Table 3. Continued

No.	AMPLIFY's Critical Cognitive Components	Cues and Strategies	Why Difficult?
v.	The inclusion of feedback and assessment through timely and constructive feedback, formative and summative assessments, varied assessment methods, alignment with learning objectives, and continuous improvement	**Cues:** 1. Timely and Constructive Feedback: • Lack of timely feedback, leading to potential gaps in understanding. • Absence of constructive feedback hindering students' learning and growth.	
		2. Formative and Summative Assessments: • Limited use of both formative and summative assessments, impacting comprehensive evaluation.	
		3. Varied Assessment Methods: • Restriction to a single assessment format, potentially limiting a holistic understanding of students' capabilities.	
		4. Alignment with Learning Objectives: • Disconnect between assessment content and the defined learning objectives, affecting the effectiveness of evaluations.	1. Time Constraints: Providing timely and detailed feedback requires significant time and effort from educators, especially when dealing with a large number of students or complex assessments.
		5. Continuous Improvement: • Lack of a systematic approach to refining assessment strategies based on feedback and results, potentially hindering the learning experience.	2. Individual Differences: Students have diverse learning styles, abilities, and needs. Tailoring feedback and assessments to meet individual requirements can be a complex task for educators.

continued on following page

Table 3. Continued

No.	AMPLIFY's Critical Cognitive Components	Cues and Strategies	Why Difficult?
		Strategies: 1. Timely and Constructive Feedback: Provide feedback in a timely manner after assessments and ensure feedback is constructive, specific, and actionable	3. Balancing Depth and Breadth: Assessments need to strike a balance between assessing deep understanding and covering a breadth of knowledge. Designing assessments that effectively evaluate higher-order thinking skills while still providing comprehensive feedback can be a challenge.
		2. Formative and Summative Assessments: Design a balanced assessment approach that incorporates both formative assessments for ongoing feedback and summative assessments for comprehensive evaluation.	4. Engaging Students: Encouraging students to actively engage with feedback and use it to improve their learning requires fostering a positive learning environment and promoting a growth mindset.
		3. Varied Assessment Methods: Diversify assessment methods, including quizzes, projects, presentations, and practical demonstrations, to provide a more comprehensive understanding of students' skills and knowledge.	
		4. Alignment with Learning Objectives: Ensure assessments align with stated learning objectives and provide feedback that reflects intended learning goals.	
		Continuous Improvement: Establish a feedback loop for continuous improvement by analysing assessment results, gathering feedback from students, and adapting assessment strategies to enhance the overall learning experience.	

continued on following page

Table 3. Continued

No.	AMPLIFY's Critical Cognitive Components	Cues and Strategies	Why Difficult?
vi.	The incorporation of microlearning delivery methods through multimedia integration, mobile friendly design, and just in time learning	**Cues:** 1. Multimedia Integration: Limited engagement with multimedia content in the learning materials.	
		2. Mobile-Friendly Design: Challenges with accessing learning content on mobile devices.	1. Content Design and Development: Creating high-quality multimedia content requires expertise in instructional design, multimedia production, and pedagogical considerations.
		3. Just-in-Time Learning: Lack of alignment with just-in-time learning principles, resulting in a mismatch with learners' immediate needs.	2. Technological Requirements: Implementing multimedia integration and mobile-friendly design may involve complex technology requirements and compatibility issues.
		Strategies: 1. Create Engaging Multimedia Content: • Design and develop multimedia content that is visually appealing, interactive, and effectively conveys the key concepts.	3. Customization and Adaptation: Adapting the microlearning content to suit different devices, platforms, and learning contexts can be challenging and time-consuming.
		2. Mobile-Friendly Design: • Optimize learning materials for mobile devices, considering different screen sizes and functionalities. • Utilize responsive design principles to ensure seamless access to content on various mobile platforms.	4. Just-in-Time Learning Challenges: Ensuring that the just-in-time resources are relevant, up-to-date, and easily accessible requires continuous curation and maintenance.

continued on following page

Table 3. Continued

No.	AMPLIFY's Critical Cognitive Components	Cues and Strategies	Why Difficult?
		3. Just-in-Time Learning: • Curate and provide easily accessible resources that learners can refer to at the moment of need, allowing them to quickly find answers or solutions to specific problems. • Implement systems for learners to easily access specific content when needed, addressing immediate learning requirements.	
vii.	The integration of social interaction through collaborative learning, and peer feedback and review	**Cues:** 1. Limited Participation in Collaborative Activities: Observations of low engagement or interaction among students during collaborative learning tasks.	
		2. Lack of Peer Feedback Opportunities: Absence of mechanisms or instances for students to provide feedback to their peers.	
		Strategies: 1. Facilitate Collaborative Learning Activities: Design and implement structured collaborative learning tasks, projects, or discussions that encourage active participation and interaction among students.	1. Group Dynamics: Managing group dynamics and ensuring equal participation and contribution from all group members.
		2. Incorporate Peer Review Mechanisms: Integrate peer review components into assessments, assignments, or projects to provide students with opportunities to evaluate and provide feedback to their peers.	2. Assessment Validity and Reliability: Ensuring fairness and consistency in assessing group work and peer feedback, while also accounting for individual contributions.

continued on following page

Table 3. Continued

No.	AMPLIFY's Critical Cognitive Components	Cues and Strategies	Why Difficult?
		3. Utilize Online Platforms for Collaboration: Implement digital tools or online platforms that facilitate collaborative learning, allowing students to engage in discussions, group projects, or shared activities.	3. Technology and Platform Considerations: Selecting and implementing suitable online platforms or tools that support effective social interaction and peer feedback.
		4. Provide Clear Guidelines for Peer Feedback: Establish clear guidelines and criteria for peer feedback to ensure constructive and meaningful input from students.	4. Cultivating a Supportive Environment: Creating a safe and inclusive learning environment where students feel comfortable providing and receiving feedback and engaging in collaborative learning.
		5. Promote a Positive Collaborative Culture: Foster a positive and inclusive collaborative culture within the learning environment, encouraging students to actively participate, share ideas, and provide support to their peers.	
		6. Integrate Social Learning Components: Embed social learning components, such as discussion forums, group projects, or collaborative assignments, into the course structure to enhance social interaction	
viii.	Seamless integration of microlearning with curriculum through integration of learning technologies and integration with learning ecosystem	**Cues:** 1. Disjointed Learning Technologies: Identification of a lack of seamless integration between microlearning and existing learning technologies within the educational framework.	

continued on following page

Table 3. Continued

No.	AMPLIFY's Critical Cognitive Components	Cues and Strategies	Why Difficult?
		2. Absence of Learning Ecosystem Integration: Observations indicating a gap in incorporating microlearning harmoniously into the broader learning ecosystem or curriculum structure.	1. Technological Infrastructure: Ensuring the availability and compatibility of necessary technology infrastructure to support the seamless integration of microlearning across the curriculum.
		Strategies: 1. Technology Alignment and Integration: Align microlearning platforms and tools with existing learning technologies to ensure a smooth and cohesive integration.	2. Training and Support: Providing adequate training and support for educators to effectively utilize and integrate microlearning into their teaching practices.
		2. Learning Ecosystem Mapping: Conduct a comprehensive analysis of the learning ecosystem to identify opportunities for integrating microlearning effectively.	3. Resistance to Change: Overcoming resistance to change from educators or stakeholders who may be hesitant or sceptical about the integration of new approach into the curriculum.
		3. Curriculum Alignment: Align microlearning modules with the overall curriculum, ensuring they complement and enhance the broader educational objectives.	4. Cost and Resources: Managing the costs associated with the integration of microlearning and allocating resources for the necessary infrastructure, training, and ongoing support.
		4. Faculty Training and Support: Provide training and support for educators to seamlessly integrate microlearning into their teaching methods and curriculum planning.	5. Continuous Updates and Upgrades: Dealing with the rapid evolution of learning technologies and the need to keep up with updates and upgrades to ensure ongoing compatibility and effectiveness.
		5. Assessment Integration: Integrate microlearning assessments with existing evaluation methods to ensure a cohesive evaluation of student performance.	6. Pedagogical Alignment: Ensuring that the integration of microlearning is pedagogically sound and aligned with effective instructional practices, avoiding superficial or arbitrary use of technology without meaningful impact on learning outcomes.
		6. Feedback Mechanisms: Implement feedback mechanisms to gather insights from both educators and students, facilitating continuous improvement in the integration of microlearning with the curriculum.	

8.2.2 Review of Knowledge Audit for Developing Microlearning Pedagogical Practices

In conclusion, the exploration into refining pedagogical practices to optimize student learning experiences through microlearning in higher education underscores the critical importance of adopting a multi-dimensional approach. This study has meticulously examined eight fundamental elements that collectively pave the way towards creating an enriched educational landscape.

Firstly, clear and concise learning objectives establish a sense of purpose and direction in microlearning initiatives. These objectives not only guide instructional design but also ensure that learning outcomes are clearly defined and achievable. Personalization within microlearning contexts plays a pivotal role in nurturing individual growth and addressing diverse learning needs. By tailoring content and delivery methods to suit the preferences and pace of each learner, educators can enhance engagement and facilitate deeper understanding.

Secondly, effective feedback and assessment mechanisms are integral to optimizing the learning experience in microlearning environments. Timely and constructive feedback helps students track their progress, identify areas for improvement, and reinforces learning objectives. This continuous feedback loop ensures that learning remains aligned with educational goals and supports ongoing improvement in student performance.

Thirdly, microlearning delivery methods must be dynamic and adaptive to cater to the evolving needs of modern learners. Utilizing concise and focused learning units allows for flexibility in learning pace and accessibility, accommodating diverse learning styles and preferences. A structured and engaging content delivery mechanism captivates student interest and promotes active learning. Interactive elements such as quizzes, simulations, and multimedia resources not only reinforce concepts but also stimulate curiosity and critical thinking skills.

Moreover, fostering social interaction within microlearning environments enhances knowledge sharing and collaboration among students. Collaborative activities, discussion forums, and peer learning opportunities create a sense of community and encourage active participation. This collaborative approach not only deepens understanding through shared perspectives but also promotes a supportive learning environment where students feel valued and motivated to contribute.

Lastly, the seamless integration of microlearning within the broader curriculum constructs a cohesive framework of education. By aligning microlearning initiatives with existing course structures and learning objectives, educators ensure continuity and coherence in educational delivery. This integration minimizes disruptions and enhances the overall learning experience by reinforcing connections between microlearning activities and broader educational goals.

When these elements are thoughtfully integrated, they form an innovative pedagogical composition that not only enhances microlearning practices but also elevates the broader educational journey. This holistic approach fosters an environment where students' learning experiences are genuinely transformative, equipping them with essential skills and knowledge for future success in their academic and professional endeavours.

CONCLUSION

In this chapter, we explored the vital role of knowledge audits as part of Applied Cognitive Task Analysis (ACTA), focusing on their application in both electronic training for professional development and the development of microlearning pedagogical practices. The knowledge audit is a crucial step in identifying cognitive challenges and knowledge gaps within training and educational environments. By systematically evaluating these areas, knowledge audits provide critical insights that inform more effective instructional design and enhance learning outcomes.

In Section 8.1, the knowledge audit for electronic training revealed several cognitive challenges related to system design, functionality, and navigation. Experts highlighted difficulties encountered by users unfamiliar with online learning systems, including navigating non-linear content and managing system functionalities. The knowledge audit provided specific strategies to address these challenges, such as simplifying system navigation, integrating multimedia elements, and providing clearer guidance for users. These findings underscore the importance of designing user-friendly, engaging online training platforms that cater to the cognitive needs of professionals.

Section 8.2 applied the knowledge audit framework to microlearning pedagogical practices, examining how this method can improve the design and delivery of short, targeted learning modules. The audit revealed that by addressing key knowledge gaps and refining content delivery, educators can create more focused and efficient microlearning experiences. Strategies such as incorporating interactive and multimedia content, ensuring clear instructional guidance, and aligning learning modules with the specific cognitive demands of learners were identified as crucial for optimizing microlearning environments.

In conclusion, the knowledge audit serves as a powerful tool for improving both professional development training and microlearning. By systematically identifying cognitive task challenges and knowledge gaps, it enables instructional designers and educators to refine their approaches, creating more effective, learner-centred training and educational programs. As education and professional development continue to evolve, knowledge audits will remain integral to ensuring that learning systems are aligned with the needs of modern learners.

REFERENCES

Al-Areibi, I., Dickson, B. A., & Kotsopoulos, D. (2022). An analysis of synchronous and asynchronous online undergraduate motivation during the COVID pandemic. *International Journal of E-Learning & Distance Education*, 37(1). Advance online publication. DOI: 10.55667/ijede.2022.v37.i1.1213

Armah, J. K., Bervell, B., & Bonsu, N. O. (2023). Modelling the role of learner presence within the community of inquiry framework to determine online course satisfaction in distance education. *Heliyon*, 9(5). Advance online publication. DOI: 10.1016/j.heliyon.2023.e15803

Biggs, J. (2014). Constructive alignment in university teaching. *HERDSA Review of Higher Education*, 1, 5–22.

Black, P., & Wiliam, D. (2009). Developing the theory of formative assessment. *Educational Assessment, Evaluation and Accountability*, 21(1), 5–31. DOI: 10.1007/s11092-008-9068-5

Bloom, B. S. (1956) Taxonomy of Educational Objectives. In B. S., Bloom, M. D., Engelhart, E. J., Furst, W. H., Hill, & D. R. Krathwohl. (Eds.). *Handbook: The Cognitive Domain*. David McKay Publication.

Cardino Jr, J. M., & Ortega-Dela Cruz, R. A. (2020). Understanding of learning styles and teaching strategies towards improving the teaching and learning of mathematics. *LUMAT: International Journal on Math. Science and Technology Education*, 8(1), 19–43. DOI: 10.31129/LUMAT.8.1.1348

Cevallos Menéndez, I. Y., Cobeña Napa, M. A., Mendoza Moreira, M. L., & Vélez Zambrano, G. G. (2019). The importance of formative assessment in the learning teaching process. *International Journal of Social Science and Humanity*, 3(2), 238–249. DOI: 10.29332/ijssh.v3n2.322

Davis, F. D. (1986). *A technology acceptance model for empirically testing new end-user information systems: Theory and results* [Doctoral dissertation, Massachusetts Institute of Technology].

Dewey, J. (1938). *Experience and education*. Macmillan.

Diaz Redondo, R. P., Rodríguez, M., Escobar, J., & Vilas, A. F. (2021). Integrating micro-learning content in traditional e-learning platforms. *Multimedia Tools and Applications*, 80, 3121–3151. DOI: 10.1007/s11042-020-09523-z

Fidan, M. (2023). The effects of microlearning-supported flipped classroom on pre-service teachers' learning performance, motivation and engagement. *Education and Information Technologies*, 28(1), 12687–12714. DOI: 10.1007/s10639-023-11639-2

Fiock, H. (2020). Designing a community of inquiry in online courses. *International Review of Research in Open and Distance Learning*, 21(1), 135–153. DOI: 10.19173/irrodl.v20i5.3985

Flavell, J. H. (1979). Metacognition and cognitive monitoring: A new area of cognitive–developmental inquiry. *The American Psychologist*, 34(10), 906–911. DOI: 10.1037/0003-066X.34.10.906

Gacusan, J. C., Dangis, S. J., & Afalla, B. (2023). Tailoring education for alpha learners: Harnessing learning styles for maximum learning outcomes. *NURTURE*, 17(3), 302–313. DOI: 10.55951/nurture.v17i3.335

Gardner, H. (2006). *Multiple intelligences: New horizons* (Completely rev. and updated.). BasicBooks.

Garrison, D. R., Anderson, T., & Archer, W. (2000). Critical inquiry in a text-based environment: Computer conferencing in higher education model. *The Internet and Higher Education*, 2(2-3), 87–105. DOI: 10.1016/S1096-7516(00)00016-6

Germuth, A. A. (2018, December). Professional Development that Changes Teaching and Improves Learning. *Journal of Interdisciplinary Teacher Leadership*, 2(1). Advance online publication. DOI: 10.46767/kfp.2016-0025

Ghimire, N. B. (2020). Teachers' professional development in community campuses of TU: Expectation and responsibility. *Siddhajyoti Interdisciplinary Journal*, 1, 63–73. DOI: 10.3126/sij.v1i0.34921

Haleem, A., Javaid, M., Qadri, M. A., & Suman, R. (2022). Understanding the role of digital technologies in education: A review. *Sustainable Operations and Computers*, 3, 275–285. DOI: 10.1016/j.susoc.2022.05.004

Hamdoun, M. W. (2021, February 5). *Constructive Alignment Approach: Enhancing Learning and Teaching* [Paper presentation]. The Nile TESOL 2021 International Virtual Conference. Sphinx Publishing Company.

Hattie, J., & Timperley, H. (2007). The power of feedback. *Review of Educational Research*, 77(1), 81–112. DOI: 10.3102/003465430298487

Ismail, S. M., Rahul, D. R., & Patra, I.. (2022). Formative vs. summative assessment: Impacts on academic motivation, attitude toward learning, test anxiety, and self-regulation skill. *Language Testing in Asia*, 12, 40. DOI: 10.1186/s40468-022-00191-4

Jönsson, A., & Prins, F. (2019). Editorial: Transparency in Assessment—Exploring the Influence of Explicit Assessment Criteria. *Frontiers in Education*, 3. Advance online publication. DOI: 10.3389/feduc.2018.00119

Kennedy, D., Hyland, Á., & Ryan, N. (2007). *Writing and Using Learning Outcomes: A Practical Guide*. University College Cork.

Kossen, C., & Ooi, C.-Y. (2021). Trialling micro-learning design to increase engagement in online courses. *Asian Association of Open Universities Journal*, 16(3), 299–310. DOI: 10.1108/AAOUJ-09-2021-0107

Kumar, T., Soozandehfar, S. M. A., & Hashemifardnia, A.. (2023). Self vs. peer assessment activities in EFL-speaking classes: Impacts on students' self-regulated learning, critical thinking, and problem-solving skills. *Language Testing in Asia*, 13(1), 36. DOI: 10.1186/s40468-023-00251-3

Latif, L. A., Subramaniam, T. T., & Khatab, Z. A. (2020). Learner Profiling Towards Improving Learner Success. In Li, K. C., Tsang, E. Y. M., & Wong, B. T. M. (Eds.), *Innovating Education in Technology-Supported Environments*. Springer Education Innovation Series. DOI: 10.1007/978-981-15-6591-5_14

Mayer, R. E. (2014). Cognitive theory of multimedia learning. In Mayer, R. E. (Ed.), *The Cambridge handbook of multimedia learning* (pp. 43–71). Cambridge University Press. DOI: 10.1017/CBO9781139547369.005

McNeill, L., & Fitch, D. (2023). Microlearning through the Lens of Gagne's Nine Events of Instruction: A Qualitative Study. *TechTrends*, 67, 521–533. DOI: 10.1007/s11528-022-00805-x

Merrill, D. (2002). First principles of instruction. *Educational Technology Research and Development*, 50(3), 43–59. DOI: 10.1007/BF02505024

Merrill, M. D. (2013). *First Principles of Instruction: Identifying and Designing Effective, Efficient, and Engaging Instruction*. Pfeiffer.

Miller, B., & Staley, L. (2021). Importance of Feedback, Training and Media Format for Students' Reflective Practice. *International Journal of Knowledge and Learning*, 10, 31–40. DOI: 10.53615/2232-5697.10.31-40

Moges, B. (2018). The implementations and challenges of assessment practices for students' learning in public selected universities, Ethiopia. *Universal Journal of Educational Research*, 6(12), 2789–2806. DOI: 10.13189/ujer.2018.061213

Mutezo, A. T., & Maré, S. (2023). Teaching and cognitive presences: The mediating effect of social presence in a developing world context. *Cogent Education*, 10(1). Advance online publication. DOI: 10.1080/2331186X.2023.2171176

Ordu, U. B.-A. (2021). The Role of Teaching and Learning Aids/Methods in a Changing World. In *New Challenges to Education: Lessons from Around the World* (Vol. 19, pp. 210). Bulgarian Comparative Education Society.

Panke, S., Kohls, C., & Gaiser, B. (2017). Social media strategies for seamless learning: Approaches and metrics. In Şad, S. N., & Ebner, M. (Eds.), *Digital Tools for Seamless Learning* (pp. 278–296). IGI Global. DOI: 10.4018/978-1-5225-1692-7.ch014

Piaget, J. (1971). The Theory of Stages in Cognitive Development. In Green, D., Ford, M. P., & Flamer, G. B. (Eds.), *Measurement and Piaget* (pp. 1–11). McGraw-Hill.

Reniers, J., Mathany, C., Farkas, M., Pollock, H., & Husband, B. C. (2022). Uncharted territory: Curriculum mapping multiple majors simultaneously. *The Canadian Journal for the Scholarship of Teaching and Learning*, 13(1). Advance online publication. DOI: 10.5206/cjsotlrcacea.2022.1.8553

Romanenko, Y. N., Solodovnikova, E., & Maksimenko, N. (2023). Microlearning as a new method of teaching soft skills to university students. *Frontiers in Education*, 8, 1177516. DOI: 10.3389/feduc.2023.1177516

Ropohl, M., & Rönnebeck, S. (2019). Making learning effective – quantity and quality of pre-service teachers' feedback. *International Journal of Science Education*, 41(15), 2156–2176. DOI: 10.1080/09500693.2019.1663452

Ryan, R. M., & Deci, E. L. (2000). Self-determination theory and the facilitation of intrinsic motivation, social development, and well-being. *The American Psychologist*, 55(1), 68–78. DOI: 10.1037/0003-066X.55.1.68

Senge, P. (2006). *The Fifth Discipline: The art and practice of the learning organization*. Random House Books.

Shaltry, C. (2020). A new model for organizing curriculum alignment initiatives. *Advances in Physiology Education*, 44, 658–663.

Sharples, M., Taylor, J., & Vavoula, G. (2007). A theory of learning for the mobile age. In Andrews, R., & Haythornthwaite, C. (Eds.), *The SAGE Handbook of E-Learning Research* (pp. 221–247). Sage Publication.

Singh, J., Singh, L., & Matthees, B. (2022). Establishing social, cognitive, and teaching presence in online learning—A panacea in COVID-19 pandemic, post vaccine and post pandemic times. *Journal of Educational Technology Systems*, 51(1), 568–585. DOI: 10.1177/00472395221095169

Stanley, D., Coman, S., Murdoch, D., & Stanley, K. (2020). Writing exceptional (specific, student and criterion-focused) rubrics for nursing studies. *Nurse Education in Practice*, 49, 102851. DOI: 10.1016/j.nepr.2020.102851

Stufflebeam, D. L. (2003). The CIPP model for evaluation. In Kellaghan, T., & Stufflebeam, D. L. (Eds.), *Evaluation models: Viewpoints on educational and human services evaluation* (pp. 279–317). Springer Netherlands. DOI: 10.1007/978-94-010-0309-4_4

Susilana, R., Dewi, L., Rullyana, G., Hadiapurwa, A., & Khaerunnisa, N. (2022). Can microlearning strategy assist students' online learning? *Cakrawala Pendidikan Jurnal Ilmiah Pendidikan*, 41(2), 437–451. DOI: 10.21831/cp.v41i2.43387

Sweller, J. (2011). Cognitive load theory. *Psychology of Learning and Motivation*, 55, 37–76. DOI: 10.1016/B978-0-12-387691-1.00002-8

Trespalacios, J., & Uribe-Florez, L. J. (2020). Developing online sense of community: Graduate students' experiences and perceptions. *Turkish Online Journal of Distance Education-TOJDE*, 21(1), 57–72.

Valzado, R. B., Subaldo, C. B., Diacor, R. S., & Rellon, J. (2023). Dimensions of pedagogical skills of teachers in the new normal: An exploratory sequential design. *Southeast Asian Journal of Multidisciplinary Studies, 3*(2).

Vygotsky, L. S. (1978). *Mind in Society: The Development of Higher Psychological Processes*. Harvard University Press.

Wang, , TTowey, , DNg, , R.YGill, , A.S., (2021). Towards post-pandemic transformative teaching and learning: Case studies of microlearning implementations in two post-secondary educational institutions. *SN Computer Science*, 2(4), 271–279. DOI: 10.1007/s42979-021-00663-z

Wang, L. (2023). The benefits of a well-calibrated micro-learning ecosystem. UNESCO.

Yin, H., & Chai, C. (2020). Catering for Learner Diversity: Issues and Trends. *ECNU Review of Education*, 3(4), 607–609. DOI: 10.1177/2096531120927620

Zarestky, J., Bigler, M., Brazile, M., Lopes, T., & Bangerth, W. (2022). Reflective Writing Supports Metacognition and Self-regulation in Graduate Computational Science and Engineering. *Computers and Education Open*, 3, 100085. DOI: 10.1016/j.caeo.2022.100085

Zimmerman, B. J., & Schunk, D. H. (2011). Self-regulated learning and performance: An introduction and an overview. In Zimmerman, B. J., & Schunk, D. H. (Eds.), *Handbook of self-regulation of learning and performance* (pp. 1–12). Routledge.

Chapter 9
Simulation

ABSTRACT

This chapter examines the use of simulation techniques in Applied Cognitive Task Analysis (ACTA) to improve educational practices. It is divided into two sections. The first, "Simulation of Electronic Training for Professional Development," explores how simulation models real-world scenarios to enhance electronic training, offering hands-on learning that improves skill acquisition and retention. This section shows how simulation-based training provides an immersive experience that bridges theory and practice for professional development. The second section, "Simulation for Microlearning Pedagogical Practices," discusses how simulations create dynamic, adaptive learning environments in higher education. These simulations help deliver personalized, relevant content, fostering engagement and better learning outcomes. The chapter underscores simulation's importance in optimizing educational strategies for both electronic training and microlearning.

INTRODUCTION

Simulation serves as the third critical step in the Applied Cognitive Task Analysis (ACTA) framework. By replicating real-world tasks and environments, simulations provide a valuable platform for understanding how individuals perform under varying conditions. This method not only allows for the observation of cognitive processes in action but also helps identify potential areas of improvement in task performance, instructional design, and training programs. In educational settings, simulations are particularly useful for bridging the gap between theoretical knowledge and practical application, offering learners an immersive experience that enhances skill development and retention.

This chapter focuses on the application of simulation in two distinct educational contexts: electronic training for professional development and microlearning pedagogical practices.

DOI: 10.4018/979-8-3693-7733-8.ch009

Section 9.1 explores the use of simulation in electronic training systems for professional development. Simulations in this context allow professionals to engage with complex tasks in a risk-free environment, mirroring the challenges they may face in real-world scenarios. By simulating tasks such as decision-making, system navigation, and problem-solving, these training modules provide learners with hands-on experience while minimizing the risks associated with real-time errors. Simulation in professional development serves as an invaluable tool for reinforcing learned skills and enhancing performance in high-stakes environments.

Section 9.2 examines how simulations can be integrated into microlearning pedagogical practices. Microlearning focuses on delivering content in short, focused bursts, and incorporating simulations into this framework allows learners to immediately apply knowledge in practical scenarios. This section will explore how simulations tailored for microlearning environments can help students engage more deeply with content, improve retention, and foster the application of theoretical knowledge in real-world contexts.

Throughout this chapter, we will discuss the role of simulation as a dynamic method in the ACTA framework, emphasizing its potential to enhance both professional development and higher education. By creating realistic, immersive learning environments, simulations can significantly improve learner engagement, understanding, and skill acquisition.

9.1 Simulation of Electronic Training System for Professional Development

A simulation interview will be conducted towards the experts in order to discover their experiences and opinions of TaSPoD system. Experts are allowed to interact and experience the simulation.

9.1.1 Simulation Interview Table for System Design

Generally, all the experts agreed that the challenging elements were contributed by four main components in TaSPoD design; system functionality design, course content design and downloadable content, as well as assessments design.

All experts agreed that the system has a proper navigation and guidance. However, they stated that the system will be complicated to access if there's no proper flow to navigate to all subsequent pages. This will lead to a not user-friendly system and further will sway users away from completing the training course. Difficulty in navigating the training system is also affected by improper lesson plan which is not designed in generic perspective.

As for course content, the experts believed that trainers need to determine appropriate instructional activities and align the lessons from different users' perspective. There will be possibility that the learning process is inaccurate if the course content design is not match with the system design. Additionally, this issue will also affect assessment design. The findings are consistent with the research conducted by Rafique & Aziz (2019) which highlights a significant connection between pedagogical content knowledge and instructional design. Regardless of their teaching methods, educators tend to employ a variety of instructional strategies to optimize students' learning time and encourage active engagement in educational activities.

Additionally, experts recommend incorporating communication tools such as chat programs and video conferencing into TaSPoD, rather than relying solely on traditional online forums and emails. This aligns with findings by Silva, Mendes, & Gomes (2020), which emphasize that communication tools are a vital component of collaboration in e-learning environments.

In addition to the provided content, the training system includes a feature that allows users to download content or documents. However, experts highlighted the importance of assessing the usability of downloadable materials, as well as the need for clear guidelines to support the downloading process. This functionality is essential to ensure ongoing engagement with online learning platforms. Saintika, Astiti, Kusuma, & Muhammad (2021) also recommended that institutions implement policies to optimize the use of e-learning systems.

The simulation process also identified that users would face difficulties in designing and determining the appropriateness of summative assessment. This potential error may be due to the inability of the assessments to address learning outcomes of the course and the inability to fulfil the needs of assessments according to different level of users. HLEs might find difficulties to complete any assigned assessments when there is no justification is provided for each of the assessments.

Data Presentation

Table 1 exhibit the identified challenging tasks in TaSPoD design, showcasing the four identified components (Actions, Situation Assessments, Critical Cues, Potential Errors). This table provides a visual summary of the insights obtained from the simulation interviews, offering a detailed overview of the framework's flexibility, effectiveness, and potential obstacles within the context of higher education.

Table 1. Simulation interview table

Major Challenging Tasks	Actions	Situation Assessment	Critical Cues	Potential Errors
System Functionality design	Proper navigation and guidance with pedagogical based learning design. System should be designed from different user's perspective.	It is complicated if there are no proper flow to navigate the lesson and a course outline page (anchor page) to link to all subsequent pages. The new HLE's find difficulty to navigate the lesson plan if not designed in a generic perspective	Interaction between TaSPoD and HLEs.	Screen designs are not user-friendly and poorly guided and navigated as it is if mouse roll over option provided. User views are not clear.
Course content design	To find suitability between pedagogical model with instructional activities and aligning the lessons from different users' perspective Verify and validate course content design match with system design.	The language and information should be comprehensive and easy to follow. Learning process and assessment for HLEs is inaccurate due to flow of the content	Interaction between course and HLEs. HLEs might be asked wrong question at the wrong module or unit. Grade given might not really access learner's learning path.	TaSPoD should focus on structure the learning systematically for different users. Mismatch content and assessment will cause demotivation among learners
Downloadable Content	To find the usability of downloadable content documents.	Provide proper guidelines for download and software requirements, if necessary, to complete the downloading process.	Enhance the functionality of accessing documents.	Document templates should be properly design and self-explanatory. HLEs might need to install supported software to open the document
Assessment design	To evaluate appropriateness of formative and summative assessment design. Verify and validate assessment based on content module. Design assessment strategies for different level of users.	Difficult for HLEs to follow the guidance and unable to complete the assigned tasks / assessments, for different level of the users. Inappropriate assessment for certain topics. No transparent justification for the given assessment	Enhance understanding of subject matter and able to apply and integrate the knowledge into the task / assessment given. Too many assessments with high difficulty might discourage HLEs from continuing the course	Assessment did not address the learning outcomes of the course. Assessment did not take into account the level of difficulty. HLEs might not be able to proceed to the next level if he did not meet the assessment requirements

9.1.2 Simulation Interview Table for System Prototype

In the simulation interview technique, experts were presented with the TaSPoD prototype. Username and password were provided to each of the experts to access the online training system. The experts have 'user privilege' that allow them to use all the available functions and features in the system.

Generally, all the experts agreed that the system's challenging tasks were contributed by the three main components of TaSPoD; system functionality, course content and assessments design. Majority of the experts also acknowledged the need of having real-time assistance besides the live session.

The experts concurred that the system can be navigated from one page to another perfectly. However, they stated that the system has insufficient instructions or guidelines to assist HLEs resume to their last accessed content from the previous login. This would restrain the benefit of using electronic training at anytime and anywhere. In addition, this will induce users to forget the previous content they have accessed before. Poorly developed navigation would cause learners to take longer time and feel frustrated to continue using e-learning (Nordin, Singh, & Mansor, 2020).

As for course content, the experts believed that trainers need to determine appropriate instructional activities; as well as utilizing varieties of multimedia elements to deliver the course content. Absence of these elements will diminish enthusiasm in using the electronic training system, which will lead to incompletion of training session. It is important to draw learners' attention and ignite trainees' excitement in completing the course (Mokhtar, Adnan, Shazali, & Ahmad, 2020). Thus, satisfy one of the electronic training benefits for independent learning.

The simulation process also identified that trainers would face difficulties in designing and determining the appropriateness of summative assessment. This potential error may be due to trainer's inability to address the course learning outcomes, as well as inability to fulfil different level of users. On the other hand, experts also pointed out the needs of having a feature that automatically assign participants into group when groupwork is required.

Although live chat sessions are scheduled in the electronic training system to attend to participants' queries, experts have confidence that assistance is required in the chatroom beyond the scheduled sessions. Experts believed its necessary for the electronic training system to provide immediate answers to participants' queries.

Data Presentation

Table 2 presented the identified challenging tasks in TaSPoD prototype, showcasing the four identified components (Actions, Situation Assessments, Critical Cues, Potential Errors). This table provides a visual summary of the insights obtained from the simulation interviews, offering a detailed overview of the framework's flexibility, effectiveness, and potential obstacles within the context of higher education.

Table 2. Simulation interview table

Major Challenging Tasks	Actions	Situation Assessment	Critical Cues	Potential Errors
System Functionality	Proper guidance to navigate from last-accessed page that was last accessed	It is challenging because users need to be able to access to the page that was last accessed from the previous login.	Users are not able to start back on the beginning page after returning.	Users will have the tendency to forget the previous content materials.
Course content	To find suitability of the instructional activities. To be able to backtrack of topics. User-friendly icons to assist the downloading action. HLE are able to download complete course content.	Use of more multimedia elements is required. Allow downloadable course content for HLE that wanted to review the course content, instead of going through all the course contents step by step	Less usage of multimedia elements will lead to lack of excitement. Scrolling to a very long page (when there lots of text) Detailed module outline may be added (consist of instructor details, weekly topics, assessment etc.)	Lack of excitement will lead to incomplete of course. Avoid scrolling to a very long page for a more stimulating learning process
Assessment design	To evaluate appropriateness of summative assessment design. Assign participants into group automatically when groupwork is required.	Difficult to HLEs to follow the guidance and unable to complete the assessments. Design assessment strategies for different level of users Group work or discussion is possible to be conducted as one of the assessments.	Enhance understanding of subject matter and able to apply and integrate the knowledge into the task / assessment given Participant list - creating group when groupwork is needed	Inability to address assessment to the learning outcomes of the course and also assessment strategies for different level of users. Assessment did not fulfil different level of users.
Real-time assistance	To provide assistant in chatroom when live chat is over.	Chat bots are needed in chatrooms	In Chatrooms if there is no user available that the chat bots should be available.	Immediate answer was not fulfilled when the HLE needed them.

9.1.3 Review of Simulation of Electronic Training System for Professional Development

In this section, the simulation interview process provided valuable insights into the challenges encountered in the design and functionality of the TaSPoD system for electronic training. By simulating real-world usage scenarios, the experts identified several critical areas in which the system and its prototype could be improved to enhance the learning experience for higher learner educators (HLEs).

For the system design, the simulation interviews revealed challenges related to system functionality, course content, and assessment design. Experts pointed out that proper navigation and user-friendly design were essential to ensuring that users could easily access and complete the training without frustration. Issues such as unclear navigation flows, poorly designed screen layouts, and the complexity of content retrieval posed significant obstacles to user engagement. Experts also emphasized the need for better alignment between course content and system functionality, particularly to support a smooth learning process and reduce cognitive load.

In terms of downloadable content, experts agreed that clear guidelines and improved usability features would enhance the overall experience for users, making it easier to access and interact with supplementary materials. Additionally, the assessment design was flagged as a key area where improvements were necessary, particularly in ensuring that assessments were aligned with course learning outcomes and catered to different levels of learners.

The system prototype simulation interviews further identified several key areas for improvement, including the need for real-time assistance and clear instructions to help users resume training from their last-accessed content. The lack of multimedia elements and groupwork features in the system was also highlighted as a potential issue in maintaining user engagement. Experts suggested that a more interactive, multimedia-rich approach, along with better assessment strategies, would improve the system's ability to meet the diverse needs of learners.

Overall, the simulation interviews provided a comprehensive understanding of the strengths and weaknesses of the TaSPoD system, offering critical insights into how the system could be improved for better professional development outcomes. By addressing the challenges identified during the simulation, the system can be optimized to offer a more intuitive, engaging, and effective learning experience.

9.2 Simulation for Microlearning Pedagogical Practices

9.2.1 Application of Simulation Interview Technique in the Evaluation Phase

In the fourth and final phase of this study, the focus shifts to the Evaluation of the Microlearning Educational Experience (AMPLIFY) Instructional Framework, addressing Research Question 4: How do experts perceive the usability, practicality, and feasibility of the microlearning framework's design principles for integration into higher education? This phase utilises the third and final methodologies from Applied Cognitive Task Analysis (ACTA) - the Simulation Interview and Cognitive Demand Table techniques.

Simulation Interview

The Simulation Interview, a key methodology in this phase, involved subject matter experts engaging in a practical and scenario-based assessment of the micro-learning framework, 'AMPLIFY.' This immersive experience aimed to replicate the real-world application of AMPLIFY in a classroom setting. By collaborating with experts, the simulation interviews facilitated a comprehensive exploration of the framework's functionality, effectiveness, and potential areas for improvement. The goal was to gain valuable insights into the practical application of AMPLIFY and its alignment with the cognitive processes of both educators and students in higher education.

Evaluation Criteria

The evaluation criteria considered during the simulation interviews revolve around four key components:

- Actions: Analysis of actions related to the framework's implementation, providing insights into specific strategies employed to seamlessly incorporate microlearning into the learning environment.
- Situation Assessments: A thorough examination of the current state and context of higher learning classrooms, shedding light on factors that may facilitate or hinder the successful integration of microlearning.
- Critical Cues: Identification of significant indicators offering insights into the level of success achieved during the implementation process or highlighting potential challenges that need addressing.

- Potential Errors: Examination of potential barriers or obstacles that might impede the smooth integration of the microlearning framework.

Data Presentation

The findings from the Simulation Interviews are presented in Table 4.11, showcasing the four identified components (Actions, Situation Assessments, Critical Cues, Potential Errors). This table serves as a visual representation of the insights gained from the simulation interviews, offering a comprehensive overview of the framework's adaptability, effectiveness, and potential challenges in the higher education context.

Implications for Implementation

The information gathered from the Simulation Interviews forms a crucial foundation for the evaluation phase. It guides decision-making on how effectively the AMPLIFY framework aligns with experts' opinions and its potential for successful implementation in higher education. The insights gained during the simulation interviews inform educators and stakeholders about the feasibility, practicality, and usability of integrating AMPLIFY into higher learning classrooms.

Conclusion of Evaluation Phase

Concluding the Evaluation Phase, the Simulation Interview Table Technique proves instrumental in assessing the microlearning framework's integration into higher education. By incorporating real-world scenarios and engaging experts in practical assessments, the study gains valuable perspectives that contribute to the refinement and potential widespread adoption of the AMPLIFY framework.

Insights and Findings

The findings from the simulation interview focused on practical, scenario-based assessments of the 'AMPLIFY' microlearning framework. This involved an in-depth examination of how the framework operates within a real educational setting. The purpose of the simulation was to replicate the use of AMPLIFY in a classroom environment, allowing experts to evaluate its effectiveness and identify any potential challenges or areas for improvement. Through these simulated interviews, the research obtained valuable insights into the practical application of AMPLIFY and its alignment with the cognitive processes of both educators and students in higher education.

i. Key Element 1: Establishing Clear and Concise Learning Objectives Aligned with Students' Needs and Educational Goals

In crafting effective microlearning experiences, it is crucial to establish clear and concise learning objectives that resonate with the learners' needs and aspirations. This process involves several critical steps. Begin by conducting a thorough needs analysis to grasp the learners' requirements, prior knowledge, and preferences, thereby customizing the objectives to meet their specific needs (Irfadila & Noprika, 2022). Collaboration with subject matter experts (SMEs) is vital to ensure the accuracy and relevance of curriculum objectives, as SMEs contribute expert domain knowledge, skills, and behaviours that are critical for developing effective learning experiences (Gottler, 2023). These objectives should be well-defined, measurable, and attainable, providing focus and direction. Furthermore, they ought to align seamlessly with the overarching educational objectives of the course or program.

Conducting a comprehensive situation assessment is crucial for the successful design of microlearning experiences. Evaluating the learning environment, context, and target audience helps ensure that the learning objectives are relevant and aligned with learners' real-world needs. Additionally, assessing the availability of resources—such as technology and support—is necessary to facilitate the smooth implementation of microlearning activities that effectively align with the intended learning outcomes. Avoiding pitfalls, such as vague or ambiguous objectives, is critical, as they can confuse learners and diminish the effectiveness of the learning activities. Misalignment between learning objectives and learners' needs or a lack of synchronization between microlearning tasks and overarching objectives can prevent students from achieving the desired outcomes. Furthermore, failing to break down complex objectives can overwhelm learners, hindering their ability to absorb information efficiently. Involving subject matter experts (SMEs) in the development process is also essential to avoid content inaccuracies and inadequacies, which can create gaps between the objectives and the actual learning outcomes. By adhering to these guidelines, conducting thorough assessments, and being mindful of potential errors, clear and concise learning objectives can be developed. This ensures that microlearning activities are tailored to learners' needs and goals, resulting in an impactful and engaging learning experience.

ii. Key Element 2: Incorporating Personalization Through Learner-Centric Microlearning Activities and Customized Content

To incorporate personalization in microlearning, several essential steps must be taken. Firstly, conducting learner profiling is crucial; this involves collecting data on learners' preferences, learning styles, prior knowledge, and goals to create individualized profiles. These profiles provide a comprehensive view of each learner's cognitive knowledge, skills, competencies, and behavioural patterns, which are essential for tailoring learning experiences to their specific needs. Leveraging methods from Learning Analytics (LA) can further enhance the effectiveness of learner profiles, allowing for scalable implementation across educational systems (Barthakur et al., 2023). This information forms the foundation for the next step: developing customized content tailored to each learner's profile and specific needs. Experts highlight the importance of examining individual students' multiple intelligences and profiles to develop personalized content that aligns with their preferences, learning styles, prior knowledge, and goals. Another critical step is implementing adaptive learning paths, which ensure learners receive individualized learning journeys based on their progress and performance. Adaptive learning (AL) tools, such as CogBooks®, have been shown to significantly improve student achievement, whether in online or face-to-face settings. Recent research has demonstrated that integrating AL into university courses, particularly in flexible, interactive modalities, leads to better learning outcomes and increased student satisfaction (Contrino et al., 2024). Finally, promoting self-directed learning is vital, fostering learner autonomy and allowing them to explore content at their own pace, catering to diverse learning preferences. Self-directed learning, as emphasized in the communities of inquiry framework, enables students to take responsibility for their own educational journey, providing students with opportunities to engage in authentic tasks, apply their own learning strategies, and construct knowledge in a social context (Bosch & Laubscher, 2022).

An effective situation assessment is also vital for personalization efforts. Experts emphasize the need to consider factors such as age, educational background, skill level, and learning goals to develop comprehensive learner profiles. Moreover, assessing the relevance and suitability of customized content is crucial to ensure it effectively addresses specific learner needs and interests.

However, several potential errors can undermine the effectiveness of personalization in microlearning design. Disregarding learner preferences and needs can result in disengagement and decreased motivation to participate in personalized activities. To prevent this, personalization efforts should focus on accommodating individual learner preferences. Additionally, offering a limited variety of personalized content options can restrict learners' exposure to different perspectives and topics, diminishing the effectiveness of personalized learning. Continuous personalization, which involves regularly adapting content and activities to evolving learner needs, is essential to overcome these challenges. Without this, learners may feel disengaged and unmotivated due to content that does not align with their current learning journey. By addressing these potential errors, personalization efforts can be optimized to create a truly engaging and effective personalized microlearning experience.

iii. Key Element 3: Incorporating Feedback and Assessment for Effective Microlearning Experiences

To ensure an effective assessment process in microlearning, educators can implement several key actions. Experts emphasize the importance of providing timely and constructive feedback to learners, identifying areas of strength and improvement, and ensuring that feedback is actionable to encourage learner progress. Constructive feedback, guided by principles such as Vygotsky's Zone of Proximal Development and formative assessment, plays a crucial role in enhancing learning outcomes by helping students understand their performance and how to improve. By customizing feedback to individual learner needs and utilizing technology for efficient delivery, educators can create a more supportive and effective learning environment (Kumar & Kumar, 2024). Integrating formative assessments throughout the learning process allows educators to gauge learners' understanding and identify areas needing further attention and adaptation in instructional strategies. Conducting summative assessments is also essential, as it evaluates learners' overall mastery and achievement of learning objectives, offering a comprehensive overview of their performance. Utilizing a variety of assessment methods, such as quizzes, projects, presentations, group work, and self-assessments, caters to diverse learning styles and abilities. Additionally, aligning all assessments with the stated learning objectives provides a clear indication of learners' attainment of intended outcomes. Continuous improvement is essential, with educators regularly reviewing assessment data to identify trends and areas for enhancement, refining their teaching approaches to create an optimized learning experience.

In the situation assessment phase, experts highlight the importance of evaluating the alignment between assessments and learning objectives to ensure that assessments effectively measure the intended learning outcomes. Assessing the diversity of assessment methods is also crucial to determine whether they cater to different cognitive and affective domains of learning.

However, several potential pitfalls can undermine the effectiveness of assessment in microlearning. Experts caution against providing ambiguous or inadequate feedback, as it can leave learners unsure about their progress and how to improve. They also advise against relying too heavily on final assessments without sufficient ongoing formative assessments, which could restrict opportunities for timely intervention and instructional adjustments. Depending solely on final exams as the primary evaluation method may be restrictive and may not fully reflect students' comprehension and capabilities. Integrating more formative assessment approaches, such as creating videos, podcasts, and presentations, allows students to learn from their errors, refine their understanding, and ultimately achieve better outcomes. These concise, targeted activities enable learners to interact with the material in manageable segments, fostering active engagement and improving retention. Through micro-learning strategies, students can continuously enhance their knowledge and skills, aligning with the dynamic and adaptable nature of modern learning environments.

Experts stress the importance of avoiding assessment methods that do not match the learning objectives, as these can fail to provide a comprehensive gauge of learners' progress. They point out the frustration and difficulties students encounter when taught using certain methods or objectives, only to be assessed with a different, often repetitive or singular, approach. This disconnect between teaching and assessment methods can result in misalignment and confusion, ultimately impeding students' ability to showcase their true understanding and abilities. By following these principles—conducting thorough situation assessments, paying attention to critical indicators, and steering clear of potential pitfalls—educators can establish a robust and effective assessment process that fosters learners' growth and success.

iv. Key Element 4: Integrating Structured and Engaging Content Delivery in Microlearning

To ensure a structured and engaging delivery of microlearning content, experts stressed several crucial steps. They underscored the need to develop concise, well-organized microlearning materials that align meticulously with the learning objectives, making it easier for learners to grasp essential concepts. Breaking down content into manageable segments was highlighted as key to enhancing retention and avoiding cognitive overload. Liu (2024) demonstrated that high segmentation in multimedia learning environments significantly reduces cognitive load while improving vocabulary acquisition, retention, and reading comprehension. These findings suggest that segmenting instructional content can enhance learning efficiency and promote a deeper understanding of the material, making it an essential consideration for educators designing digital learning experiences. Additionally, organizing content in a logical sequence, where each unit builds upon the previous one, was emphasized to promote a smooth learning flow. Experts also emphasized the integration of interactive elements such as quizzes, simulations, and activities to actively engage learners and reinforce their understanding of the content. In the context of micro-learning, incorporating such interactive features has been shown to enhance learner motivation and retention. A study on the design of a micro-learning application for teaching thermography demonstrated that learners respond positively to these elements, finding them more engaging and motivating compared to traditional learning methods (Nilsson, 2021).

During the situation assessment phase, experts emphasized the criticality of evaluating content clarity, highlighting its coherence and effectiveness in conveying information clearly and without ambiguity. They also underscored the importance of assessing how content is chunked, ensuring that the units are appropriately sized to enhance effective learning outcomes. Experts stressed the importance of avoiding potential pitfalls in microlearning content delivery. Educators, as emphasized, should be vigilant against content overload, as presenting excessive information at once can overwhelm learners and impede their understanding. Ensuring a logical progression is crucial, as highlighted by experts; disjointed content sequencing can confuse learners and disrupt their learning process. Similarly, experts emphasized the importance of breaking down complex concepts into smaller, manageable points or objectives. This approach allows learners to grasp fundamental concepts quickly and then gradually connect these smaller pieces to develop a comprehensive understanding. By structuring learning in this way, educators promote a coherent and structured learning experience, enabling learners to build their knowledge step by step in a logical and interconnected manner.

To sustain engagement, educators must incorporate interactive elements, as experts emphasize that inadequate interactivity can result in passive learning and reduced motivation. Furthermore, experts stress the importance of aligning content directly with learning objectives to avoid wasting time and effort for both learners and educators. By skilfully integrating these strategies—conducting thorough situation assessments, heeding critical indicators, and avoiding common pitfalls—educators can effectively integrate structured and engaging content delivery into their instructional design, as highlighted by experts. This approach enhances learners' comprehension, promotes active participation, and fosters a positive and impactful learning experience.

v. Key Element 5: Incorporating Effective Microlearning Delivery Methods

To ensure effective delivery of microlearning methods, experts stressed several critical actions. They emphasized the integration of multimedia, advocating for the inclusion of diverse elements like videos, audio, infographics, and interactive simulations. According to their insights, these elements enhance the learning experience and cater to the varied preferences of learners, promoting better engagement and understanding in microlearning-supported environments. Fidan (2023) demonstrated that incorporating multimedia in microlearning not only increases motivation and performance but also allows for flexible, self-paced learning that is accessible anytime and anywhere. Additionally, experts highlighted the importance of designing content that is mobile-friendly. They emphasized the need to ensure that microlearning materials are not only accessible but also optimized for use on mobile devices. This approach enables learners to engage conveniently and flexibly with the content, facilitating learning at any time and from any location. Moreover, experts underscored the value of just-in-time learning, emphasizing that providing learners with immediate access to relevant and practical information empowers them to apply knowledge promptly in real-life situations, a principle they strongly endorsed.

During the situation assessment, experts emphasized the criticality of evaluating the impact of multimedia elements. They underscored the importance of assessing how these elements influence learners' engagement, comprehension, and retention of knowledge. Additionally, experts stressed the need to thoroughly evaluate the accessibility and usability of microlearning content across different mobile devices. They highlighted the significance of ensuring that learners can navigate and interact with the content seamlessly on their preferred devices. Furthermore, experts emphasized the importance of assessing the effectiveness of just-in-time learning. They highlighted the need to evaluate how well this approach meets learners' immediate needs and addresses their challenges, considering this aspect crucial in instructional design. To prevent potential mistakes, experts warned against overwhelming learners with an excess of multimedia elements. They emphasized that while multimedia can enhance learning, overusing these elements can distract learners and detract from focusing on key learning objectives.

Additionally, experts emphasized the risk of developing content that is not accessible on mobile devices. They stressed that neglecting to optimize microlearning content for mobile usage could restrict learners' accessibility and flexibility, which they identified as a significant concern. Finally, experts warned against offering just-in-time learning content that lacks relevance or direct applicability to learners' immediate needs. They highlighted that such content could result in disengagement and reduced effectiveness, stressing the importance of ensuring that content is highly pertinent and immediately beneficial to learners. Moreover, experts emphasized that addressing current, in-demand skills further enhance the alignment of learning with learners' needs and real-world applications.

By carefully implementing these strategies—conducting thorough situation assessments, paying attention to critical indicators, and avoiding common pitfalls—educators can effectively integrate microlearning delivery methods into their instructional design, as experts emphasize. This approach, as highlighted by experts, not only enhances learners' engagement but also provides flexible learning opportunities and supports the practical application of knowledge in real-world contexts.

vi. Key Element 6: Integrating Interactive Microlearning Experiences

To ensure an effective microlearning experience, experts emphasized several crucial actions. They highlighted the importance of incorporating authentic sources, such as real-life case studies, interviews, and practical scenarios, into microlearning content. These authentic sources provide learners with relevant and practical examples that resonate with real-world applications. According to de Vries et al.

(2018), microlearning is particularly effective in promoting self-directed learning and ownership of the learning process, as it allows participants to engage with content that is directly applicable to their professional or educational contexts. Experts also stressed the critical importance of promoting active learning strategies. They pointed out that engaging learners actively with the content through discussions, problem-solving exercises, and hands-on projects fosters a deeper understanding and application of knowledge. According to Sulaiman & Abdullah (2024), the use of active learning techniques not only enhances academic achievement but also supports the development of critical thinking, motivation, and inclusivity, making it a key approach for improving student engagement and learning outcomes in higher education. Additionally, experts underscored the value of creating interactive activities. They noted that developing interactive elements such as quizzes, simulations, and gamified content enhances learner engagement and creates an immersive learning experience.

In line with situation assessments, experts emphasized the importance of evaluating learners' needs and preferences. They noted that aligning interactive learning experiences with individual learning styles and preferences ensures optimal engagement and effectiveness. Additionally, experts stressed the need to assess the availability and accessibility of technological resources, highlighting that the effective delivery of interactive content depends on factors such as device compatibility and internet connectivity, which they considered crucial. Furthermore, experts underscored the relevance of interactive learning experiences, emphasizing the necessity of ensuring that these experiences directly contribute to the learning objectives and meaningfully enhance overall learning outcomes.

Experts cautioned against potential technology-related errors, highlighting that technical glitches or malfunctioning interactive elements can disrupt the learning experience. They emphasized the necessity for seamless technological integration to avoid these issues. Experts also warned against the potential pitfall of overwhelming complexity. They stressed that presenting learners with too many interactive elements or complex activities can cause cognitive overload and reduce learning effectiveness. Experts highlighted the importance of carefully selecting and prioritizing learning methods to prevent overwhelming learners with excessive options or intricate tasks. Striking a balance between offering engaging activities and avoiding cognitive overload is crucial when designing microlearning experiences, ensuring learners can engage effectively without being overwhelmed.

Moreover, experts emphasized the importance of aligning interactive learning experiences with learning objectives. They stressed that a failure to establish this alignment could lead to disengagement and a perceived lack of relevance for learners. Lastly, experts cautioned against inadequate assessment during interactive activities. They highlighted the necessity of robustly evaluating learners' performance in these activities to provide meaningful feedback and drive continuous improvement, a principle they regarded as crucial.

By carefully considering these strategies, conducting thorough situation assessments, paying attention to critical indicators, and proactively avoiding potential mistakes, educators can successfully integrate interactive learning experiences into microlearning to enrich learners' educational journeys and overall learning outcomes, as experts have noted. This approach, as highlighted by experts, results in more engaging and impactful learning experiences that promote a deeper understanding and practical application of knowledge.

vii. Key Element 7: Integrating Social Interaction Through Collaborative Learning and Peer Feedback

To create an engaging microlearning experience, experts emphasized several important actions. They highlighted the significance of incorporating collaborative learning activities, such as designing and facilitating group projects, discussions, and teamwork exercises that foster social interaction and collaboration among learners. These activities not only enhance engagement but also deepen understanding through shared insights and perspectives. According to Sidgi (2024), universities should prioritize creating collaborative learning environments, as these spaces encourage positive group dynamics and promote the exchange of ideas, ultimately improving academic achievement and classroom involvement. Experts also stressed the critical importance of promoting peer feedback and review. They emphasized cultivating a culture where learners actively provide constructive feedback and review each other's work, which enhances social engagement and improves comprehension through diverse viewpoints. According to Kuyyogsuy (2019), peer feedback fosters critical thinking, autonomy, and social interaction among learners, making it a valuable strategy for improving both academic performance and collaborative learning environments.

In terms of assessing the situation, experts stressed the importance of evaluating learner interests and preferences regarding participation in social interactions and collaborative activities. They emphasized aligning the collaborative approach with learners' motivations and past experiences, considering this aspect crucial. Additionally, experts emphasized the significance of understanding group dynamics. They highlighted that learners have diverse backgrounds and personalities, underscoring the importance of fostering balanced and effective group dynamics for successful collaborative learning.

Addressing potential pitfalls, experts warned against uneven participation in collaborative activities. They highlighted the risk of certain learners dominating group interactions, which can diminish engagement and limit learning opportunities for others—an issue they emphasized. Additionally, experts noted the potential for ineffective communication. They stressed the importance of clear communication and guidance to foster meaningful collaboration and maintain the effectiveness of social interactions. Finally, experts warned about the risks of a deficient feedback culture. They emphasized that if learners are not open to feedback or lack the ability to offer constructive criticism, the effectiveness of peer review could be undermined—a concern they highlighted. Experts stressed that feedback involves not only critiquing the recipient but also fostering a learning process for the giver. They underscored the importance of articulating perspectives thoughtfully and substantively.

By following these steps—conducting a comprehensive situation assessment, paying attention to critical indicators, and actively preventing potential pitfalls—educators can successfully incorporate social interaction into microlearning through collaborative learning and peer feedback, in line with expert advice. This method promotes increased engagement, better retention of knowledge, and improved collaborative skills among learners, as emphasized by experts, resulting in enhanced learning outcomes.

viii. Key Element 8: Seamless integration with curriculum through integration of learning technologies and integration with learning ecosystem

To ensure the smooth integration of microlearning into the curriculum, experts emphasized several key actions. They stressed the importance of curriculum mapping as a foundational step, emphasizing the need to align microlearning activities with the existing curriculum. This alignment ensures consistency and relevance throughout the learning process, creating a cohesive educational experience for learners. According to Alam and Benaida (2022), curriculum mapping helps institutions accurately estimate learning outcomes, maximize resource use, and enhance overall educational performance by connecting disciplines to academic outcomes. Additionally, experts underscored the critical role of selecting appropriate learning technologies. They highlighted the importance of choosing tools and technologies that support microlearning delivery and enhance the learning experience. Haleem et al. (2022) emphasized how digital technologies have revolutionized the education system, enabling more interactive, efficient, and accessible learning environments, which are essential for meeting the goals of inclusive and equitable quality education in line with the United Nations' sustainable development agenda. Furthermore, these experts emphasized the value of integrating microlearning content into the institution's Learning Management System (LMS), emphasizing the importance of easy access and user-friendly navigation for learners.

Experts highlighted the importance of incorporating gamification elements into microlearning. They emphasized that gamification significantly enhances learner engagement and motivation, creating a more dynamic and interactive learning environment. According to experts, this approach introduces elements of enjoyment and competition that positively influence learner participation. For example, they suggested using quick quiz games where students compete, earn points for correct answers, and track their progress on a leaderboard. These gamified elements aim to make the learning experience more engaging and motivating for learners.

In the context of situation assessment, experts emphasized the necessity of evaluating the institution's technology infrastructure and capabilities. They highlighted that a thorough review of the technology landscape is essential to determine the feasibility of integrating microlearning into the curriculum. Additionally, experts pointed out the importance of assessing faculty readiness, stressing that understanding faculty members' willingness and preparedness to adopt and incorporate microlearning strategies into their teaching practices is crucial for successful implementation. Furthermore, experts underscored the need to consider learners' device accessibility, noting that ensuring compatibility with various devices, including smartphones, tablets, and laptops, is vital for effectively reaching all learners.

Regarding potential errors, experts warned about technological barriers. They stressed that insufficient technology support or outdated systems can obstruct the successful integration of microlearning into the curriculum, potentially diminishing its effectiveness. Experts highlighted the necessity of a platform that enables learners to monitor their progress and completion of specific microlearning content. Experts also underscored the importance of having adequate technology support and systems in place to ensure the seamless integration of microlearning into the curriculum.

Additionally, experts noted the potential challenge posed by a lack of faculty training. They emphasized that inadequate training for faculty on using learning technologies and incorporating microlearning can result in ineffective implementation and diminished outcomes. In addition, experts highlighted the risk of overwhelming integration. They stressed that poorly integrated microlearning activities could burden learners with excessive workloads or disrupt the learning flow, negatively affecting their engagement and motivation. Experts pointed out the importance of ensuring that microlearning activities complement and enhance the existing curriculum, rather than creating an additional and overwhelming workload. They emphasized that microlearning should contribute to higher-level curriculum goals, ensuring seamless integration.

Lastly, experts pointed out the potential issue of inconsistent content quality. They warned that without proper quality control measures, the effectiveness of microlearning content could vary across different modules. The experts emphasized the importance of maintaining a high standard for content quality in microlearning to ensure its impact and effectiveness. They stressed the need for a regular review process to keep the content up-to-date, aligned with the curriculum, and relevant. Without these measures, microlearning content might not be uniformly effective. They further highlighted that adhering to the fundamental elements in AMPLIFY not only ensures seamless integration of microlearning with the curriculum but also establishes a standard for quality.

By carefully implementing these actions, conducting a thorough situation assessment, paying attention to critical cues, and proactively avoiding potential errors, educators can effectively integrate microlearning into the curriculum, creating a comprehensive and effective learning environment. This strategy, aligned with experts' insights, promotes a seamless educational experience that enhances learner engagement and academic success.

Table 3 illustrates the simulation interview technique for effectively addressing and assessing challenging tasks within the AMPLIFY framework. This technique involves a systematic approach to identifying critical actions, assessing situations, recognizing key cues, and anticipating potential errors. It serves as a comprehensive guide for educators and instructional designers to enhance microlearning experiences by aligning learning objectives, personalizing content, providing timely feedback, and integrating interactive and collaborative elements. By employing this technique, the table ensures a structured and effective implementation of microlearning strategies, ultimately leading to improved learner engagement and outcomes.

Table 3. Simulation interview table

AMPLIFY Challenging Tasks	Actions	Situation Assessments	Critical Cues	Potential Errors
1. The establishment of clear and concise learning objectives and alignment of microlearning activities with needs and goals.	1. Conduct a Needs Analysis: Assess the learners' needs, prior knowledge, and learning preferences to identify the specific learning objectives that align with their requirements.	1. Contextual Relevance: Evaluate the learning environment, context, and target audience to ensure the learning objectives are relevant and applicable to the learners' real-world needs.	1. Student Engagement: Monitor the level of engagement during microlearning activities as a cue to the effectiveness of alignment.	1. Vague Learning Objectives: Unclear or ambiguous learning objectives may lead to confusion among learners and hinder the effectiveness of the microlearning activities.
	2. Collaborate with Subject Matter Experts (SMEs): Work closely with SMEs to gain insights into the subject matter and ensure the learning objectives are accurate, relevant, and aligned with the curriculum.	2. Resource Availability: Assess the availability of resources, technology, and support needed to implement the microlearning activities aligned with the learning objectives.	2. Learning Outcomes: Evaluate whether the desired learning outcomes are achieved, providing a cue for the effectiveness of objectives and alignment.	2. Misalignment with Learners' Needs: If the learning objectives are not aligned with the learners' needs and goals, the microlearning activities may fail to engage and motivate the learners.
	3. Develop Clear Learning Objectives: Create well-defined, measurable, and achievable learning objectives that clearly specify what learners are expected to achieve after completing the microlearning activities.			3. Lack of Alignment: If microlearning activities are not aligned with the established objectives, students may not achieve the desired outcomes.
	4. Align Learning Objectives with Overall Goals: Ensure that the learning objectives align with the broader educational goals and objectives of the course or program.			4. Overwhelming Content: Failing to break down complex learning objectives into manageable microlearning activities can overwhelm learners and hinder their ability to absorb the information effectively.

continued on following page

Table 3. Continued

AMPLIFY Challenging Tasks	Actions	Situation Assessments	Critical Cues	Potential Errors
	5. Break Down Learning Objectives: Divide complex learning objectives into smaller, manageable chunks that can be effectively addressed through microlearning activities.			5. Lack of SME Input: Not involving SMEs in the process of developing learning objectives can result in inaccuracies and inadequacies in the content, leading to a gap between the objectives and the actual learning outcomes.
2. The incorporation of personalization through learner-centric activities and customized content.	1. Conduct Learner Profiling: Gather data on learners' preferences, learning styles, prior knowledge, and goals to create personalized profiles.	1. Learner Profiles: Create comprehensive learner profiles by considering factors such as age, educational background, skill level, and learning goals.	1. Individual Progress: Monitor the progress of individual learners to gauge the effectiveness of personalized content and activities.	1. Ignoring Learner Preferences: Failing to consider learners' preferences and needs can result in disengagement and reduced motivation to participate in the personalized activities.
	2. Create Customized Content: Develop tailored learning materials and resources that align with learners' profiles and specific needs.	2. Content Relevance: Evaluate the relevance and appropriateness of the customized content to ensure it addresses specific learner needs and interests.	2. Engagement Levels: Track the level of engagement with customized content and activities as an indicator of their relevance and interest.	2. Insufficient Content Variety: Failing to offer a diverse range of personalized content options may limit learners' exposure to different perspectives and topics.
	3. Offer Adaptive Learning Paths: Implement adaptive learning strategies to provide individualized learning pathways based on learners' progress and performance.		3. Learning Retention: Assess how well learners retain information from personalized content compared to generic content.	3. Continuous Personalization: Failing to adapt content and activities to evolving learner needs

continued on following page

Table 3. Continued

AMPLIFY Challenging Tasks	Actions	Situation Assessments	Critical Cues	Potential Errors
	4. Encourage Self-Directed Learning: Promote learner autonomy and self-directed exploration of content to cater to diverse learning preferences.			
3. The inclusion of feedback and assessment through timely and constructive feedback, formative and summative assessments, varied assessment methods, alignment with learning objectives, and continuous improvement.	1. Timely and Constructive Feedback: Provide timely feedback to learners on their performance, highlighting areas of strength and improvement. Ensure that the feedback is constructive and actionable, encouraging learners to make progress.	1. Learning Objectives Alignment: Evaluate the alignment between assessments and learning objectives to verify that the assessments effectively measure the intended learning outcomes.	1. Learning Progression: • Monitor the progression of student learning through formative assessments to identify areas that may require additional attention. • Assess the overall achievement and mastery demonstrated in summative assessments.	1. Inadequate Feedback: Providing vague or insufficient feedback can leave learners uncertain about their progress and how to improve.
	2. Formative Assessments: Integrate formative assessments throughout the learning process to gauge learners' understanding and progress. Use formative assessments to identify areas that need further attention and adaptation in instructional strategies.	2. Assessment Variety: Assess the diversity of assessment methods to ascertain if they cater to the different cognitive and affective domains of learning.	2. Feedback Uptake: • Observe how students incorporate feedback into their learning and whether it leads to improvement. • Gauge the responsiveness of students to different types of feedback.	2. Overemphasis on Summative Assessment: Overreliance on summative assessments without sufficient formative assessments may limit opportunities for timely intervention and adjustments in instruction.

continued on following page

Table 3. Continued

AMPLIFY Challenging Tasks	Actions	Situation Assessments	Critical Cues	Potential Errors
	3. Summative Assessments: Conduct summative assessments to evaluate learners' overall mastery of the material and achievement of learning objectives. Summative assessments provide a comprehensive overview of learners' performance.		3. Alignment Check: ● Regularly check the alignment between assessments and learning objectives to ensure coherence.	3. Mismatched Assessment Methods: Using assessment methods that do not align with the learning objectives may fail to provide a comprehensive understanding of learners' progress.
	4. Varied Assessment Methods: Employ a variety of assessment methods, such as quizzes, projects, presentations, group work, and self-assessments, to cater to different learning styles and abilities.		4. Evaluate whether assessments cover a representative sample of the learning content.	
	5. Alignment with Learning Objectives: Ensure that all assessments are aligned with the stated learning objectives, providing a clear indication of learners' attainment of the intended outcomes.			
	6. Continuous Improvement: Regularly review assessment data to identify trends and areas for improvement. Use this information to refine teaching approaches and enhance the microlearning experience.			

continued on following page

Table 3. Continued

AMPLIFY Challenging Tasks	Actions	Situation Assessments	Critical Cues	Potential Errors
4. The integration of structured and engaging content delivery through clear and focused content, bite-sized units, logical progression, and interactivity and reinforcement.	1. Clear and Focused Content: Develop content that is concise, well-organized, and directly aligned with the learning objectives, ensuring learners can easily grasp the key concepts.	1. Content Clarity: Assess the clarity and coherence of the content to ensure that it conveys information effectively without ambiguity.	1. Engagement Levels: ● Monitor the engagement levels of learners during content delivery to gauge its effectiveness. ● Observe participation in interactive elements as a cue for the engagement of the learning audience.	1. Content Overload: Presenting too much information at once can overwhelm learners and hinder their ability to retain and comprehend the material.
	2. Bite-sized Units: Divide the content into small, manageable units to facilitate better retention and prevent cognitive overload.	2. Content Chunking: Evaluate how the content is divided into chunks and assess if the units are appropriately sized for effective learning.	2. Comprehension and Retention: ● Assess how well learners comprehend and retain information presented in bite-sized units. ● Use quizzes and assessments to measure the retention of key concepts over time.	2. Disjointed Progression: Lack of a logical progression can confuse learners and disrupt their learning experience.
	3. Logical Progression: Arrange the content in a logical sequence, ensuring that each unit builds upon the previous one, creating a smooth flow of learning.			3. Insufficient Interactivity: Failing to include interactive elements may lead to passive learning, reducing learners' engagement and motivation.

continued on following page

Table 3. Continued

AMPLIFY Challenging Tasks	Actions	Situation Assessments	Critical Cues	Potential Errors
	4. Interactivity and Reinforcement: Incorporate interactive elements, such as quizzes, simulations, and activities, to engage learners actively and reinforce their understanding of the content.			4. Unaligned Content: Content that is not directly related to the learning objectives can lead to wasted time and effort for both learners and educators.
5. The incorporation of microlearning delivery methods through multimedia integration, mobile-friendly design, and just-in-time learning.	1. Multimedia Integration: Integrate multimedia elements such as videos, audio, infographics, and interactive simulations to enhance the learning experience and cater to diverse learning preferences.	1. Multimedia Effectiveness: Assess the impact of multimedia elements on learners' engagement, comprehension, and knowledge retention.	1. Access Patterns: • Monitor how frequently and consistently learners access microlearning content on mobile devices. • Track the types of devices used to ensure compatibility with mobile-friendly design.	1. Overwhelming Multimedia: Excessive use of multimedia elements can distract learners and impede their focus on key learning points.
	2. Mobile-Friendly Design: Ensure that the microlearning content is accessible and optimized for mobile devices, allowing learners to engage with the material anytime and anywhere.	2. Mobile Accessibility: Evaluate the accessibility and user-friendliness of the microlearning content on various mobile devices.	2. Engagement with Multimedia: • Assess the level of engagement with multimedia elements to determine their effectiveness. • Look for critical cues such as completion rates, interaction with interactive features, and feedback from learners.	2. Inaccessible Design: Failing to optimize the microlearning content for mobile devices may limit learners' accessibility and flexibility.

continued on following page

Table 3. Continued

AMPLIFY Challenging Tasks	Actions	Situation Assessments	Critical Cues	Potential Errors
	3. Just-in-Time Learning: Provide learners with on-demand access to relevant and practical information, allowing them to apply knowledge immediately in real-life situations.	3. Applicability of Just-in-Time Learning: Determine the effectiveness of just-in-time learning in addressing learners' immediate needs and challenges.	3. Application of Just-in-Time Learning: • Observe whether learners apply knowledge and skills gained through just-in-time microlearning to their immediate tasks.	3. Lack of Relevance: Providing just-in-time learning that lacks direct applicability or relevance to learners' needs may lead to disengagement.
			4. Gauge the effectiveness of just-in-time learning by evaluating performance improvements in relevant activities.	
6. The integration of interactive learning experiences through authentic sources, active learning, interactive activities, and experiential learning.	1. Incorporate Authentic Sources: Integrate real-life, authentic sources such as case studies, interviews, or real-world scenarios into the microlearning content to provide learners with relevant and practical examples.	1. Learner Needs and Preferences: Assess learners' needs and preferences to ensure that the integration of interactive learning experiences aligns with their learning styles and preferences.	1. Engagement Levels: • Monitor the level of engagement during interactive learning experiences, including participation in discussions and completion of activities. • Track the time spent on interactive elements to assess their impact.	1. Technology-related Issues: Technical glitches or inadequately functioning interactive elements may hinder learners' learning experience.

continued on following page

Table 3. Continued

AMPLIFY Challenging Tasks	Actions	Situation Assessments	Critical Cues	Potential Errors
	2. Promote Active Learning: Encourage learners to actively engage with the content through activities like discussions, problem-solving exercises, or hands-on projects that require active participation.	2. Technological Resources: Evaluate the availability and accessibility of technological resources needed to deliver interactive content effectively, considering factors like device compatibility and internet connectivity.	2. Application of Knowledge: • Evaluate whether learners can apply theoretical knowledge gained from interactive experiences to practical situations. • Assess the transfer of skills and understanding from interactive activities to real-world scenarios.	2. Overwhelming Complexity: Overloading learners with too many interactive elements or complex activities can lead to cognitive overload and reduced learning effectiveness.
	3. Create Interactive Activities: Develop interactive elements like quizzes, simulations, and gamified content to enhance learner engagement and create an immersive learning experience.	3. Content Relevance: Ensure that the interactive learning experiences are relevant to the learning objectives and contribute meaningfully to the overall learning outcomes.	3. Feedback and Reflection: • Gather feedback from learners on their experiences with interactive activities.	3. Lack of Alignment: Failure to align the interactive learning experiences with the learning objectives may result in disengagement and lack of relevance for learners.
	4. Facilitate Experiential Learning: Provide opportunities for learners to apply their knowledge in practical situations through simulations, role-plays, or real-life scenarios.		4. Encourage reflective practices to help learners articulate what they have learned and how it applies to their context.	4. Inadequate Assessment: Insufficient evaluation of learners' performance during interactive activities may lead to a lack of feedback and hinder continuous improvement.

continued on following page

Table 3. Continued

AMPLIFY Challenging Tasks	Actions	Situation Assessments	Critical Cues	Potential Errors
7. The integration of social interaction through collaborative learning and peer feedback and review.	1. Implement Collaborative Learning Activities: Design and facilitate group projects, discussions, and teamwork exercises that promote social interaction and collaboration among learners.	1. Learner Interests and Preferences: Assess learners' willingness to engage in social interactions and collaborative activities based on their interests and prior experiences.	1. Collaboration Levels: • Monitor the level of collaboration among learners in group activities or projects. • Observe the quality of contributions and the effectiveness of group dynamics.	1. Unequal Participation: Some learners may dominate collaborative activities, leading to unequal engagement and reduced learning opportunities for others.
	2. Encourage Peer Feedback and Review: Foster a culture of constructive feedback where learners provide input and review each other's work, fostering social engagement and deeper understanding.	2. Group Dynamics: Consider learners' diverse backgrounds and personalities to create balanced and effective group dynamics for collaborative learning.	2. Peer Feedback Quality: • Assess the quality of feedback provided by peers in terms of specificity, relevance, and constructive criticism. • Gather feedback from learners about the helpfulness of peer reviews.	2. Ineffective Communication: Lack of clear communication and guidance can hinder meaningful collaboration and compromise the effectiveness of social interactions.
			3. Community Engagement: • Measure the engagement levels in discussion forums or collaborative platforms.	3. Inadequate Feedback Culture: If learners are not receptive to feedback or lack the skills to provide constructive criticism, the value of peer review may be diminished.
			4. Evaluate the diversity of topics discussed and the extent of knowledge sharing within the community.	

continued on following page

Table 3. Continued

AMPLIFY Challenging Tasks	Actions	Situation Assessments	Critical Cues	Potential Errors
8. Seamless integration of microlearning with the curriculum through the integration of learning technologies and integration with the learning ecosystem.	1. Curriculum Mapping: Align microlearning activities with the existing curriculum to ensure coherence and relevance in the overall learning journey..	1. Technology Infrastructure: Evaluate the institution's technology infrastructure and capabilities to determine the feasibility of integrating microlearning with the curriculum.	1. Usage Analytics: • Track user engagement and completion rates of microlearning modules. • Analyze data to identify which microlearning elements are most effective and where improvements may be needed.	1. Technological Barriers: Inadequate technology support or outdated systems can hinder the successful integration of microlearning with the curriculum.
	2. Identify Learning Technology: Select appropriate learning technologies and tools that complement microlearning delivery and enhance the learning experience	2. Faculty Readiness: Assess the readiness and willingness of faculty to adopt and integrate microlearning into their teaching practices.	2. Alignment with Curriculum Progression: • Assess how well microlearning aligns with the chronological progression of the curriculum. • Ensure that microlearning content is introduced at appropriate points in the learning journey.	2. Lack of Faculty Training: Insufficient training for faculty on using learning technologies and integrating microlearning can result in ineffective implementation.

continued on following page

Table 3. Continued

AMPLIFY Challenging Tasks	Actions	Situation Assessments	Critical Cues	Potential Errors
	3. Integrate with Learning Management System (LMS): Seamlessly integrate microlearning content and activities within the institution's LMS to provide easy access to learners.	3. Learner Device Accessibility: Consider learners' access to devices (e.g., smartphones, tablets, laptops) to ensure compatibility with microlearning content delivery.	3. Interactivity and Engagement: • Gauge the level of interactivity and engagement within microlearning modules.	3. Overwhelming Integration: Poorly integrated microlearning activities may overwhelm learners with an excessive workload or disrupt the learning flow.
	4. Incorporate Gamification Elements: Integrate gamification elements into microlearning to enhance learner engagement and motivation.		4. Look for cues such as user participation in quizzes, discussions, or other interactive elements.	4. Inconsistent Content Quality: Without proper quality control measures, microlearning content may vary in quality and effectiveness across different modules.

9.2.2 Review of Simulation for Microlearning Pedagogical Practices

In conclusion, the simulation interviews conducted using the Applied Cognitive Task Analysis (ACTA) method provided valuable insights into the practical application of the microlearning framework 'AMPLIFY' in higher education. These interviews engaged subject matter experts in scenario-based assessments, aiming to replicate classroom use and evaluate AMPLIFY's effectiveness. Through this process, significant aspects such as framework implementation actions, thorough situation assessments of classroom contexts, critical cues indicating success or challenges, and identification of potential errors were meticulously examined.

The findings from these evaluations underscore AMPLIFY's potential feasibility, usability, and practicality for integration into higher education classrooms. They serve as a critical foundation for informed decision-making by educators and stakeholders regarding the framework's adaptability and effectiveness. By aligning with expert insights and addressing identified challenges, AMPLIFY shows promise in enhancing educational experiences through microlearning strategies. This comprehensive evaluation framework supports the ongoing advancement and implementation of innovative educational practices in higher learning environments.

Additionally, the simulation interviews highlighted the importance of contextual adaptability when implementing the AMPLIFY framework across diverse learning environments. Experts noted that while the framework offers significant advantages in promoting focused, modular learning, its success is highly dependent on the alignment with specific institutional goals, student needs, and course objectives. This underscores the necessity for flexibility in both instructional design and content delivery, allowing educators to tailor the microlearning modules to fit their unique classroom settings. By incorporating feedback loops and continuous assessment, AMPLIFY can be fine-tuned to not only improve engagement but also ensure sustained learning outcomes over time. This iterative approach will be key in maximizing the framework's impact on student learning and overall educational effectiveness in higher education institutions.

CONCLUSION

In this chapter, we explored the application of simulation, the third step in the Applied Cognitive Task Analysis (ACTA) framework, to enhance both electronic training for professional development and microlearning pedagogical practices. Simulations serve as an invaluable tool in educational design by allowing for the replication of real-world tasks, providing insights into the cognitive processes and challenges that learners encounter. By engaging experts in scenario-based assessments, we were able to identify key areas for improvement in both educational approaches, ensuring they are better aligned with learners' needs and expectations.

Section 9.1 focused on the simulation of electronic training systems designed for professional development. Through the use of simulation interviews, experts assessed various aspects of system functionality, content design, navigation, and assessment strategies. The insights gained highlighted critical cognitive challenges faced by learners, such as navigating non-linear content, engaging with course materials, and the usability of assessments. These findings underscore the importance of developing intuitive, user-centred electronic training systems that address both the cognitive and practical needs of professionals in digital learning environments.

Section 9.2 examined the simulation of microlearning pedagogical practices. Through expert interviews, key cognitive tasks involved in microlearning were evaluated, and suggestions were made for enhancing the design and delivery of microlearning modules. The feedback highlighted areas for improvement, such as the integration of multimedia elements, the usability of content, and ensuring that assessments align with learning objectives. These insights from the simulation interviews provide valuable guidance for refining microlearning strategies to better meet the needs of learners in higher education.

In conclusion, the use of simulation interviews in this chapter allowed for a thorough evaluation of both electronic training and microlearning practices. By leveraging expert insights, we were able to identify cognitive challenges and propose strategies to enhance the design and effectiveness of these educational approaches. The findings from these simulations lay a strong foundation for ongoing improvements, ensuring that both professional development programs and microlearning practices are better suited to the cognitive demands of learners and educators alike.

REFERENCES

Alam, T., & Benaida, M. (2022). Smart curriculum mapping and its role in outcome-based education. *Informatica (Vilnius)*, 46(4). Advance online publication. DOI: 10.31449/inf.v46i4.3717

Barthakur, A., Dawson, S., & Kovanovic, V. (2023). Advancing learner profiles with learning analytics: A scoping review of current trends and challenges. In *Proceedings of the 13th International Learning Analytics and Knowledge Conference (LAK23)*. DOI: 10.1145/3576050.3576083

Bosch, C., & Laubscher, D. J. (2022). Promoting self-directed learning as learning presence through cooperative blended learning. *International Journal of Learning. Teaching and Educational Research*, 21(9), 18–33. DOI: 10.26803/ijlter.21.9.2

Contrino, M. F., Reyes-Millán, M., Vazquez-Villegas, P., & Membrillo-Hernández, J. (2024). Using an adaptive learning tool to improve student performance and satisfaction in online and face-to-face education for a more personalized approach. *Smart Learning Environments*, 11(1). Advance online publication. DOI: 10.1186/s40561-024-00292-y

de Vries, P., Van den Bogaard, M., & Overschie, M. (2018, September). Microlearning to support authentic learning in continuing education and for engineering students. In *SEFI2018 - Creativity, Innovation and Entrepreneurship for Engineering Education Excellence* (Conference proceedings).

Fidan, M. (2023). The effects of microlearning-supported flipped classroom on pre-service teachers' learning performance, motivation and engagement. *Education and Information Technologies*, 28, 12687–12714. DOI: 10.1007/s10639-023-11639-2

Gottler, A. (2023). Collaboration between instructional designers and subject matter experts in digital transformation projects. *Studies in Technology Enhanced Learning*, 3(2). Advance online publication. DOI: 10.21428/8c225f6e.93df9a6e

Haleem, A., Javaid, M., Qadri, M. A., & Suman, R. (2022). Understanding the role of digital technologies in education: A review. *Sustainable Operations and Computers*, 3, 275–285. DOI: 10.1016/j.susoc.2022.05.004

Irfadila, M. S., & Noprika, W. (2022). The importance of needs analysis in curriculum development language learning. In *Proceedings of the 2nd International Conference on Language Pedagogy* (pp. 224). Universitas Muhammadiyah Sumatera Barat & Universitas Adzkia.

Kumar, M. V., & Kumar, A. (2024). Incorporating constructive written feedback to enhance students' writing skills: Strategies and practices. *African Journal of Biological Sciences*, 6(14), 6341–6353. DOI: 10.33472/AFJBS.6.14.2024.6341-6353

Kuyyogsuy, S. (2019). Promoting peer feedback in developing students' English writing ability in L2 writing class. *International Education Studies*, 12(9), 76–87. DOI: 10.5539/ies.v12n9p76

Liu, D. (2024). The effects of segmentation on cognitive load, vocabulary learning and retention, and reading comprehension in a multimedia learning environment. *BMC Psychology*, 12(1). Advance online publication. DOI: 10.1186/s40359-023-01489-5

Mokhtar, S. M., Nur, A. A. A., Shazali, N. M., & Ahmad, N. (2020). Why Are Students Involved In E-Learning? A Reasoning Study At Univerisity Kebangsaan Malaysia (UKM). *International Journal of Education and Pedagogy*, 2(1), 152–159. http://myjms.moe.gov.my/index.php/ijeap

Nilsson, R. (2021). *Designing an interactive micro-learning application: A new way to learn thermography* (Examensarbete, 30 hp). Uppsala Universitet.

Nordin, H., Singh, D., & Mansor, Z. (2020). An empirical study of e-learning interface design elements for generation Z. *International Journal of Advanced Computer Science and Applications*, 11(9), 507–515. DOI: 10.14569/IJACSA.2020.0110961

Rafique, S., & Aziz, S. (2019). Relationship of Pedagogical Skills and Content Knowledge with Pedagogical Design Capacity in Higher Education of Pakistan. *Global Regional Review*, IV(IV), 85–94. DOI: 10.31703/grr.2019(iv-iv).10

Saintika, Y., Astiti, S., Kusuma, D. J. A., & Muhammad, A. W. (2021). Analysis of E-learning readiness level of public and private universities in central Java, Indonesia. *Register: Jurnal Ilmiah Teknologi Sistem Informasi*, 7(1), 15–30. DOI: 10.26594/register.v7i1.2042

Sidgi, L. F. S. (2022). The benefits of using collaborative learning strategy in higher education. *International Journal of English Literature and Social Sciences*, 7(6). Advance online publication. DOI: 10.22161/ijels

Silva, L., Mendes, A. J., & Gomes, A. (2020). Computer-supported collaborative learning in programming education: A systematic literature review. In *IEEE Global Engineering Education Conference, EDUCON* (Vol. 2020-April, pp. 1086–1095). IEEE Computer Society. DOI: 10.1109/EDUCON45650.2020.9125237

Sulaiman, A. H., & Abdullah, S. A. (2024). The use of active learning strategies to foster effective teaching in higher education institutions. *Zanco Journal of Humanity Sciences*, 28(2). Advance online publication. DOI: 10.21271/zjhs.28.2.11

Chapter 10
Cognitive Demand Table

ABSTRACT

This chapter discusses the development and use of Cognitive Demand Tables in Applied Cognitive Task Analysis (ACTA) to assess and optimize cognitive load in educational settings. It is divided into two sections. The first, "Cognitive Demand Table of Electronic Training for Professional Development," examines how cognitive demands in electronic training are identified, analyzed, and balanced to prevent overload, ensuring effective and accessible professional development. The second, "Cognitive Demand Table of Microlearning Pedagogical Practices," focuses on optimizing the cognitive load of microlearning strategies in higher education, enhancing understanding and retention. This chapter highlights the importance of managing cognitive demands to create effective, learner-centered instructional strategies.

INTRODUCTION

The Cognitive Demand Table represents the fourth step in the Applied Cognitive Task Analysis (ACTA) framework. This tool is designed to systematically assess the mental and cognitive demands that learners encounter while performing tasks within educational and training environments. By identifying and categorizing the cognitive challenges involved in specific learning activities, the Cognitive Demand Table helps educators, instructional designers, and trainers pinpoint areas where learners may struggle and develop strategies to address those challenges effectively.

In this chapter, we examine the application of the Cognitive Demand Table in two key educational settings: electronic training for professional development and microlearning pedagogical practices.

Section 10.1 focuses on creating a Cognitive Demand Table for electronic training systems aimed at professional development. This section will outline the cognitive complexities faced by professionals engaging with digital training programs, such as navigating complex systems, processing course content, and completing assess-

DOI: 10.4018/979-8-3693-7733-8.ch010

ments. By systematically mapping out these demands, the table provides valuable insights into how training programs can be tailored to better support professional learners, ensuring a balance between challenge and accessibility.

Section 10.2 applies the Cognitive Demand Table to microlearning pedagogical practices. Microlearning, with its focus on delivering content in brief, targeted units, presents unique cognitive challenges, such as maintaining focus and retaining information over short bursts of learning. The Cognitive Demand Table in this context helps educators identify the mental load required for specific microlearning tasks and offers strategies for designing modules that enhance retention and engagement while minimizing cognitive overload (Hultberg, Santandreu Calonge, & Safiullin Lee, 2018).

Throughout this chapter, we will explore how the Cognitive Demand Table serves as a powerful tool for evaluating and improving the cognitive effectiveness of educational programs. By breaking down complex tasks and assessing the mental effort required to complete them, this tool supports the creation of more effective, learner-centred instructional designs in both professional development and higher education settings.

10.1 Cognitive Demand Table of Electronic Training System for Professional Development

10.1.1 Cognitive Demand Table for System Design

The cognitive demand table provided a comprehensive view of the cognitive task difficulties and expertise experienced by experts. Generally, the difficulties discussed by experts are caused mainly by lack of experience in using online learning. Users who do not have any experience with similar online learning environment, will face a much greater difficulty assessing and navigating elements in TaSPoD. The screen view which was design with mouse rollover options could-be the cause for overfull screens and thus discourage users from completing the session. Furthermore, absence of multimedia elements in the content may cause demotivation as well.

Experts have identified potential challenges for users in terms of communication with the system, noting that the lack of real-time communication, feedback, or response mechanisms could hinder user interaction. The current system relies solely on online forums, which may not provide the immediate support users need. This concern aligns with the findings Silva, Mendes, & Gomes (2020), who emphasized that incorporating communication tools, such as real-time messaging or video conferencing, is essential for fostering collaboration among learners in e-learning environments. By integrating more interactive communication features, the system could significantly enhance user engagement and collaboration, ultimately improving the overall learning experience (Theodorio, 2024).

HLEs need to respond to the online forum and at the same time being active in completing assessments provided in the same training session. Too many assessments to be completed in any single training session will also be a discouraging factor for incompletion of the course in general and incompletion of the assessments in specific. Experts highlighted the importance of utilizing varied assessment strategies tailored to different levels of users within the TaSPoD system as learners expressed high satisfaction with the diverse assessment approaches implemented (Bahati, Fors, Hansen, Nouri, & Mukama, 2019).

The findings from this table determine the cause of the errors were noticed by experts who have experience in electronic training or electronic learning environment. Both trainers and trainees as the main users of the system are obliged to enhance their comprehension on the electronic platform to enable them to apply and integrate knowledge into tasks.

Conclusion of Evaluation Phase

Table 1 showed the major difficult cognitive elements integrated from the previous three techniques. System functionality design, course content design, and assessments design are the three difficult cognitive elements summarized in the study. The findings of this study suggest that TaSPoD, as an electronic training system, serves as a valuable platform to help educators sustain their professional development activities, particularly during unexpected circumstances such as the COVID-19 pandemic. As well as being an alternative to reach the unreach (Alasmari, 2020). A structured, guided, and supported approach to professional development delivery enhances the effectiveness, increases the efficacy and technical confidence of educators (Gong, Collins, & Amrein-Beardsley, 2022), and maximizes the benefits of the electronic training system. The findings will be a significant study towards the development of electronic training systems.

Table 1. Cognitive demand table

Difficult cognitive elements	Why difficult	Causes of common errors	Cues and strategies used	
System functionality design *Course content design* *Assessment design*	Difficult to navigate through the system and content especially for users without experience with similar online applications or Learning Management Platform and for users who don't know how to adjust their browsers. Difficult to motivate to complete the task. Difficult to be active in the online forum, assignment, etc. Difficult to communicate between TaSPoD and among HLEs. Difficult to complete the assessments.	Not well-structured guidance and navigation. TaSPoD did not prepare well about the element of formative and summative assessment. TaSPoD did not provide proper step-by-step pedagogical guidance to the HLEs. Poor online communication and feedback tools design between TaSPoD and HLEs.	***Cues:*** Screen designs will look crowded because of mouse rollover option as it might create pressure when HLE do the self-learning. Having animations and pictures would enhance the interest of the user. Navigation and placement of buttons should be self-explanatory. HLEs only can communicate via electronic forum, which are considered quite conventional and not fun. HLEs need to respond to online forums and completing assessments. Too many assessments might discourage HLEs from continuing the course.	***Strategies:*** A simple, straight-forward, user-friendly and attractive screen views should be designed in accordance to the needs of the higher learning educators (HLEs). The screen views need to be designed using graphical user interface that contains simple and appropriate objects, tabs and icons, and video upload tools. TaSPoD should embed appropriate pedagogical and instructional-based model and automate learning process in sequential manner in order to give clear navigation, guidance and engagement in learning to HLEs. There can be more interesting applications to communicate with the TaSPoD such as Chat programs, video conferencing rather than the usual online forums and emails. The downloading and uploading of the documents via PDF and word processing application to be customised within the TaSPoD.

10.1.2 Cognitive Demand Table for System Prototype

The cognitive demand table provides a comprehensive view of the cognitive task difficulties and expertise experienced by experts. This technique integrates the data obtained from the data findings from the previous techniques: task diagram, knowledge audit and simulation.

Table 2 shows four difficult cognitive elements integrated from the previous two techniques. System functionality, course content, assessments design and real-time assistance are the four difficult cognitive elements identified. These difficulties vary between trainers and trainees.

Generally, trainees would find difficulties in navigating the system, specifically to resume from the last access. Trainees will need to know effective ways of identifying hyperlinks or hypertext links that allows content navigation not according in sequence. Trainees would also need assistance while using the system that attend almost instantly to their questions and queries. Absence of these elements will potentially lead trainees to disregard some pages of content and even to the incompletion of the course.

Whereas for trainers, they will find difficulties in determining the suitability of instructional activities. Trainers need to specify and decide different constructive methods or strategies to facilitate the class, with different level of users. Apart of the instructional activities, trainers might also encounter difficulties in determining appropriate summative assessment design (Hultberg et al., 2018). Both difficult cognitive elements require trainers to ensure that course learning outcomes are being addressed.

The findings from this table determine the cause of the errors often noticed by experts who have experience in electronic training or electronic learning. Both trainers and trainees obliged to enhance their comprehension on the electronic platform to enable them to apply and integrate knowledge into tasks.

Table 2. Cognitive demand table

Difficult cognitive elements	Why difficult	Common errors	Cues and strategies used
System Functionality *Course Content* *Assessment design* *Real-time assistance*	Difficult to navigate throughout the system, specifically to continue after previous login. Difficult to determine the suitability of the instructional activities. Difficult to evaluate appropriateness of summative assessment design. Difficult to get real-time assistance while using the system.	There will be tendency for HLEs to omit or disregards some pages. Lack of excitement will lead to incomplete of course. Assessment did not address the learning outcomes of the course and did not fulfil different level of users. Immediate answer might not be fulfilled when the HLE needed them.	***Cues:*** Users are not able to start back on the beginning page after returning. Less usage of multimedia elements will lead to lack of excitement. Enhance understanding of subject matter and able to apply and integrate the knowledge into the task / assessment given. ***Strategies:*** Enable a log to examine or inspect hypertext link that being assessed. Trainers tend to use only text, images and videos to deliver the course content. Instructions should be provided to inform achievement to move to next lesson/level. Usage of different assessments strategies for different level of users.

10.1.3 Review of Cognitive Demand Table of Electronic Training for Professional Development

The Cognitive Demand Table has provided a detailed analysis of the cognitive challenges faced by both trainers and trainees within the TaSPoD system. This table allowed us to comprehensively map out the difficult cognitive elements across system functionality design, course content design, and assessment design, as well as real-time assistance in the system prototype. The integration of insights from the task diagram, knowledge audit, and simulation techniques enabled a deeper understanding of how users interact with the system and where the cognitive bottlenecks occur.

For the system design, the primary cognitive difficulties stemmed from a lack of experience with similar online learning platforms and inadequate navigation cues. The findings emphasized the importance of designing user-friendly, visually intuitive systems that provide clear, step-by-step guidance. Enhancing the use of multimedia elements and simplifying navigation are key strategies to ensure learners are engaged and motivated to complete the training.

In the system prototype, additional difficulties emerged, including challenges with resuming content from previous sessions and the absence of immediate real-time assistance. Trainers also faced significant challenges in tailoring instructional activities and designing assessments that appropriately matched learning outcomes for different levels of users. These findings indicate the need for more structured content delivery and better-designed assessment strategies to enhance the learning experience (Tai et al., 2023).

Overall, the Cognitive Demand Table offers actionable insights for improving both system design and user experience. Addressing these cognitive challenges will lead to a more efficient, engaging, and supportive environment for users of the TaSPoD system, ultimately enhancing the effectiveness of electronic training for professional development. By implementing the strategies outlined in the table, the system can better meet the diverse cognitive needs of its users, ensuring a smoother, more productive learning process.

10.2 Cognitive Demand Table of Microlearning Pedagogical Practices

10.2.1 Application of Cognitive Demand Table in the Evaluation Phase

Cognitive Demand Table

The Cognitive Demand Table serves as a pivotal tool in the evaluation of the AMPLIFY framework, playing a crucial role in consolidating insights from various ACTA techniques, including task diagrams, knowledge audits, and simulations. This structured table offers a comprehensive overview of the cognitive challenges and levels of expertise encountered by experts during their engagement with the framework.

Role in Evaluation

The significance of the Cognitive Demand Table lies in its ability to provide insightful and detailed information about the cognitive demands associated with the implementation of the AMPLIFY framework. This tool acts as a bridge between theory and practice, shedding light on the practicality, usability, and feasibility of the framework.

Comprehensive Data Consolidation

The Cognitive Demand Table consolidates data from task diagrams, knowledge audits, and simulations, offering a holistic understanding of the cognitive complexities inherent in the AMPLIFY framework. It provides educators and researchers with a structured representation of the cognitive elements associated with the framework's pedagogical practices.

Practicality, Usability, and Feasibility

By analysing the cognitive elements presented in the table, educators and researchers gain insights into potential challenges that both educators and learners might face when interacting with the framework. This understanding contributes to refining the framework, ensuring it remains accessible, relevant, and effective across diverse educational contexts.

i. Practicality: The table informs the practicality of the framework by highlighting cognitive complexities involved in implementing microlearning strategies. This awareness allows educators to fine-tune their approach, aligning content and activities with learners' cognitive capacities and challenges.

ii. Usability: Potential cognitive challenges impacting learners' engagement and comprehension are identified, informing the design of user-friendly experiences. Educators can create content that caters to learners' cognitive abilities, enhancing the overall usability of the AMPLIFY framework.
iii. Feasibility: The table contributes to assessing the feasibility of the framework by identifying cognitive elements that may present difficulties for learners. This proactive approach enables educators to address challenges during development and implementation, ensuring a smoother and more effective learning experience.

Validation and Acceptance

The Cognitive Demand Table serves as a validation tool, indicating the effectiveness of the AMPLIFY framework based on experts' experiences. When experts engage with the framework and contribute to the data informing the table, their assessments serve as a form of validation. The table showcases that experts can effectively navigate the framework, suggesting that its pedagogical practices are well-structured and tailored to optimise microlearning experiences.

Iterative Refinement

The insights gained from the Cognitive Demand Table contribute to an iterative refinement process for the AMPLIFY framework. Educators and researchers can use this information to make informed adjustments, ensuring that the framework evolves to align seamlessly with learners' cognitive capacities, thereby maximising its impact on enhancing microlearning experiences.

Conclusion of Evaluation Phase

In conclusion, the Cognitive Demand Table stands as a critical component in the evaluation phase, offering a nuanced understanding of the cognitive demands associated with the AMPLIFY framework. By bridging theory and practice, this tool guides educators and researchers in refining the framework for optimal practicality, usability, and feasibility, ultimately enhancing the microlearning experiences in higher education.

Table 3, presented below, encapsulates the Cognitive Demand Table, providing a visual representation of the cognitive challenges and strategies identified in the evaluation of the AMPLIFY framework.

Table 3. Cognitive demand table

Difficult Cognitive Elements	Why Difficult	Common Errors	Cues and Strategies	
1. The establishment of clear and concise learning objectives and alignment of microlearning activities with needs and goals.	1. Subject Complexity: Some subject matters may be inherently complex, requiring careful consideration to define clear and focused learning objectives that adequately cover the required knowledge and skills.	1. Vague Learning Objectives: Unclear or ambiguous learning objectives may lead to confusion among learners and hinder the effectiveness of the microlearning activities.	**Cues:** 1. Confusion or lack of understanding regarding learning objectives.	**Strategies:** 1. Conducting a thorough analysis of learner needs, including their prior knowledge, skills, and aspirations.
	2. Varying Learner Needs: Designing learning objectives that align with the diverse needs and goals of learners can be challenging, as different learners may have different backgrounds, prior knowledge, and skill levels.	2. Misalignment with Learners' Needs: If the learning objectives are not aligned with the learners' needs and goals, the microlearning activities may fail to engage and motivate the learners.	2. Mismatch between stated objectives and learners' needs.	2. Aligning the learning objectives with the overall goals and objectives of the course, program, or curriculum.
	3. Curriculum Alignment: Ensuring that learning objectives align with the broader curriculum or program goals requires careful coordination and consideration of the overall learning outcomes.	3. Lack of Alignment: If microlearning activities are not aligned with the established objectives, students may not achieve the desired outcomes.	3. Inconsistent alignment of learning objectives with overarching goals.	3. Incorporating input from stakeholders, such as students, subject matter experts, and industry professionals, to ensure relevance and alignment with real-world needs.

continued on following page

Table 3. Continued

Difficult Cognitive Elements	Why Difficult	Common Errors	Cues and Strategies	
	4. Emerging Needs and Changing Goals: Educational contexts are dynamic, and needs and goals may evolve over time. Designing learning objectives that can adapt to these changing needs and goals requires ongoing reflection and adjustment.	4. Overwhelming Content: Failing to break down complex learning objectives into manageable microlearning activities can overwhelm learners and hinder their ability to absorb the information effectively.	4. Varied levels of engagement or achievement related to learning objectives.	Regularly reviewing and updating learning objectives to ensure they remain aligned with evolving needs and goals.
		5. Lack of SME Input: Not involving SMEs in the process of developing learning objectives can result in inaccuracies and inadequacies in the content, leading to a gap between the objectives and the actual learning outcomes.		
2. The incorporation of personalization through learner-centric activities and customized content.	1. Resource Intensiveness: Developing personalized content and activities for each learner requires significant time, effort, and resources.	1. Ignoring Learner Preferences: Failing to consider learners' preferences and needs can result in disengagement and reduced motivation to participate in the personalized activities.	**Cues:** 1. Lack of learner engagement in activities tailored to their preferences.	**Strategies:** 1. Developing learner profiles by gathering information about preferences, interests, and prior knowledge.

continued on following page

Table 3. Continued

Difficult Cognitive Elements	Why Difficult	Common Errors	Cues and Strategies	
	2. Scalability: Ensuring personalization at scale, especially in large learning environments, can be challenging, as it may require sophisticated technology and instructional design approaches.	2. Insufficient Content Variety: Failing to offer a diverse range of personalized content options may limit learners' exposure to different perspectives and topics.	2. Limited participation in personalized learning paths.	2. Designing adaptive learning paths that adjust based on learners' progress and provide personalized recommendations and content.
	3. Data Collection and Analysis: Gathering and analysing learner data to inform personalization strategies	3. Continuous Personalization: Failing to adapt content and activities to evolving learner needs	3. Low utilization of varied content formats by learners.	3. Offering content in different formats (text, audio, video, interactive modules) to accommodate diverse learning styles.
	4. Pedagogical Expertise: Designing effective personalized learning experiences requires expertise in instructional design and understanding of learners' cognitive processes and learning needs.		4. Limited choices or flexibility in learner selections.	4. Allowing learners to have choices in selecting activities or topics of interest and providing options for pacing and sequencing.
	5. Technological Infrastructure: Implementing personalized learning experiences often relies on robust and adaptive learning platforms, which may require investment in technology and infrastructure.		5. Absence of mechanisms for feedback and reflection.	5. Providing timely and meaningful feedback, encouraging self-reflection, and enabling learners to track their progress and set personal goals.

continued on following page

Table 3. Continued

Difficult Cognitive Elements	Why Difficult	Common Errors	Cues and Strategies	
3. The inclusion of feedback and assessment through timely and constructive feedback, formative and summative assessments, varied assessment methods, alignment with learning objectives, and continuous improvement.	1. Time Constraints: Providing timely and detailed feedback requires significant time and effort from educators, especially when dealing with a large number of students or complex assessments.	1. Inadequate Feedback: Providing vague or insufficient feedback can leave learners uncertain about their progress and how to improve.	**Cues:** 1. Timely and Constructive Feedback: • Lack of timely feedback, leading to potential gaps in understanding. • Absence of constructive feedback hindering students' learning and growth.	**Strategies:** 1. Timely and Constructive Feedback: Provide feedback in a timely manner after assessments and ensure feedback is constructive, specific, and actionable.
	2. Individual Differences: Students have diverse learning styles, abilities, and needs. Tailoring feedback and assessments to meet individual requirements can be a complex task for educators.	2. Overemphasis on Summative Assessment: Overreliance on summative assessments without sufficient formative assessments may limit opportunities for timely intervention and adjustments in instruction.	2. Formative and Summative Assessments: • Limited use of both formative and summative assessments, impacting comprehensive evaluation.	2. Formative and Summative Assessments: Design a balanced assessment approach that incorporates both formative assessments for ongoing feedback and summative assessments for comprehensive evaluation.
	3. Balancing Depth and Breadth: Assessments need to strike a balance between assessing deep understanding and covering a breadth of knowledge. Designing assessments that effectively evaluate higher-order thinking skills while still providing comprehensive feedback can be a challenge.	3. Mismatched Assessment Methods: Using assessment methods that do not align with the learning objectives may fail to provide a comprehensive understanding of learners' progress.	3. Varied Assessment Methods: • Restriction to a single assessment format, potentially limiting a holistic understanding of students' capabilities.	3. Varied Assessment Methods: Diversify assessment methods, including quizzes, projects, presentations, and practical demonstrations, to provide a more comprehensive understanding of students' skills and knowledge.

continued on following page

Table 3. Continued

Difficult Cognitive Elements	Why Difficult	Common Errors	Cues and Strategies	
	4. Engaging Students: Encouraging students to actively engage with feedback and use it to improve their learning requires fostering a positive learning environment and promoting a growth mindset.		4. Alignment with Learning Objectives: • Disconnect between assessment content and the defined learning objectives, affecting the effectiveness of evaluations.	4. Alignment with Learning Objectives: Ensure assessments align with stated learning objectives and provide feedback that reflects intended learning goals.
			5. Continuous Improvement: • Lack of a systematic approach to refining assessment strategies based on feedback and results, potentially hindering the learning experience.	5. Continuous Improvement: Establish a feedback loop for continuous improvement by analysing assessment results, gathering feedback from students, and adapting assessment strategies to enhance the overall learning experience.
4. The integration of structured and engaging content delivery through clear and focused content, bite-sized units, logical progression, and interactivity and reinforcement.	1. Balancing depth and brevity: Ensuring that the content is comprehensive enough to cover the necessary knowledge and skills while keeping it concise and focused can be challenging.	1. Content Overload: Presenting too much information at once can overwhelm learners and hinder their ability to retain and comprehend the material.	**Cues:** 1. Clear and Focused Content: Lack of clarity in content organization and delivery.	**Strategies:** 1. Clear and Focused Content: • Conduct a thorough content analysis and mapping to determine the most essential and relevant information to be included.
	2. Maintaining coherence: Creating a seamless flow and logical progression of content requires careful planning and organization.	2. Disjointed Progression: Lack of a logical progression can confuse learners and disrupt their learning experience.	2. Bite-Sized Units: Disengagement during lesson.	2. Bite-Sized Units: • Breaking down the content into smaller, manageable chunks or modules.

continued on following page

Table 3. Continued

Difficult Cognitive Elements	Why Difficult	Common Errors	Cues and Strategies	
	3. Designing effective interactive elements: Designing interactive activities and incorporating appropriate technology tools to enhance engagement and learning effectiveness can be complex.	3. Insufficient Interactivity: Failing to include interactive elements may lead to passive learning, reducing learners' engagement and motivation.	3. Logical Progression: Difficulty in following the content.	3. Logical Progression: • Plan and organize the learning materials in a progressive manner, building upon previously covered concepts and skills.
	4. Providing meaningful reinforcement: Designing reinforcement activities and assessments that effectively reinforce learning and address individual learner needs can be demanding.	4. Unaligned Content: Content that is not directly related to the learning objectives can lead to wasted time and effort for both learners and educators.	4. Interactivity and Reinforcement: Limited engagement during class activities.	4. Interactivity and Reinforcement: • Incorporate interactive components such as quizzes, simulations, discussions, and multimedia to enhance learner engagement and participation. • Design activities, exercises, and assessments that allow learners to apply their knowledge and skills, receive feedback, and reinforce their understanding
5. The incorporation of microlearning delivery methods through multimedia integration, mobile-friendly design, and just-in-time learning	1. Content Design and Development: Creating high-quality multimedia content requires expertise in instructional design, multimedia production, and pedagogical considerations.	1. Overwhelming Multimedia: Excessive use of multimedia elements can distract learners and impede their focus on key learning points.	**Cues:** 1. Multimedia Integration: Limited engagement with multimedia content in the learning materials.	**Strategies:** 1. Create Engaging Multimedia Content: • Design and develop multimedia content that is visually appealing, interactive, and effectively conveys the key concepts.

continued on following page

Table 3. Continued

Difficult Cognitive Elements	Why Difficult	Common Errors	Cues and Strategies	
	2. Technological Requirements: Implementing multimedia integration and mobile-friendly design may involve complex technology requirements and compatibility issues.	2. Inaccessible Design: Failing to optimize the microlearning content for mobile devices may limit learners' accessibility and flexibility.	2. Mobile-Friendly Design: Challenges with accessing learning content on mobile devices.	2. Mobile-Friendly Design: • Optimize learning materials for mobile devices, considering different screen sizes and functionalities. • Utilize responsive design principles to ensure seamless access to content on various mobile platforms.
	3. Customization and Adaptation: Adapting the microlearning content to suit different devices, platforms, and learning contexts can be challenging and time-consuming.	3. Lack of Relevance: Providing just-in-time learning that lacks direct applicability or relevance to learners' needs may lead to disengagement.	3. Just-in-Time Learning: Lack of alignment with just-in-time learning principles, resulting in a mismatch with learners' immediate needs.	3. Just-in-Time Learning: • Curate and provide easily accessible resources that learners can refer to at the moment of need, allowing them to quickly find answers or solutions to specific problems. • Implement systems for learners to easily access specific content when needed, addressing immediate learning requirements.
	4. Just-in-Time Learning Challenges: Ensuring that the just-in-time resources are relevant, up-to-date, and easily accessible requires continuous curation and maintenance.			

continued on following page

Table 3. Continued

Difficult Cognitive Elements	Why Difficult	Common Errors	Cues and Strategies	
6. The integration of interactive learning experiences through authentic sources, active learning, interactive activities, and experiential learning.	1. Resource Constraints: Developing or sourcing authentic and relevant materials can be time-consuming and may require access to specific resources, expertise, or technologies.	1. Technology-related Issues: Technical glitches or inadequately functioning interactive elements may hinder learners' learning experience.	**Cues:** 1. Authentic Sources: Limited engagement in activities that involve authentic sources.	**Strategies:** 1. Authentic Sources: Identifying and curating authentic sources, ensuring their alignment with learning objectives, and integrating them into the learning materials and activities to enhance relevance and authenticity.
	2. Technology and Infrastructure: Implementing interactive activities and experiential learning may depend on the availability and accessibility of suitable technologies, platforms, and infrastructure to support such experiences.	2. Overwhelming Complexity: Overloading learners with too many interactive elements or complex activities can lead to cognitive overload and reduced learning effectiveness.	2. Active Learning: Inactive participation in learning tasks.	2. Active Learning: Designing and facilitating activities that promote learner engagement, critical thinking, problem-solving, and collaboration. It may include group discussions, hands-on experiments, role plays, or project-based learning.
	3. Curriculum Alignment: Ensuring that the interactive microlearning experiences align with the learning objectives and curriculum can be complex, as it requires careful planning and coordination across different topics, modules, or subjects.	3. Lack of Alignment: Failure to align the interactive learning experiences with the learning objectives may result in disengagement and lack of relevance for learners.	3. Interactive Activities: Low involvement in class activities.	3. Interactive Activities: Designing and implementing interactive elements such as quizzes, simulations, games, and multimedia presentations to enhance learner engagement, motivation, and knowledge retention.

continued on following page

Table 3. Continued

Difficult Cognitive Elements	Why Difficult	Common Errors	Cues and Strategies	
	4. Assessment and Feedback: Designing effective assessments that measure learner progress and provide timely feedback can be demanding, particularly in the context of interactive and experiential learning, where traditional assessment methods may need to be adapted.	4. Inadequate Assessment: Insufficient evaluation of learners' performance during interactive activities may lead to a lack of feedback and hinder continuous improvement.	4. Experiential Learning: Lack of opportunities for learners to directly apply theoretical knowledge in real or simulated environments.	4. Experiential Learning: Designing opportunities for learners to actively apply their knowledge and skills in real or simulated environments, providing hands-on experiences, reflection activities, and feedback to facilitate deeper learning.
	5. Instructor Competence: Facilitating interactive learning experiences requires instructors to possess the necessary pedagogical skills, knowledge of technology tools, and the ability to create a supportive learning environment that fosters active participation and engagement.			
7. The integration of social interaction through collaborative learning and peer feedback and review.	1. Group Dynamics: Managing group dynamics and ensuring equal participation and contribution from all group members.	1. Unequal Participation: Some learners may dominate collaborative activities, leading to unequal engagement and reduced learning opportunities for others.	**Cues:** 1. Limited Participation in Collaborative Activities: Observations of low engagement or interaction among students during collaborative learning tasks.	**Strategies:** 1. Facilitate Collaborative Learning Activities: Design and implement structured collaborative learning tasks, projects, or discussions that encourage active participation and interaction among students.

continued on following page

Table 3. Continued

Difficult Cognitive Elements	Why Difficult	Common Errors	Cues and Strategies	
	2. Assessment Validity and Reliability: Ensuring fairness and consistency in assessing group work and peer feedback, while also accounting for individual contributions.	2. Ineffective Communication: Lack of clear communication and guidance can hinder meaningful collaboration and compromise the effectiveness of social interactions.	2. Lack of Peer Feedback Opportunities: Absence of mechanisms or instances for students to provide feedback to their peers.	2. Incorporate Peer Review Mechanisms: Integrate peer review components into assessments, assignments, or projects to provide students with opportunities to evaluate and provide feedback to their peers.
	3. Technology and Platform Considerations: Selecting and implementing suitable online platforms or tools that support effective social interaction and peer feedback.	3. Inadequate Feedback Culture: If learners are not receptive to feedback or lack the skills to provide constructive criticism, the value of peer review may be diminished.		3. Utilize Online Platforms for Collaboration: Implement digital tools or online platforms that facilitate collaborative learning, allowing students to engage in discussions, group projects, or shared activities.
	4. Cultivating a Supportive Environment: Creating a safe and inclusive learning environment where students feel comfortable providing and receiving feedback and engaging in collaborative learning.			4. Provide Clear Guidelines for Peer Feedback: Establish clear guidelines and criteria for peer feedback to ensure constructive and meaningful input from students.
				5. Promote a Positive Collaborative Culture: Foster a positive and inclusive collaborative culture within the learning environment, encouraging students to actively participate, share ideas, and provide support to their peers.

continued on following page

Table 3. Continued

Difficult Cognitive Elements	Why Difficult	Common Errors	Cues and Strategies	
				6. Integrate Social Learning Components: Embed social learning components, such as discussion forums, group projects, or collaborative assignments, into the course structure to enhance social interaction.
8. Seamless integration with the curriculum through the integration of learning technologies and integration with the learning ecosystem.	1. Technological Infrastructure: Ensuring the availability and compatibility of necessary technology infrastructure to support the seamless integration of learning technologies across the curriculum.	1. Technological Barriers: Inadequate technology support or outdated systems can hinder the successful integration of microlearning with the curriculum.	**Cues:** 1. Disjointed Learning Technologies: Identification of a lack of seamless integration between microlearning and existing learning technologies within the educational framework.	**Strategies:** 1. Technology Alignment and Integration: Align microlearning platforms and tools with existing learning technologies to ensure a smooth and cohesive integration.
	2. Training and Support: Providing adequate training and support for educators to effectively utilize and integrate learning technologies into their teaching practices.	2. Lack of Faculty Training: Insufficient training for faculty on using learning technologies and integrating microlearning can result in ineffective implementation.	2. Absence of Learning Ecosystem Integration: Observations indicating a gap in incorporating microlearning harmoniously into the broader learning ecosystem or curriculum structure.	2. Learning Ecosystem Mapping: Conduct a comprehensive analysis of the learning ecosystem to identify opportunities for integrating microlearning effectively.

continued on following page

Table 3. Continued

Difficult Cognitive Elements	Why Difficult	Common Errors	Cues and Strategies	
	3. Resistance to Change: Overcoming resistance to change from educators or stakeholders who may be hesitant or sceptical about the integration of new technologies into the curriculum.	3. Overwhelming Integration: Poorly integrated microlearning activities may overwhelm learners with an excessive workload or disrupt the learning flow.		3. Curriculum Alignment: Align microlearning modules with the overall curriculum, ensuring they complement and enhance the broader educational objectives.
	4. Cost and Resources: Managing the costs associated with the integration of learning technologies and allocating resources for the necessary infrastructure, training, and ongoing support.	4. Inconsistent Content Quality: Without proper quality control measures, microlearning content may vary in quality and effectiveness across different modules.		4. Faculty Training and Support: Provide training and support for educators to seamlessly integrate microlearning into their teaching methods and curriculum planning.
	5. Continuous Updates and Upgrades: Dealing with the rapid evolution of learning technologies and the need to keep up with updates and upgrades to ensure ongoing compatibility and effectiveness.			5. Assessment Integration: Integrate microlearning assessments with existing evaluation methods to ensure a cohesive evaluation of student performance.
	6. Pedagogical Alignment: Ensuring that the integration of learning technologies is pedagogically sound and aligned with effective instructional practices, avoiding superficial or arbitrary use of technology without meaningful impact on learning outcomes.			6. Feedback Mechanisms: Implement feedback mechanisms to gather insights from both educators and students, facilitating continuous improvement in the integration of microlearning with the curriculum.

10.2.2 Findings from the Cognitive Demand Table

i. Challenges and Strategies in Establishing and Aligning Microlearning Objectives

Establishing clear and concise learning objectives and aligning microlearning activities with learners' needs and goals presents significant cognitive challenges. The complexity of some subjects makes it difficult to define focused learning objectives that adequately cover required knowledge and skills while remaining comprehensible. Additionally, learners' diverse backgrounds, prior knowledge, and skill levels add to the challenge of designing objectives that meet everyone's needs and goals. Aligning these objectives with broader curriculum or program goals requires careful coordination, and educational contexts' dynamic nature necessitates adaptable objectives that can evolve over time. Involving subject matter experts (SMEs) in developing learning objectives is crucial to avoid inaccuracies and gaps between the objectives and actual learning outcomes.

Common errors in this process include vague learning objectives that confuse learners and hinder microlearning activities' effectiveness, misalignment with learners' needs and goals, and lack of alignment between microlearning activities and established objectives. Failing to break down complex learning objectives into manageable microlearning activities can also overwhelm learners, affecting their ability to absorb information effectively. Cues indicating these errors include learners' confusion or lack of understanding about learning objectives, a mismatch between objectives and learners' needs, inconsistencies between objectives and overarching goals, and varied engagement or achievement levels related to learning objectives.

To address these challenges, a thorough analysis of learners' needs, including their prior knowledge, skills, and aspirations, is essential. Learning objectives should align with the overall goals of the course, program, or curriculum, and stakeholder input from students, SMEs, and industry professionals should be incorporated to ensure relevance and alignment with real-world needs. Regularly reviewing and updating learning objectives to align with evolving needs and goals is also necessary. These findings highlight the complexity of establishing and aligning learning objectives within a microlearning framework and underscore the importance of strategic and iterative approaches. By recognizing common errors and implementing targeted strategies, educators can enhance the clarity and effectiveness of learning objectives, improving the microlearning experience for diverse learners. Involving SMEs and maintaining a dynamic review process are essential for ensuring that learning objectives remain relevant and aligned with educational goals and learners' needs.

ii. Challenges and Strategies for Effective Personalization in Learner-Centric Activities and Customized Content

The findings regarding the incorporation of personalization through learner-centric activities and customized content reveal several significant challenges and strategies essential for effective implementation. Developing personalized content and activities for each learner is resource-intensive, requiring substantial time, effort, and resources. Ensuring scalability in large learning environments further complicates personalization, often necessitating sophisticated technology and instructional design approaches. Gathering and analysing learner data to inform personalization strategies demands robust data collection and analysis mechanisms, while designing effective personalized learning experiences requires pedagogical expertise in instructional design and an understanding of learners' cognitive processes and needs (Gunawardena, Bishop, & Aviruppola, 2024).

Common errors in the personalization process include ignoring learner preferences, which can lead to disengagement and reduced motivation. Additionally, offering insufficient content variety may limit learners' exposure to diverse perspectives and topics, while failing to continuously adapt content and activities to evolving learner needs can hinder the personalization effort's effectiveness.

Cues indicating personalization challenges include a lack of learner engagement in tailored activities, limited participation in personalized learning paths, low utilization of varied content formats, limited choices or flexibility in learner selections, and the absence of mechanisms for feedback and reflection. Strategies to address these challenges involve developing learner profiles by gathering information about preferences, interests, and prior knowledge. Designing adaptive learning paths that adjust based on learners' progress, offering content in different formats to accommodate diverse learning styles, and allowing learners to choose activities or topics of interest while providing options for pacing and sequencing are critical . Additionally, providing timely and meaningful feedback, encouraging self-reflection, and enabling learners to track their progress and set personal goals are essential strategies for successful personalization.

Overall, while the incorporation of personalization in learner-centric activities and customized content presents significant challenges, careful consideration of common errors and strategic implementation can enhance engagement and effectiveness, ultimately leading to a more tailored and impactful learning experience (Theodorio, 2024).

iii. Challenges and Strategies for Effective Feedback and Assessment in Microlearning

The incorporation of feedback and assessment in microlearning through timely and constructive feedback, formative and summative assessments, varied assessment methods, alignment with learning objectives, and continuous improvement presents several cognitive challenges (Muhammad Younis, Parveen Imdad, Aneela Atta Ur Rahman, 2021). Providing detailed and timely feedback requires significant time and effort from educators, especially when handling a large number of students or complex assessments. Additionally, students' diverse learning styles, abilities, and needs necessitate tailored feedback and assessments, making it a complex task for educators (Shemshack & Spector, 2020). Designing assessments that balance depth and breadth to evaluate higher-order thinking skills while providing comprehensive feedback is another challenge.

Common errors in this process include providing inadequate feedback that leaves learners uncertain about their progress, overemphasis on summative assessments without sufficient formative assessments, and using mismatched assessment methods that do not align with learning objectives. These issues can result in limited opportunities for timely intervention and adjustments in instruction, as well as a lack of comprehensive understanding of learners' progress.

Several cues indicate these challenges. For timely and constructive feedback, the lack of timely feedback can lead to gaps in understanding, and the absence of constructive feedback can hinder students' learning and growth. Limited use of both formative and summative assessments can impact comprehensive evaluation. Restriction to a single assessment format may limit a holistic understanding of students' capabilities, and a disconnect between assessment content and defined learning objectives can affect the effectiveness of evaluations. Additionally, encouraging students to actively engage with feedback and use it to improve their learning requires fostering a positive learning environment and promoting a growth mindset. The lack of a systematic approach to refining assessment strategies based on feedback and results can potentially hinder the learning experience.

To address these challenges, educators can implement several strategies. Providing feedback in a timely manner after assessments and ensuring it is constructive, specific, and actionable can help. Designing a balanced assessment approach that incorporates both formative assessments for ongoing feedback and summative assessments for comprehensive evaluation is crucial. Diversifying assessment methods, including quizzes, projects, presentations, and practical demonstrations, can provide a more comprehensive understanding of students' skills and knowledge. Ensuring assessments align with stated learning objectives and providing feedback that reflects intended learning goals is essential. Finally, establishing a feedback loop for continuous improvement by analysing assessment results, gathering feedback from students, and adapting assessment strategies can enhance the overall learning experience.

iv. Challenges and Strategies for Integrating Structured and Engaging Content Delivery in Microlearning

The integration of structured and engaging content delivery in microlearning, characterized by clear and focused content, bite-sized units, logical progression, and interactivity and reinforcement, presents several cognitive challenges. Balancing the depth and brevity of content is a significant difficulty, as it requires ensuring the content is comprehensive enough to cover the necessary knowledge and skills while keeping it concise and focused. Maintaining coherence and creating a seamless flow and logical progression of content requires careful planning and organization.

Common errors in this process include content overload, where presenting too much information at once can overwhelm learners and hinder their ability to retain and comprehend the material. Another issue is disjointed progression, where a lack of logical progression can confuse learners and disrupt their learning experience.

Several cues indicate these challenges. For clear and focused content, the lack of clarity in content organization and delivery can be problematic. Disengagement during lessons can signal issues with bite-sized units, and difficulty in following the content can indicate problems with logical progression. Limited engagement during class activities suggests insufficient interactivity and reinforcement.

Designing effective interactive elements and incorporating appropriate technology tools to enhance engagement and learning effectiveness can be complex. Providing meaningful reinforcement through activities and assessments that effectively reinforce learning and address individual learner needs can also be demanding. Insufficient interactivity can lead to passive learning, reducing learners' engagement and motivation, and unaligned content that is not directly related to the learning objectives can lead to wasted time and effort for both learners and educators.

To address these challenges, several strategies can be implemented. For clear and focused content, conducting a thorough content analysis and mapping to determine the most essential and relevant information to include is crucial. Breaking down the content into smaller, manageable chunks or modules can help with bite-sized units. Planning and organizing the learning materials in a progressive manner, building upon previously covered concepts and skills, can address issues with logical progression. Incorporating interactive components such as quizzes, simulations, discussions, and multimedia can enhance learner engagement and participation. Designing activities, exercises, and assessments that allow learners to apply their knowledge and skills, receive feedback, and reinforce their understanding is essential for effective interactivity and reinforcement.

v. Challenges and Strategies for Incorporating Microlearning Delivery Methods

The incorporation of microlearning delivery methods through multimedia integration, mobile-friendly design, and just-in-time learning presents several cognitive challenges. Creating high-quality multimedia content necessitates expertise in instructional design, multimedia production, and pedagogical considerations, making content design and development a significant difficulty. Technological requirements pose another challenge, as implementing multimedia integration and mobile-friendly design often involves complex technology requirements and compatibility issues. Customizing and adapting microlearning content to suit different devices, platforms, and learning contexts is also challenging and time-consuming.

Common errors in this process include overwhelming multimedia, where excessive use of multimedia elements can distract learners and impede their focus on key learning points. Inaccessible design, where the failure to optimize microlearning content for mobile devices, can limit learners' accessibility and flexibility. Additionally, providing just-in-time learning that lacks direct applicability or relevance to learners' needs may lead to disengagement.

Several cues indicate these challenges. Limited engagement with multimedia content in learning materials suggests problems with multimedia integration. Challenges with accessing learning content on mobile devices point to issues with mobile-friendly design. A lack of alignment with just-in-time learning principles, resulting in a mismatch with learners' immediate needs, indicates difficulties in just-in-time learning. Ensuring that just-in-time resources are relevant, up-to-date, and easily accessible requires continuous curation and maintenance. This presents a unique challenge as it demands ongoing efforts to keep resources pertinent and useful for learners.

To address these challenges, several strategies can be implemented. To create engaging multimedia content, design and develop multimedia elements that are visually appealing, interactive, and effectively convey key concepts. For mobile-friendly design, optimize learning materials for mobile devices by considering different screen sizes and functionalities and utilizing responsive design principles to ensure seamless access to content on various mobile platforms. For just-in-time learning, curate and provide easily accessible resources that learners can refer to at the moment of need, allowing them to quickly find answers or solutions to specific problems. Implement systems for learners to easily access specific content when needed, addressing immediate learning requirements.

vi. Challenges and Strategies for Integrating Interactive Learning Experiences

Integrating interactive learning experiences through authentic sources, active learning, interactive activities, and experiential learning presents several cognitive challenges. Resource constraints pose a significant difficulty, as developing or sourcing authentic and relevant materials can be time-consuming and may require access to specific resources, expertise, or technologies. Technology and infrastructure also present challenges, as implementing interactive activities and experiential learning depends on the availability and accessibility of suitable technologies, platforms, and infrastructure to support such experiences (Haleem, Javaid, Qadri, & Suman, 2022). Curriculum alignment further complicates the process, requiring careful planning and coordination across different topics, modules, or subjects to ensure that the interactive microlearning experiences align with the learning objectives and curriculum (Wijngaards-de Meij & Merx, 2018).

Common errors in this process include technology-related issues, such as technical glitches or inadequately functioning interactive elements that may hinder learners' learning experience. Overwhelming complexity, where learners are overloaded with too many interactive elements or complex activities, can lead to cognitive overload and reduced learning effectiveness. Lack of alignment, where interactive learning experiences fail to align with the learning objectives, can result in disengagement and a lack of relevance for learners.

Several cues indicate these challenges. Limited engagement in activities involving authentic sources suggests problems with using authentic sources. Inactive participation in learning tasks indicates issues with active learning. Low involvement in class activities points to difficulties with interactive activities. A lack of opportunities for learners to directly apply theoretical knowledge in real or simulated environments suggests problems with experiential learning. (Tai et al., 2023) and feedback are also challenging in this context. Designing effective assessments that measure learner progress and provide timely feedback is demanding, particularly in the context of interactive and experiential learning, where traditional assessment methods may need to be adapted. Instructor competence is another challenge, as facilitating interactive learning experiences requires instructors to possess the necessary pedagogical skills, knowledge of technology tools, and the ability to create a supportive learning environment that fosters active participation and engagement.

To address these challenges, several strategies can be implemented. For authentic sources, identify and curate authentic sources, ensuring their alignment with learning objectives and integrating them into the learning materials and activities to enhance relevance and authenticity. For active learning, design and facilitate activities that promote learner engagement, critical thinking, problem-solving, and collaboration, such as group discussions, hands-on experiments, role plays, or project-based learning (Merta, Ratminingsih, & Budasi, 2023). For interactive activities, design and implement interactive elements such as quizzes, simulations, games, and multimedia presentations to enhance learner engagement, motivation, and knowledge retention. For experiential learning, design opportunities for learners to actively apply their knowledge and skills in real or simulated environments, providing hands-on experiences, reflection activities, and feedback to facilitate deeper learning. Additionally, ensure that assessments are adequate by evaluating learners' performance during interactive activities and providing timely feedback to support continuous improvement.

vii. Challenges and Strategies for Integrating Social Interaction through Collaborative Learning and Peer Feedback

Integrating social interaction through collaborative learning and peer feedback presents several cognitive challenges. Managing group dynamics is a significant difficulty, as ensuring equal participation and contribution from all group members can be complex. Assessment validity and reliability are also challenging, requiring fairness and consistency in assessing group work and peer feedback while accounting for individual contributions. Additionally, technology and platform considerations, such as selecting and implementing suitable online platforms or tools that support effective social interaction and peer feedback, are crucial. Cultivating a supportive environment where students feel comfortable providing and receiving feedback and engaging in collaborative learning is essential but can be demanding.

Common errors in this process include unequal participation, where some learners may dominate collaborative activities, leading to reduced learning opportunities for others. Ineffective communication can hinder meaningful collaboration and compromise the effectiveness of social interactions. An inadequate feedback culture, where learners are not receptive to feedback or lack the skills to provide constructive criticism, can diminish the value of peer review.

Several cues indicate these challenges. Limited participation in collaborative activities suggests low engagement or interaction among students during collaborative learning tasks. The absence of mechanisms or instances for students to provide feedback to their peers indicates a lack of peer feedback opportunities.

Strategies to address these challenges include facilitating collaborative learning activities by designing and implementing structured collaborative tasks, projects, or discussions that encourage active participation and interaction among students. Incorporating peer review mechanisms into assessments, assignments, or projects provides students with opportunities to evaluate and provide feedback to their peers. Utilizing online platforms for collaboration, such as digital tools or online platforms that facilitate collaborative learning, allows students to engage in discussions, group projects, or shared activities. Providing clear guidelines for peer feedback ensures constructive and meaningful input from students. Promoting a positive collaborative culture within the learning environment encourages students to actively participate, share ideas, and provide support to their peers. Integrating social learning components, such as discussion forums, group projects, or collaborative assignments, into the course structure enhances social interaction.

viii. Challenges and Strategies for Seamless Integration with the Curriculum

Achieving seamless integration with the curriculum through learning technologies poses several cognitive challenges. Ensuring technological infrastructure is up to par is crucial, as it supports the compatibility and availability needed for effective integration across the curriculum. Providing adequate training and support for educators is also essential to empower them in utilizing these technologies effectively. Overcoming resistance to change from educators or stakeholders hesitant about new technologies and managing associated costs are additional complexities.

Common errors include technological barriers such as inadequate support systems or outdated technology hindering microlearning integration. Insufficient faculty training in using learning technologies and integrating microlearning can lead to ineffective implementation. Poorly integrated microlearning activities may overwhelm learners or disrupt the learning flow, while inconsistent content quality across modules may impact effectiveness. Observable cues indicating these challenges include disjointed learning technologies and gaps in integrating microlearning harmoniously into the broader learning ecosystem or curriculum structure.

Strategies to address these challenges involve aligning microlearning platforms with existing learning technologies to ensure cohesive integration. Conducting a thorough analysis of the learning ecosystem helps identify opportunities for effective microlearning integration. Continuous updates and upgrades are necessary to keep pace with evolving technologies and maintain compatibility. Ensuring pedagogical alignment by integrating learning technologies in ways that enhance instructional practices and learning outcomes is crucial. Further strategies include aligning microlearning modules with the curriculum, supporting faculty through training and ongoing support, integrating microlearning assessments with existing evaluation methods, and implementing feedback mechanisms for continuous improvement. By addressing these cognitive demands and employing strategic approaches, educational institutions can effectively integrate microlearning technologies into their curricula, enhancing learning experiences and outcomes for students.

10.2.3 Review of Cognitive Demand Table of Microlearning Pedagogical Practices

In navigating the complexities of cognitive demands in microlearning integration, educators and instructional designers face multifaceted challenges and opportunities. The cognitive demand table illuminates critical areas where careful planning, strategic implementation, and continuous refinement are essential to optimizing learning experiences.

Clarity and Alignment of Learning Objectives:

Effective microlearning hinges on clearly defined objectives that align seamlessly with learner needs and curriculum goals. Challenges such as vague objectives or misalignment with learner expectations underscore the importance of rigorous planning and stakeholder collaboration. Strategies involving thorough needs analysis and stakeholder input ensure that microlearning activities remain relevant and impactful.

Personalization and Learner-Centric Approaches:

Tailoring learning experiences to individual learner preferences and needs presents challenges in scalability and resource allocation. The integration of diverse content formats and adaptive technologies, while ensuring pedagogical soundness, is crucial for sustained engagement and effective knowledge acquisition. Strategies focused on robust learner profiling and adaptive learning paths foster personalized learning journeys that resonate with diverse learner cohorts.

Feedback and Assessment Integration:

Timely and constructive feedback, coupled with varied assessment methods aligned with learning objectives, is pivotal for gauging learner progress and fostering continuous improvement. Challenges such as time constraints and the need for balanced assessment approaches underscore the importance of pedagogical expertise and technological support. Implementing comprehensive feedback mechanisms and diverse assessment strategies enhances the efficacy of microlearning initiatives, promoting deeper learning outcomes.

Structured and Engaging Content Delivery:

The design and delivery of content in bite-sized, logically sequenced formats require careful attention to maintain coherence and learner engagement. Challenges such as content overload or disjointed progression highlight the need for interactive elements and reinforcement activities. Strategies focusing on interactive design principles and content curation ensure that microlearning modules are both engaging and conducive to effective knowledge retention.

Multimedia Integration and Just-in-Time Learning:

Integrating multimedia elements and facilitating just-in-time learning necessitate robust technological infrastructure and pedagogical agility. Challenges such as overwhelming multimedia or inaccessible design underscore the importance of user-centric design and responsive technology platforms. Strategies focusing on engaging multimedia development and accessible learning formats ensure that microlearning remains relevant and accessible across diverse learning environments.

Interactive Learning Experiences and Social Interaction:

Promoting active learning through interactive experiences and fostering social interaction requires careful facilitation and supportive learning environments. Challenges such as technological constraints or ineffective communication underscore the need for collaborative learning frameworks and clear communication channels. Strategies emphasizing collaborative project design and peer feedback mechanisms enrich microlearning environments, fostering a culture of active engagement and knowledge sharing.

Integration with Learning Technologies and Ecosystem:

Seamlessly integrating microlearning with existing learning technologies and broader educational ecosystems demands strategic alignment and ongoing adaptation. Challenges such as technological barriers or resistance to change highlight the importance of comprehensive technology integration and faculty support. Strategies focusing on pedagogical alignment and continuous professional development empower educators to leverage microlearning effectively within evolving educational landscapes.

In conclusion, navigating the cognitive demands of microlearning integration requires a holistic approach that balances innovation with pedagogical rigor, ensuring that learning experiences are not only impactful but also adaptive to the diverse needs of learners and educational contexts. By addressing these challenges through informed strategies and collaborative efforts, educators can unlock the full potential of microlearning to enhance learning outcomes and prepare learners for success in the digital age.

CONCLUSION

This chapter explored the application of the Cognitive Demand Table, the fourth step in the Applied Cognitive Task Analysis (ACTA) framework, in two key educational contexts: electronic training for professional development and microlearning pedagogical practices. By systematically analyzing the cognitive challenges faced by users in each environment, the Cognitive Demand Table provided valuable insights into how system functionality, content design, assessments, and other instructional components can be improved to better meet learners' needs.

In Section 10.1, the Cognitive Demand Table for Electronic Training for Professional Development identified key cognitive difficulties in navigating system functionality, content alignment, and assessment design. These challenges were primarily attributed to the lack of familiarity with online learning platforms and insufficient guidance within the training system. The findings emphasized the need for user-friendly navigation, multimedia integration, and real-time assistance to create a more supportive and engaging training environment. By addressing these cognitive demands, electronic training systems can be optimized for both trainers and trainees, enhancing the overall learning experience and improving knowledge retention.

Section 10.2 focused on the Cognitive Demand Table for Microlearning Pedagogical Practices, revealing the importance of clarity, personalization, and interactive content delivery in microlearning environments. The analysis highlighted the challenges of maintaining learner engagement, aligning content with learning objectives, and providing timely feedback. Addressing these challenges through well-designed microlearning modules, adaptive learning paths, and effective multimedia integration can significantly improve learners' cognitive engagement and overall learning outcomes. By embedding these strategies into microlearning practices, educators can create flexible, accessible, and impactful learning experiences.

In conclusion, the Cognitive Demand Table serves as a critical tool for identifying and addressing cognitive challenges in both electronic training and microlearning. By focusing on the cognitive load and usability of instructional systems, this method allows educators and instructional designers to refine their approaches, ensuring that learners are fully supported and engaged in their educational journey. Ultimately, the insights gained from this analysis pave the way for more effective, learner-centred systems that align with the demands of modern education and professional development.

REFERENCES

Alasmari, T. M. (2020). Can mobile learning technology close the gap caused by gender segregation in the Saudi educational institutions? *Journal of Information Technology Education*, 19, 655–670. DOI: 10.28945/4634

Bahati, B., Fors, U., Hansen, P., Nouri, J., & Mukama, E. (2019). Measuring learner satisfaction with formative e-assessment strategies. *International Journal of Emerging Technologies in Learning*, 14(7), 61–79. DOI: 10.3991/ijet.v14i07.9120

Gong, B., Collins, C., & Amrein-Beardsley, A. (2022). An International Professional Development Collaboration: Supporting Teachers in the Kingdom of Saudi Arabia through an Immersion Program in the United States. [GESPR]. *Gulf Education and Social Policy Review*, 2(2), 115–143. DOI: 10.18502/gespr.v2i1.10045

Gunawardena, M., Bishop, P., & Aviruppola, K. (2024). Personalized learning: The simple, the complicated, the complex and the chaotic. *Teaching and Teacher Education, 139*(December 2023), 104429. DOI: 10.1016/j.tate.2023.104429

Haleem, A., Javaid, M., Qadri, M. A., & Suman, R. (2022). Understanding the role of digital technologies in education: A review. *Sustainable Operations and Computers*, 3(May), 275–285. DOI: 10.1016/j.susoc.2022.05.004

Hultberg, P., Santandreu Calonge, D., & Safiullin Lee, A. E. (2018). Promoting Long-lasting Learning Through Instructional Design. *The Journal of Scholarship of Teaching and Learning*, 18(3), 26–43. DOI: 10.14434/josotl.v18i3.23179

Merta, L. W. S., Ratminingsih, N. M., & Budasi, I. G. (2023). The Integration of Technology in English Language Teaching to Stimulate Students' Critical Thinking. *Language Circle: Journal of Language and Literature*, 17(2), 333–341. DOI: 10.15294/lc.v17i2.39097

Shemshack, A., & Spector, J. M. (2020). A systematic literature review of personalized learning terms. *Smart Learning Environments*, 7(1). Advance online publication. DOI: 10.1186/s40561-020-00140-9

Silva, L., Mendes, A. J., & Gomes, A. (2020). Computer-supported collaborative learning in programming education: A systematic literature review. In *IEEE Global Engineering Education Conference, EDUCON* (Vol. 2020-April, pp. 1086–1095). IEEE Computer Society. DOI: 10.1109/EDUCON45650.2020.9125237

Tai, J. H. M., Dollinger, M., Ajjawi, R., Jorre de St Jorre, T., Krattli, S., McCarthy, D., & Prezioso, D. (2023). Designing assessment for inclusion: An exploration of diverse students' assessment experiences. *Assessment & Evaluation in Higher Education*, 48(3), 403–417. DOI: 10.1080/02602938.2022.2082373

Theodorio, A. O. (2024). Examining the support required by educators for successful technology integration in teacher professional development program. *Cogent Education*, 11(1). Advance online publication. DOI: 10.1080/2331186X.2023.2298607

Wijngaards-de Meij, L., & Merx, S. (2018). Improving curriculum alignment and achieving learning goals by making the curriculum visible. *The International Journal for Academic Development*, 23(3), 219–231. DOI: 10.1080/1360144X.2018.1462187

Younis, M., Imdad, P., & Aneela, A. U. R. (2021). Effects of Constructive and Timely Feedback on Academic Performance of Students. *Pakistan Journal of Educational Research*, 4(4), 670–677. DOI: 10.52337/pjer.v4i4.464

Section 4
Advanced Topics and Future Directions

Chapter 11
Innovative Approaches in ACTA

ABSTRACT

This chapter explores advancements in Applied Cognitive Task Analysis (ACTA) driven by emerging technologies and methodologies. It examines how artificial intelligence (AI) and virtual reality (VR) are transforming ACTA practices by automating data analysis, optimizing task performance assessments, and personalizing learning experiences. The chapter highlights AI's role in enhancing ACTA and VR's application in simulating complex cognitive tasks and providing immersive training environments. It also looks at future trends and tools that could further advance ACTA. By providing insights into these innovations, the chapter discusses their implications for educational and professional settings, emphasizing how ACTA can evolve in a technology-driven world.

INTRODUCTION

In this chapter, we delve into the innovative approaches currently shaping the landscape of Applied Cognitive Task Analysis (ACTA). As the field evolves, the integration of new technologies, such as artificial intelligence (AI) and virtual reality (VR), plays a crucial role in enhancing both the effectiveness and applicability of ACTA. These technologies are enabling deeper insights into cognitive processes, allowing for more accurate modelling and analysis of complex tasks across a variety of settings. With AI's ability to process vast amounts of data and VR's immersive environments, ACTA is becoming a more powerful tool for improving decision-making, training, and task performance.

In addition to exploring these technological advancements, we will examine future trends and emerging tools in cognitive task analysis. As new methodologies and techniques develop, they offer the potential to further revolutionize how cognitive

DOI: 10.4018/979-8-3693-7733-8.ch011

tasks are studied and applied. This chapter will provide a comprehensive outlook on the implications of these advancements, focusing on their potential impact in education and other sectors. By analysing these future trends, we aim to highlight how ACTA will continue to evolve and adapt, offering innovative solutions to both current and future challenges.

11.1 Integration of New Technologies: Artificial Intelligence and Virtual Reality

The incorporation of Artificial Intelligence (AI) and Virtual Reality (VR) into Applied Cognitive Task Analysis (ACTA) methodologies marks a substantial advancement in the field. These cutting-edge technologies offer more immersive, precise, and adaptive approaches to capturing and evaluating cognitive processes, thereby optimizing task performance and learning outcomes. By providing enhanced data analysis, personalized learning experiences, and real-time monitoring and feedback, AI enhances the precision and efficacy of ACTA.

11.1.1 Artificial Intelligence (AI)

AI has the potential to revolutionize the way cognitive tasks are analysed and optimized within the ACTA framework. This integration offers several methodological advantages:

- **Enhanced Data Analysis**:
    - o **Efficiency in Processing**: AI algorithms can process and analyse vast amounts of data much faster and more efficiently than traditional methods. ACTA typically involves the detailed breakdown and examination of cognitive tasks, which can generate substantial amounts of data. AI can handle this data efficiently, identifying patterns and insights that might be overlooked by human analysts due to the sheer volume and complexity of the information. For example, in an educational context, machine learning models can analyse students' interactions with learning materials to identify common challenges and areas for improvement.
    - o **Predictive Capabilities**: AI's machine learning models are particularly adept at predicting and suggesting improvements in task performance based on historical data. In the context of ACTA, this means AI can predict potential cognitive bottlenecks or areas where learners might struggle, allowing for pre-emptive adjustments to instructional strategies. A study by Romero and Ventura (2020) demonstrated how AI algorithms could predict student performance and provide targeted interventions to

enhance learning outcomes. This predictive capability can be integrated into ACTA to continually refine and optimize cognitive task analyses.

- **Personalized Learning:**
  - o **Customization:** AI can tailor educational content to the specific cognitive needs and learning styles of individual students. This personalization is crucial for ACTA, as it ensures that the cognitive tasks analysed are directly relevant to each learner's unique context. AI-driven customization involves providing adaptive learning pathways and customized feedback, which significantly enhances the learning experience and efficacy of the instructional design. For instance, adaptive learning systems can modify the difficulty of tasks in real-time based on learner performance, ensuring optimal cognitive load and engagement.
  - o **Improved Outcomes:** Personalized learning powered by AI has been shown to improve academic performance and student engagement. In the context of ACTA, personalized feedback can help learners understand and navigate their cognitive processes more effectively, leading to better retention and application of knowledge. A study by Kasepalu et al. (2022) discusses how AI supports teacher pedagogical actions in collaborative learning environments. By enhancing feedback and learning adaptation, AI can personalize the learning experience. AI's role in providing tailored feedback based on individual student performance and interactions allows for more personalized instructional approaches. This helps teachers to address the unique needs of each student, thereby facilitating a more personalized learning experience. Similarly, the work of Fu et al. (2020) investigates AI-enabled automatic scoring applications and their impact on learners' continuous learning intentions. These applications can provide personalized feedback based on students' performance, helping them understand their strengths and weaknesses. The empirical evidence from China shows how such AI tools can support continuous learning by providing timely and personalized feedback, which encourages learners to stay engaged and motivated.
- **Real-Time Monitoring and Feedback**:
  - o **Dynamic Interaction**: AI-powered systems can monitor cognitive load and task performance in real-time, allowing for immediate feedback to learners and instructors. This dynamic interaction is particularly valuable in ACTA, where real-time data can provide insights into how learners engage with cognitive tasks and where they encounter difficulties. AI systems can offer instant feedback, helping learners adjust their strategies and improve performance on the fly.

- **Balancing Cognitive Demand**: Real-time monitoring helps in managing cognitive load, ensuring that learners are neither overwhelmed nor under-challenged. This is a critical aspect of ACTA, as understanding and managing cognitive load is essential for effective task analysis and instructional design. AI-driven real-time feedback systems enable educators to maintain an optimal balance of cognitive demand, enhancing overall learning outcomes. Research by Zheng et al. (2024) demonstrate the effectiveness of AI-driven real-time feedback systems in achieving this balance. Their study explored an AI-enabled feedback and feed-forward approach in online collaborative learning, revealing that it significantly enhances collaborative knowledge building, coregulated behaviours, and group performance. By offering timely recommendations and actionable insights, these systems help educators maintain an optimal cognitive demand, thereby improving learning outcomes and facilitating more effective online collaboration.

In conclusion, the integration of AI into ACTA methodologies offers a transformative approach to analysing and optimizing cognitive tasks. By leveraging the capabilities of AI for enhanced data analysis, personalized learning, and real-time monitoring, ACTA can significantly improve the precision and efficacy of cognitive task analysis. The incorporation of these advanced technologies not only addresses current challenges in cognitive task analysis but also sets the stage for future innovations and improvements in the field.

11.1.2 Virtual Reality (VR)

- **Immersive Training Environments**

Virtual Reality (VR) offers transformative opportunities for training and cognitive task analysis, aligning closely with the principles of Applied Cognitive Task Analysis (ACTA). VR creates highly immersive training environments where users can engage in realistic simulations of tasks, thereby providing a rich context for understanding and analysing cognitive processes. This immersion allows for a detailed examination of task performance in controlled yet lifelike settings, which is essential for identifying the cognitive demands and strategies required for complex tasks.

The integration of VR with ACTA principles is particularly powerful because it allows for a nuanced exploration of both the physical and cognitive aspects of task performance. ACTA focuses on dissecting the cognitive processes involved in complex tasks, such as decision-making, problem-solving, and skill execution. VR environments can be tailored to replicate these tasks with high fidelity, providing a platform to study cognitive functions in contexts that closely resemble real-world scenarios.

For instance, research by Zhou et al. (2023) introduced an innovative method of evaluating cognitive processing speed using an immersive virtual street environment embedded with a Stroop task, known as VR-Street. This study illustrates the synergy between VR and ACTA by engaging participants in a dual-task scenario where they navigate a virtual street while performing a Stroop task. The VR-Street environment allowed researchers to capture comprehensive behavioural and performance data, offering insights into how participants manage cognitive resources under realistic conditions.

The study's findings revealed significant correlations between the features of the Stroop task and cognitive test results, suggesting that the VR-Street environment effectively mobilizes participants' cognitive resources. The integration of behavioural data with Stroop task features in a machine learning model demonstrated high accuracy in estimating processing speed, underscoring VR's potential for precise and quantitative cognitive assessment.

This approach not only enhances our understanding of cognitive processing in realistic scenarios but also provides a valuable tool for clinicians in diagnosing and evaluating cognitive functions in daily life contexts. By using VR to simulate real-world tasks and incorporating ACTA principles, researchers and practitioners can gain deeper insights into the cognitive demands of complex tasks and develop more effective training and assessment methods.

In summary, VR's immersive training environments, when combined with ACTA, offer a powerful means of studying and improving cognitive task performance. VR allows for the creation of detailed, lifelike simulations that facilitate a comprehensive analysis of cognitive processes, while ACTA ensures that these simulations address the specific cognitive demands associated with complex tasks. This integration enhances both training and assessment, providing valuable tools for improving cognitive performance and understanding in real-world contexts.

- **Safe and Controlled Conditions**

Moreover, VR enables the simulation of high-risk or complex scenarios in a safe and controlled manner, making it invaluable in fields such as medicine, aviation, and military training. This aspect of VR is particularly beneficial when real-world training poses significant risks or is impractical. Studies such as those by Holuša et al. (2023) demonstrate how VR can effectively enhance training and safety in industrial contexts, such as the raw materials industry. Their research highlights that VR provides an immersive and interactive learning environment, allowing employees to engage with realistic simulations and scenarios without the inherent risks of real-world operations.

Incorporating Applied Cognitive Task Analysis (ACTA) into VR training enhances its effectiveness by focusing on the cognitive demands of tasks performed in high-risk environments. ACTA aims to identify and analyse the cognitive processes involved in complex tasks, such as decision-making, problem-solving, and skill execution. By applying ACTA to VR simulations, developers can create more accurate and relevant training scenarios that align closely with the real-world cognitive challenges faced by users.

Holuša et al. (2023) found that VR-based training improves practical skills, information retention, and critical soft skills like communication and teamwork. Integrating ACTA into the development of these VR scenarios ensures that the simulations not only replicate physical tasks but also address the cognitive aspects of these tasks. For instance, VR simulations informed by ACTA can be designed to challenge users' decision-making abilities and problem-solving skills under realistic conditions, enhancing the overall training experience.

By aligning VR training with the insights gained from ACTA, training programs can better prepare individuals for real-world challenges. This approach allows for the creation of scenarios that are not only physically realistic but also cognitively demanding, providing a more comprehensive training experience. The integration of ACTA with VR also facilitates more effective feedback and assessment, as it allows trainers to evaluate both the cognitive and physical performance of trainees.

In summary, combining VR with ACTA offers a powerful approach to training in high-risk environments. VR provides a safe, controlled space for simulation, while ACTA ensures that these simulations accurately reflect the cognitive demands of real-world tasks. This combination enhances the effectiveness of training programs by addressing both the physical and cognitive aspects of complex scenarios, ultimately improving preparedness and performance in high-risk situations.

- **Interactive and Engaging Learning**

Additionally, VR enhances engagement and motivation through interactive and visually stimulating learning experiences. This heightened engagement fosters deeper learning and better retention of information, as evidenced by recent findings from Vats and Joshi (2023). Their study provides a comprehensive exploration of how VR technologies can transform educational settings by creating immersive environments that captivate and actively involve students.

Vats and Joshi (2023) highlight the potential of VR to offer hands-on, interactive learning experiences that simulate real-world scenarios, ranging from historical landmarks to complex scientific experiments. Such immersive simulations are not only engaging but also enable students to interact with content in a way that traditional methods often cannot. This interactivity is crucial for fostering an active learning environment, where students can engage deeply with the material, explore scenarios in detail, and apply their knowledge in realistic contexts.

The study also reveals that VR-based training modules significantly enhance learner engagement and knowledge retention compared to conventional teaching methods. By integrating these interactive elements, VR supports the principles of Applied Cognitive Task Analysis (ACTA). ACTA focuses on understanding the cognitive demands and strategies required for complex tasks. VR environments can be designed to replicate these tasks with high fidelity, providing a detailed and dynamic context where cognitive processes can be observed and analysed.

For instance, VR simulations can be tailored to present tasks that involve decision-making, problem-solving, and critical thinking, allowing researchers and educators to study how learners manage cognitive demands in real-time. The interactive nature of VR facilitates a more nuanced analysis of cognitive functions, such as attention, memory, and skill acquisition. By offering a controlled yet lifelike setting, VR enables a detailed examination of how learners approach and perform complex tasks, aligning well with the goals of ACTA.

In summary, Vats and Joshi (2023) demonstrate that VR's immersive and interactive capabilities not only enhance engagement and retention but also provide valuable insights into cognitive processes. By creating dynamic learning environments, VR supports the ACTA framework by addressing the cognitive demands of complex tasks and improving overall training outcomes. This integration of VR with ACTA principles ensures that learning experiences are both engaging and effective, leading to more profound and practical educational benefits.

11.2 Future Trends and Emerging Tools in Cognitive Task Analysis

As we look ahead, several emerging trends and tools are poised to further transform cognitive task analysis, making it more effective and accessible.

11.2.1 Wearable Technology

- **Biosensors and Monitoring Devices:**

Wearable technology embedded with biosensors is becoming increasingly sophisticated, offering new ways to monitor and analyse physiological indicators. These devices can track metrics such as heart rate, skin conductivity, and brain activity, providing a more comprehensive understanding of cognitive load and task performance. For example, recent studies have demonstrated that wearable biosensors can capture real-time data on cognitive workload and stress levels, which can be crucial for optimizing task design and training programs. A narrative review by Iarlori et al. (2024) offers a comprehensive analysis of various techniques for collecting and integrating physiological data through wearable sensors, highlighting their effectiveness in monitoring cognitive workload in human-machine interaction (HMI) scenarios.

The study by Iarlori et al. (2024) underscores the potential of wearable sensors to provide a multimodal approach to cognitive workload assessment by integrating multiple physiological signals. This approach aligns with the principles of Applied Cognitive Task Analysis (ACTA), which focuses on understanding and modelling cognitive processes involved in task performance. By incorporating real-time physiological data, ACTA can gain deeper insights into cognitive workload and task execution, leading to more informed task design and training interventions. The integration of AI methods in data analysis, as discussed in the study, further enhances the capability of ACTA to model and optimize cognitive workload, addressing challenges in standardization and applicability across diverse contexts.

- **Augmented Reality (AR):**

Augmented Reality (AR) technology is making significant strides in enhancing cognitive task analysis. Wearable AR devices, such as smart glasses, can overlay digital information onto the physical environment, providing real-time guidance and support. This can improve situational awareness and facilitate more effective learning and task execution by integrating contextual information and interactive elements directly into the user's field of view.

Recent research by Woodward and Ruiz (2022) highlights the critical role of AR in improving situational awareness, particularly in safety-critical domains. Their systematic literature review examines how information is currently presented in AR systems and identifies design considerations for optimizing situational awareness. Woodward and Ruiz (2022) point out that while AR technology has demonstrated potential in enhancing situational awareness, prior studies have often focused more on the general applicability of AR rather than on the specifics of information design and presentation. The review underscores the importance of designing effective information displays in AR headsets to enhance user situational awareness, which is vital for improving cognitive task performance.

This research is particularly relevant to Applied Cognitive Task Analysis (ACTA), which focuses on understanding and modelling cognitive processes involved in task performance. ACTA aims to provide detailed insights into how cognitive tasks are executed and how they can be optimized. By improving situational awareness through better AR information design, ACTA can benefit from more accurate and contextually relevant data, leading to a deeper understanding of cognitive workload and task execution. The integration of AR technology, as highlighted by Woodward and Ruiz (2022), supports ACTA by offering real-time, context-sensitive information that can enhance cognitive task analysis and lead to more effective task design and training. This alignment between AR advancements and ACTA principles demonstrates how emerging technologies can enrich the analysis and optimization of cognitive tasks, leading to improved performance and reduced error rates in complex environments.

11.2.2 Big Data and Analytics

- **Comprehensive Data Collection:**

The advent of big data and analytics has revolutionized the field of cognitive task analysis (CTA) by enabling the collection and analysis of vast amounts of data from diverse sources. This includes biometric data, task performance metrics, and user interactions, which together provide a more comprehensive understanding of cognitive processes.

Applied Cognitive Task Analysis (ACTA) benefits from these advancements as it relies on detailed insights into cognitive workload to enhance task performance analysis. Recent research, including the work by Chakladar and Roy (2023), has demonstrated how integrating various physiological measures—such as Electroencephalography (EEG), Functional Magnetic Resonance Imaging (fMRI), and eye activity—with machine learning techniques can significantly improve cognitive workload estimation. This approach aligns with ACTA's goal of understanding cognitive demands in complex tasks by providing a more detailed and accurate picture of how cognitive resources are allocated during task performance. By combining these diverse data sources, researchers achieve a multidimensional view of cognitive workload, uncovering deeper insights into cognitive processes and improving CTA effectiveness.

Applied Cognitive Task Analysis (ACTA) focuses on understanding and modelling cognitive processes involved in task performance, with the aim of optimizing task design and improving training programs. The integration of big data aligns with ACTA principles by providing detailed insights into cognitive workload and task execution. Comprehensive data collection supports ACTA by offering a richer, more nuanced view of cognitive processes. The ability to analyse data from multiple sources, such as biometric data, task performance metrics, and user interactions, allows for a better understanding of how cognitive load fluctuates with task demands. This leads to more accurate models of cognitive task performance, enhancing the effectiveness of task analysis and training programs.

- **Predictive Analytics:**

Advanced analytics and predictive modelling further enhance CTA by enabling the anticipation of potential issues and the identification of opportunities for task performance optimization. Predictive analytics involves analysing historical data to forecast future performance and detect areas needing improvement.

The study by Ahmad, Keller, Robb, and Lohan (2023) on cognitive load detection exemplifies this by proposing a framework that uses physiological data from the eyes and heart to predict cognitive load levels with high accuracy. By employing classifiers such as Random Forest and Naive Bayes, which achieved accuracies of 91.66% and 85.83% respectively, the study demonstrates the potential of predictive analytics in monitoring user cognitive load in real time. This approach allows for efficient workload management and the reduction of human error in applications like automation, robotics, and aerospace. The ability to accurately predict low, medium, and high cognitive load levels based on physiological measures aligns with the goals of predictive analytics in CTA, offering valuable insights for optimizing task performance and improving user experience.

The study by Aksu, Çakıt, and Dağdeviren (2024) exemplifies this by using EEG and eye tracking data to classify mental workload levels during n-back tasks through machine learning algorithms. The concurrent application of these techniques, along with the NASA-Task Load Index (TLX) subjective rating scale, allowed for the accurate prediction of mental workload levels with a 76.59% accuracy using 34 selected features. This study highlights the value of predictive analytics in real-time monitoring and classification of mental workload, contributing to the optimization of task performance and the enhancement of user experience. The ability to predict mental workload levels based on physiological and performance measures aligns with the goals of predictive analytics in CTA, providing critical insights for improving task design and training programs.

Predictive analytics enhances ACTA by providing tools for forecasting cognitive workload and identifying potential issues before they affect task performance. This proactive capability aligns with ACTA's goal of optimizing task design and training interventions based on detailed cognitive insights. By leveraging predictive models, ACTA can be applied more effectively to anticipate challenges, refine task designs, and tailor training programs to individual needs.

The integration of big data and analytics into CTA supports the development of more sophisticated models of cognitive processes and task performance. By combining comprehensive data collection with predictive analytics, ACTA can benefit from deeper insights and more proactive approaches to task optimization. Recent studies highlight the effectiveness of these advancements in enhancing our understanding of cognitive dynamics and improving task performance, demonstrating the value of big data and analytics in the field of cognitive task analysis.

11.2.3 Collaborative and Social Learning Platforms

- **Online Collaboration Tools:**

Online collaboration tools are revolutionizing the field of cognitive task analysis (CTA) by enhancing communication and knowledge sharing among learners and instructors. These platforms facilitate the exchange of insights, feedback, and best practices, which are crucial for refining and optimizing cognitive tasks.

For example, a study by Du et al. (2022) examined the impact of online collaborative problem-solving (CPS) on cognitive task performance. The research utilized brain-to-brain synchrony (BS) to measure interpersonal interaction and common attention among participants during CPS activities. The study found that higher levels of BS at the problem-understanding stage significantly correlated with better task performance, highlighting the effectiveness of collaborative platforms in enhancing cognitive engagement and task outcomes.

Another relevant study by Zhang et al. (2022) focused on optimizing online collaborative learning by considering cognitive load. The research utilized learning analytics to assess learners' states and proposed an optimized mechanism for improving collaborative performance without increasing cognitive load. The results demonstrated significant improvements in collaboration and knowledge acquisition, underscoring the importance of data-driven approaches in refining collaborative learning environments.

Applied Cognitive Task Analysis (ACTA) aims to understand and model the cognitive processes involved in task performance. Online collaboration tools align with ACTA by facilitating the detailed documentation and analysis of these processes. Through collaborative platforms, instructors and learners can share their experiences and strategies, contributing to a richer understanding of cognitive tasks. This collective knowledge can be used to refine cognitive models and develop more effective training programs.

- **Crowdsourced Data and Insights:**

Leveraging the collective intelligence of diverse user groups through crowdsourcing can provide valuable insights and enhance the robustness of cognitive task analysis methodologies. Crowdsourced data allows for the collection of a wide range of experiences and perspectives, which can be instrumental in identifying common patterns and unique challenges in cognitive tasks.

Applied Cognitive Task Analysis (ACTA) is a method used to capture the detailed cognitive processes involved in complex tasks. By integrating crowdsourced data with ACTA, researchers can enrich their understanding of cognitive tasks through diverse and comprehensive datasets. Crowdsourcing can supplement ACTA by providing a broader range of input, making the analysis more robust and generalizable.

In the realm of online collaborative learning, crowdsourcing data through learning analytics has proven to be a powerful approach. According to Zhang et al. (2022), research on online collaborative learning has increasingly adopted learning analytics to ascertain learners' states and promote collaborative performance. This method involves collecting data from a large number of learners, which is then analysed to understand and optimize collaborative learning processes.

Zhang et al. (2022) highlight the potential of integrating collaborative cognitive load theory with system dynamics modelling methods to investigate the transformation of collaborative states in online learning environments. By conducting a quasi-experiment with 91 college students, the study analysed how these students' collaborative states evolved and proposed an optimized mechanism to improve online collaboration. The findings from Zhang et al. (2022) demonstrate that students who utilized the optimized mechanism exhibited significant improvements in collaboration and knowledge acquisition. Importantly, this optimized approach did not add to the cognitive load of the students, indicating its efficiency and effectiveness. These results underscore the value of using crowdsourced data to inform and enhance collaborative learning strategies.

Sharma et al. (2023) exemplifies this approach by examining student engagement in collaborative online discussions. Their study employs social network analysis, discourse analysis, and a machine learning model to analyse data from multiple students, providing a comprehensive view of behavioural and cognitive engagement at both individual and group levels. This method of crowdsourcing data aligns with Applied Cognitive Task Analysis (ACTA), which focuses on understanding the cognitive processes involved in complex tasks. By integrating crowdsourced data with ACTA methodologies, Sharma et al. (2023) demonstrate how detailed patterns of engagement and interaction can be revealed, offering deeper insights into cognitive processes and enhancing the effectiveness of educational interventions. Their approach highlights the potential of combining diverse data sources and analytical methods to enrich ACTA and improve learning outcomes.

Integrating ACTA with crowdsourced data, as demonstrated by Zhang et al. (2022) and Sharma et al. (2023), can provide a more nuanced understanding of collaborative tasks. ACTA focuses on the cognitive processes and knowledge required to perform tasks, while crowdsourcing offers a wealth of data from diverse participants, enhancing the reliability and applicability of the findings. By using crowdsourced data in ACTA, researchers can identify cognitive patterns and challenges that may not be apparent through traditional methods alone.

In conclusion, these studies illustrate the significant benefits of utilizing crowdsourced data and learning analytics in online collaborative learning. By combining these approaches with ACTA, researchers can gain a deeper and more comprehensive understanding of cognitive processes, leading to the development of optimized mechanisms that enhance learning outcomes without increasing cognitive load.

Collaborative and social learning platforms significantly enhance the cognitive task analysis process by facilitating communication, knowledge sharing, and the aggregation of diverse insights. Online collaboration tools support the sharing of best practices and feedback, while crowdsourced data provides a broad base of information that enhances the robustness of cognitive task models. Integrating these platforms with Applied Cognitive Task Analysis (ACTA) leads to more accurate and comprehensive understanding of cognitive processes, ultimately improving task design and training effectiveness.

11.2.4 Neuroscientific Advances

- **Brain-Computer Interfaces (BCIs):**

Brain-Computer Interfaces (BCIs) are sophisticated systems that facilitate direct communication between the brain and external devices. By interpreting neural signals, BCIs offer a window into real-time brain activity and cognitive processes.

The study by Gallegos Ayala et al. (2023) exemplifies this application by introducing a novel algorithm for assessing mental workload through the extraction of frontal theta oscillations from EEG recordings. Their approach aims to address the challenge of task-dependent neural signals by providing a more accurate classification of mental workload across different cognitive tasks. The new algorithm achieved impressive performance, with binary classification rates of 92.00% and 92.35% for differentiating between low or high workload and no workload conditions, surpassing previous methods. Although it did not excel in distinguishing between high and low workload levels, it represents a significant advancement in real-time mental workload assessment. This development enhances Applied Cognitive Task Analysis (ACTA) by offering precise, objective data on cognitive states, which can refine task designs and instructional methods to better align with cognitive processing patterns. Integrating BCI-based insights from studies like Gallegos Ayala et al. (2023) can lead to more personalized and effective educational and training interventions.

- **Neurofeedback:**

Neurofeedback is a technique that provides real-time feedback to individuals about their brain activity, enabling them to learn how to regulate and optimize their cognitive functions. By observing their own brainwave patterns, individuals can develop strategies to enhance focus, reduce cognitive overload, and improve performance on specific tasks.

The study by Loriette et al. (2021) reviews the methodological principles of neurofeedback, its behavioural benefits, and the challenges it faces. Neurofeedback can complement Applied Cognitive Task Analysis (ACTA) by offering a dynamic approach to understanding and improving cognitive task performance. While ACTA focuses on breaking down tasks into cognitive components and understanding cognitive demands, neurofeedback provides a tool for real-time intervention and adjustment. For instance, if ACTA identifies a task component that is particularly cognitively demanding, neurofeedback can help individuals develop strategies to manage cognitive load more effectively. This approach allows for immediate adjustments and personal optimization, making it especially useful in training environments where cognitive efficiency is crucial. Integrating neurofeedback with ACTA can enhance the ability to monitor and improve cognitive processes on-the-fly, leading to more effective and personalized interventions.

Integrating neuroscientific advances like BCIs and neurofeedback into ACTA not only provides deeper insights into cognitive processes but also offers practical tools for enhancing cognitive task performance. These technologies enable a more comprehensive understanding of how cognitive tasks are managed and optimized, leading to more effective and personalized approaches in education and training.

In conclusion, the integration of AI, VR, and other emerging technologies is paving the way for more sophisticated and effective cognitive task analysis methodologies. As these tools continue to evolve, they hold the promise of unlocking new levels of understanding and optimization in cognitive processes, ultimately enhancing learning and performance across various domains.

CONCLUSION

In this chapter, we explored the innovative approaches that are reshaping the landscape of Applied Cognitive Task Analysis (ACTA). The integration of Artificial Intelligence (AI) and Virtual Reality (VR) represents significant advancements in how cognitive processes are analysed and optimized, particularly in high-stakes fields such as education, healthcare, and industrial training. These technologies not only enhance the precision and depth of ACTA but also offer real-time feedback, immersive learning experiences, and personalized interventions, making cognitive task analysis more dynamic and responsive to individual needs.

The role of AI in ACTA, particularly through its capabilities in data analysis, predictive modelling, and personalized learning, allows for the more efficient processing of vast amounts of cognitive data and the development of targeted strategies to optimize performance. As demonstrated in various studies, AI has been instrumental in providing real-time, adaptive feedback to learners, predicting cognitive bottlenecks, and improving educational outcomes by customizing learning paths. Similarly, VR has emerged as a powerful tool for creating immersive training environments that closely mimic real-world conditions. This integration enables a deeper exploration of cognitive tasks in safe and controlled settings, where users can engage with realistic scenarios and receive targeted feedback on their performance.

Beyond AI and VR, we also looked at emerging trends such as wearable technology, big data analytics, brain-computer interfaces (BCIs), and neurofeedback. These advancements are pushing the boundaries of cognitive task analysis by offering new ways to capture real-time data on cognitive load and task performance, providing actionable insights that enhance the effectiveness of ACTA. Wearable biosensors and augmented reality (AR) further contribute to improving task performance and situational awareness, allowing for a more comprehensive understanding of how cognitive processes unfold in dynamic environments.

As we move forward, the integration of these emerging technologies with ACTA methodologies promises to revolutionize cognitive task analysis, enabling deeper insights into human cognition and more effective task optimization. The inclusion of collaborative platforms and social learning tools further enhances the capacity for shared knowledge and problem-solving, making cognitive analysis more inclusive and reflective of real-world contexts. The use of crowdsourced data and collaborative learning platforms aligns well with ACTA's goals of optimizing performance and improving learning outcomes, particularly in team-based environments.

In conclusion, innovative approaches such as AI, VR, and wearable technologies are transforming the way cognitive tasks are analysed and addressed in ACTA. These advancements open up new possibilities for refining cognitive performance, enhancing training programs, and improving real-world applications across sectors. As these tools continue to evolve, they hold the potential to advance ACTA's precision, scalability, and relevance, making it a critical framework for future educational, industrial, and professional applications.

REFERENCES

Ahmad, M. I., Keller, I., & Robb, D. A. (2023). A framework to estimate cognitive load using physiological data. *Personal and Ubiquitous Computing*, 27(4), 2027–2041. DOI: 10.1007/s00779-020-01455-7

Aksu, Ş. H., Çakıt, E., & Dağdeviren, M. (2024). Mental workload assessment using machine learning techniques based on EEG and eye tracking data. *Applied Sciences (Basel, Switzerland)*, 14(6), 2282. DOI: 10.3390/app14062282

Chakladar, D., & Roy, P. P. (2023). Cognitive workload estimation using physiological measures: A review. *Cognitive Neurodynamics*. Advance online publication. DOI: 10.1007/s11571-023-10051-3

Du, X., Zhang, L., & Hung, J. L. (2022). Understand group interaction and cognitive state in online collaborative problem solving: Leveraging brain-to-brain synchrony data. *International Journal of Educational Technology in Higher Education*, 19, 52. DOI: 10.1186/s41239-022-00356-4

Fu, S., Gu, H., & Yang, B. (2020). The affordances of AI-enabled automatic scoring applications on learners' continuous learning intention: An empirical study in China. *British Journal of Educational Technology*, 51(5). Advance online publication. DOI: 10.1111/bjet.12995

Gallegos Ayala, G. I., Haslacher, D., Krol, L. R., Soekadar, S. R., & Zander, T. O. (2023). Assessment of mental workload across cognitive tasks using a passive brain-computer interface based on mean negative theta-band amplitudes. *Frontiers in Neuroergonomics*, 4, 1233722. DOI: 10.3389/fnrgo.2023.1233722

Holuša, V., Vaněk, M., Beneš, F., Švub, J., & Staša, P. (2023). Virtual reality as a tool for sustainable training and education of employees in industrial enterprises. *Sustainability*, 15(17), 12886. DOI: 10.3390/su151712886

Iarlori, S., Perpetuini, D., Tritto, M., & Monteriù, A. (2024). An overview of approaches and methods for the cognitive workload estimation in human–machine interaction scenarios through wearables sensors. *BioMedInformatics*, 4(2), 1155–1173. DOI: 10.3390/biomedinformatics4020064

Kasepalu, R., Prieto, L. P., Ley, T., & Chejara, P. (2022). Teacher artificial intelligence-supported pedagogical actions in collaborative learning coregulation: A Wizard-of-Oz study. *Frontiers in Education*, 7. Advance online publication. DOI: 10.3389/feduc.2022.736194

Loriette, C., Ziane, C., & Ben Hamed, S. (2021). Neurofeedback for cognitive enhancement and intervention and brain plasticity. *Revue Neurologique*, 177(9), 1133–1144. DOI: 10.1016/j.neurol.2021.08.004

Romero, C., & Ventura, S. (2020). Educational data mining and learning analytics: An updated survey. *Wiley Interdisciplinary Reviews. Data Mining and Knowledge Discovery*, 10(3), e1355. DOI: 10.1002/widm.1355

Sharma, P., Akgun, M., & Li, Q. (2023). Understanding student interaction and cognitive engagement in online discussions using social network and discourse analyses. *Educational Technology Research and Development*. Advance online publication. DOI: 10.1007/s11423-023-10261-w

Vats, S., & Joshi, R. (2024). The impact of virtual reality in education: A comprehensive research study. In S. K. Sharma, Y. K. Dwivedi, B. Metri, B. Lal, & A. Elbanna (Eds.), *Transfer, diffusion and adoption of next-generation digital technologies* (Vol. 699, pp. 137-150). IFIP Advances in Information and Communication Technology. Springer. DOI: 10.1007/978-3-031-50204-0_11

Woodward, J., & Ruiz, J. (2022). Analytic review of using augmented reality for situational awareness. *IEEE Transactions on Visualization and Computer Graphics, PP*, (99), 1–1. DOI: 10.1109/TVCG.2022.3141585

Zhang, L., Wang, X., He, T., & Han, Z. (2022). A data-driven optimized mechanism for improving online collaborative learning: Taking cognitive load into account. *International Journal of Environmental Research and Public Health*, 19(12), 6984. DOI: 10.3390/ijerph19126984

Zheng, L., Fan, Y., Chen, B., Huang, Z., Gao, L., & Long, M. (2024). An AI-enabled feedback-feedforward approach to promoting online collaborative learning. *Education and Information Technologies*, 29, 11385–11406. DOI: 10.1007/s10639-023-12292-5

Zhou, Y., Zhao, Y., Xiang, Z., Yan, Z., Shu, L., Xu, X., Zhang, L., & Tian, X. (2023). A dual-task-embedded virtual reality system for intelligent quantitative assessment of cognitive processing speed. *Frontiers in Human Neuroscience*, 17. Advance online publication. DOI: 10.3389/fnhum.2023.1158650

Chapter 12
Interdisciplinary Contributions to ACTA

ABSTRACT

This chapter explores the interdisciplinary dimensions of Applied Cognitive Task Analysis (ACTA), highlighting its integration with cognitive science, psychology, and organizational studies. It examines how ACTA benefits from advancements in these fields, showcasing its versatile applications across healthcare, education, and technology. By exploring these cross-sectoral applications, the chapter illustrates the mutual benefits of ACTA's collaboration with diverse disciplines, enhancing its effectiveness and impact. Specific examples demonstrate how ACTA methodologies address complex problems, contributing to scientific knowledge while gaining insights from various research perspectives. This comprehensive overview emphasizes how interdisciplinary contributions enrich ACTA and offers insights into future research and practical applications.

INTRODUCTION

Applied Cognitive Task Analysis (ACTA) has emerged as a powerful methodology for dissecting and optimizing cognitive processes across diverse fields. At its core, ACTA thrives on interdisciplinary collaboration, drawing from cognitive psychology, neuroscience, educational science, human-computer interaction, and other domains to enhance its analytical capabilities and practical applications. This chapter delves into the synergies between ACTA and various scientific and social science disciplines, highlighting how these interactions enrich ACTA's methodologies and expand its impact.

The convergence of disciplines enables ACTA to address complex cognitive challenges with innovative and tailored solutions. By integrating insights from neuroscience, psychology, and instructional design, ACTA provides a robust frame-

DOI: 10.4018/979-8-3693-7733-8.ch012

work for improving performance in high-stakes environments such as healthcare, military operations, and industrial settings. Furthermore, the inclusion of emerging technologies like artificial intelligence (AI) and virtual reality (VR) underscores ACTA's adaptability and relevance in modern contexts.

In this chapter, we explore the interdisciplinary contributions that shape ACTA, focusing on its integration with cutting-edge research areas such as brain-computer interfaces (BCIs), cognitive load theory, behavioural analysis, and instructional design. The chapter also examines cross-sectoral applications, showcasing how ACTA informs curriculum development, healthcare training, industrial workflows, and military simulations. These examples demonstrate ACTA's versatility and its transformative potential to enhance cognitive performance and system design across diverse domains.

12.1 Synergies With Science and Social Science Research Areas

Applied Cognitive Task Analysis (ACTA) stands at the intersection of multiple disciplines, benefiting from and contributing to both science and social science research areas. This chapter explores these synergies, highlighting how interdisciplinary collaboration enhances the scope and impact of ACTA. By integrating insights from cognitive psychology, human factors engineering, instructional design, and domain-specific expertise, ACTA provides a comprehensive framework for analysing complex tasks and improving performance in a variety of settings.

Moreover, ACTA's methodologies contribute to advancing technological fields like artificial intelligence and virtual reality, while also drawing on the social sciences to better understand human behaviour, teamwork, and decision-making in complex environments. As technology and work environments evolve, ACTA continues to adapt and expand through collaboration across disciplines, ensuring its relevance in addressing increasingly intricate cognitive tasks.

In this chapter, we explore how the convergence of diverse fields enriches ACTA, strengthening its methodologies and widening its applicability across industries. The synergies created through interdisciplinary collaboration not only enhance ACTA's analytical capabilities but also provide innovative solutions to challenges in both high-stakes and everyday environments.

12.1.1 Neuroscience

1. Integration with Brain-Computer Interfaces (BCIs):

Neuroscientific research has significantly advanced our understanding of brain activity during cognitive tasks. Brain-Computer Interfaces (BCIs) provide a novel approach to monitoring and interpreting neural signals in real time. By offering direct insights into the brain's functioning, BCIs allow for a deeper exploration of cognitive processes. Recent studies have demonstrated the potential of BCIs to enhance Applied Cognitive Task Analysis (ACTA).

- **Real-Time Monitoring and Cognitive Workload:**

A study by Zakeri et al. (2023) explores the use of Brain-Computer Interfaces (BCIs) for real-time cognitive workload assessment in the context of human-robot collaboration. The researchers utilized electroencephalography (EEG) and functional near-infrared spectroscopy (fNIRS) to monitor brain activity and track hemodynamic responses during various cognitive tasks, providing objective data on cognitive strain and efficiency. This information is critical for understanding the impact of task complexity, cobot movement speed, and cobot payload capacity on the mental stress experienced by factory workers. By integrating BCI data, this study offers valuable insights that can be incorporated into Applied Cognitive Task Analysis (ACTA) methodologies. Such integration helps refine task analysis and design, ensuring tasks are tailored to the cognitive capabilities and limitations of individuals working in smart factory environments. The study highlights the potential of physiological measures to provide unbiased and authentic estimates of stress, which can enhance the accuracy and relevance of ACTA by offering real-time, objective data on cognitive workload. This approach is particularly useful in optimizing human-robot interactions, thereby improving productivity and safety in Industry 5.0 settings.

- **Enhancing Task Design:**

Another study by Blitz and Barfield (2023) investigates how BCI feedback can improve task performance. The findings highlight that BCIs can inform Applied Cognitive Task Analysis (ACTA) by revealing the neural mechanisms underlying task execution, leading to more effective and optimized task designs. One prominent example of BCI technology involves devices that create a direct communication pathway between the brain's electrical activity and external devices, traditionally aimed at restoring sensory mechanisms or muscular movements. However, recent developments have explored BCIs for augmenting the brain's ability to form and access memories. For instance, technologies like the prototype artificial hippocampus developed by Theodore Berger and colleagues aim to restore long-term memory formation typically lost after the onset of Alzheimer's disease. Future variants may enhance or supplement normal cognitive functions, potentially even implanting memories of events never directly experienced by the individual. This kind of memory-enhancing BCI can significantly impact task design by providing insights into the brain regions activated during specific tasks. Understanding these neural mechanisms allows ACTA to develop strategies that minimize cognitive load and enhance performance. Additionally, by augmenting memory capacity and the ability to create or share memories, BCIs can lead to innovative approaches in designing tasks that align with an individual's cognitive capabilities, ultimately improving task efficiency and effectiveness.

By incorporating BCI insights, ACTA can achieve a more nuanced understanding of cognitive processes. This integration leads to better task design and optimization, ultimately improving efficiency and effectiveness in various applications.

2. Neurofeedback:

Neurofeedback is a technique that provides real-time feedback on brain activity, enabling individuals to learn how to regulate and optimize their cognitive performance. This dynamic feedback mechanism has shown promise in enhancing task execution and learning outcomes, making it a valuable complement to ACTA.

- **Optimizing Cognitive Performance**:

A study by Dessy et al. (2018) investigates the effectiveness of neurofeedback training for cognitive enhancement. The research demonstrates that neurofeedback can help individuals develop strategies to improve focus, reduce cognitive overload, and enhance task performance. In the context of non-invasive modulation of human physiological activity, both biofeedback (BFB) and neurofeedback (NFB) have shown promise in optimizing cognitive, emotional, and behavioural functioning. A comprehensive review of the state of the art in BFB and NFB studies highlights the potential of these training techniques in non-clinical settings. Despite the varying terminology, diverse methodological designs, and differences in neurofeedback frequency ranges, the evidence suggests that neurofeedback can be an effective tool for cognitive performance enhancement. Integrating neurofeedback into Applied Cognitive Task Analysis (ACTA) can provide real-time adjustments and personalized support, ensuring tasks are tailored to optimize cognitive task execution. By leveraging neurofeedback, ACTA can facilitate the development of strategies that enhance focus and reduce cognitive overload, ultimately leading to improved task performance.

- **Enhancing Learning Outcomes:**

A study by Martin et al. (2020) explores the integration of neurofeedback with adaptive learning systems. The research highlights that neurofeedback can offer personalized insights into cognitive processes, enabling real-time adjustments to learning tasks. This approach aligns with Applied Cognitive Task Analysis (ACTA)'s goal of understanding and improving cognitive task performance, leading to enhanced learning outcomes and more effective educational interventions. In a systematic review of adaptive learning research, 61 articles were analysed to describe the current state of research and identify gaps in the literature. The review revealed that the majority of studies were conducted in higher education, with a focus on design, development, implementation, and evaluation of adaptive learning strategies. By leveraging neurofeedback within adaptive learning systems, educators can gain a deeper understanding of learners' cognitive states and provide tailored support that enhances learning efficiency. This synergy between neurofeedback and adaptive learning can facilitate personalized growth, ensuring that instructional strategies are dynamically adjusted to meet the individual needs of learners, thereby optimizing educational outcomes.

By incorporating neurofeedback into ACTA, researchers and practitioners can provide dynamic feedback and support to individuals, helping them optimize their cognitive performance. This integration enhances task execution and learning outcomes, making neurofeedback a valuable tool in various educational and training contexts.

12.1.2 Psychology

1. Cognitive Load Theory:

Cognitive Load Theory (CLT) is a psychological framework that focuses on the mental effort required to process information during task performance. It provides valuable insights into how cognitive resources are allocated and how tasks can be designed to minimize cognitive overload. ACTA can effectively leverage CLT principles to enhance task design, ensuring that tasks are cognitively manageable and efficient.

- **Understanding Cognitive Overload:**

Recent research by Klepsch & Seufert (2020) emphasizes the importance of designing tasks that align with cognitive load principles, highlighting various types of cognitive load—intrinsic, extraneous, and germane—and their impact on task performance. The study reviews the use of questionnaires to measure these types of cognitive load and demonstrates their prognostic validity across different learning contexts. It reveals that variations in task difficulty reflect intrinsic cognitive load, design variations reflect extraneous cognitive load, and strategies to enhance deeper learning reflect germane cognitive load. By applying these principles, Applied Cognitive Task Analysis (ACTA) can develop tasks that minimize extraneous load (irrelevant cognitive processing) and optimize intrinsic load (task complexity). This approach leads to improved accuracy and efficiency in task execution, ensuring that tasks are designed to effectively manage cognitive demands and enhance overall performance.

- **Designing Effective Learning Materials:**

A study by Castro-Alonso et al. (2021) demonstrates how Cognitive Load Theory (CLT) can be used to design instructional materials that facilitate learning. The research focuses on five key strategies for addressing problematic instructional designs in multimedia learning: (a) the multimedia principle (using visualizations to complement text), (b) the split-attention effect (integrating text with visualizations), (c) the redundancy effect (removing nonessential information), (d) the signalling principle (highlighting essential information), and (e) the transient information effect (segmenting content). These strategies are shown to help manage cognitive load, improving learning outcomes. The study also explores how these strategies can be employed by both instructors and learners, revealing mixed results depending on learner expertise. For novice learners, instructor-managed strategies may be more effective, whereas expert learners might benefit more from self-managed approaches. Integrating these insights into ACTA allows for the development of tasks that are not only cognitively appropriate but also enhance learning outcomes by effectively applying these principles to instructional design.

By incorporating CLT principles into ACTA, researchers and practitioners can create tasks that are better aligned with cognitive capabilities, reducing cognitive overload and improving overall task performance.

2. Behavioural Analysis:

Behavioural analysis involves examining the observable behaviours and strategies employed by individuals during task performance. Integrating behavioural analysis methods into ACTA allows for a deeper understanding of cognitive strategies and provides insights into how tasks are approached and managed.

- **Examining Cognitive Strategies**:

A recent study by Laurence and Macedo (2023) investigates cognitive strategies used in matrix-reasoning tasks, presenting a comprehensive review of the advances made in this area over the past decade and a half. The study explores various methods and measures used to analyse these strategies, highlighting both the progress and limitations in the field. By reviewing the diverse applications and resulting knowledge, the study provides insights into how cognitive strategies can be effectively measured and understood. This research is relevant to ACTA as it underscores the importance of analysing cognitive strategies to gain a deeper understanding of task demands. By incorporating these insights into ACTA, researchers can better comprehend the underlying cognitive processes and decision-making strategies individuals use during complex tasks, leading to more informed and effective task design and analysis.

- **Improving Task Design:**

A study by McGee and Crowley-Koch (2019) examines how behavioural systems analysis can enhance the implementation of large-scale organizational change initiatives. It compares two approaches: one that implements changes first and then adjusts systems, and another that begins with identifying and implementing necessary system changes before rolling out the initiative. The latter approach, although slower initially, results in a more adaptive and proactive organization in the long run. This focus on behavioural systems analysis highlights how understanding and mapping out system variables can streamline and support change initiatives more effectively. This is relevant to task design as it demonstrates how behavioural analysis methods can be applied to optimize task structures. By analysing how individuals interact with tasks and identifying common strategies, researchers using Applied Cognitive Task Analysis (ACTA) can refine task design, ensuring it better meets user needs and improves overall efficiency.

Integrating behavioural analysis into ACTA enables a more detailed examination of cognitive processes and task performance. This approach provides valuable insights into how tasks are managed and allows for the development of more effective and targeted interventions.

12.1.3 Educational Science

1. Learning Theories:

Educational research has significantly contributed to the development and refinement of Applied Cognitive Task Analysis (ACTA) through various learning theories. These theories provide a framework for understanding how individuals acquire, process, and retain information, guiding the development of instructional strategies that align with cognitive capabilities and enhance learning outcomes.

- **Constructivist Learning Theory:**

A study by Zajda (2021) explores the application of constructivist principles in creating effective learning environments. Constructivism posits that meaningful knowledge and critical thinking are actively constructed through cognitive, cultural, emotional, and social engagement. The study discusses how constructivist pedagogy can be leveraged to enhance learning by considering students' characteristics, developmental stages, cultural diversity, and motivational factors. By applying these principles, ACTA (Applied Cognitive Task Analysis) can design tasks that foster active engagement, problem-solving, and critical thinking. Such tasks promote deeper cognitive processing, which aligns with ACTA's goal of understanding and optimizing cognitive task performance. Constructivist approaches emphasize the importance of creating learning experiences that encourage students to build their own understanding through direct engagement and reflection, thus supporting more effective and meaningful learning outcomes.

- **Cognitive Load Theory:**

As previously discussed, Cognitive Load Theory (CLT) offers insights into how cognitive resources are utilized during learning. A study by Skulmowski and Xudelves (2022) delves into Cognitive Load Theory (CLT) and its influence on educational psychology. The research addresses the challenge of extraneous cognitive load in digital and online learning environments, such as interactive media, immersion, disfluency, realism, and redundancy. It proposes integrating aspects of constructive alignment to balance cognitive load, considering both the positive and negative effects of design factors on cognitive resources. By aligning cognitive load with desired learning outcomes, this approach helps to formulate theory-guided hypotheses and practical strategies for leveraging emerging technologies while minimizing extraneous drawbacks. Incorporating CLT principles into Applied Cognitive Task Analysis (ACTA) allows for the creation of tasks and learning materials that are aligned with cognitive capacities, reducing unnecessary cognitive strain and enhancing task efficiency. This strategy ensures that tasks are designed to manage cognitive load effectively, thus improving learning outcomes and task performance.

- **Experiential Learning Theory:**

A study by Bui, Tong and Lien (2022) explores the principles of experiential learning, which emphasize learning through experience and reflection. The study examined how experiential learning activities were applied in the context of mathematics education, focusing on the impact of real-world problem-solving on students' motivation, engagement, and achievement. By conducting pedagogical experiments with sixth-grade students, the study found that experiential learning positively influenced students' attitudes and performance in mathematics. Integrating experiential learning principles into Applied Cognitive Task Analysis (ACTA) allows for the development of tasks that provide learners with opportunities to engage in real-world scenarios and reflect on their experiences. This approach makes tasks more relevant and engaging, thereby facilitating better learning and performance outcomes.

2. Instructional Design:

Insights from instructional design research are crucial for creating educational materials and tasks that are cognitively appropriate, engaging, and effective. ACTA can benefit from these principles to optimize task structures and instructional methods, ensuring that they are aligned with cognitive capabilities and learning goals.

- **Principles of Instructional Design**:

Research by Mayer (2014) explores key principles of instructional design through the lens of the cognitive theory of multimedia learning. This theory emphasizes the importance of designing multimedia instructional messages that align with how the human mind processes information, minimizing extraneous cognitive load and enhancing the processing of essential information. By applying these principles, ACTA can create educational tasks and materials that are more effective and engaging, thereby improving overall learning outcomes.

- **Adaptive Learning Systems:**

The systematic review by Martin et al. (2022) synthesizes research on adaptive learning, focusing on various aspects like instructional design phases, adaptive strategies, and technologies. This review aligns with the principles discussed in the study on adaptive learning systems. By incorporating adaptive learning principles into ACTA, tasks can be developed that dynamically respond to individual learners' needs, enhancing the personalization and effectiveness of educational interventions. The focus on adaptive feedback, navigation, and learning styles in the review further supports the development of personalized and effective educational tasks within ACTA.

- **Motivational Design:**

The study by Vu et al. (2022) examines the reciprocal relationship between learners' motivation and academic achievement, highlighting the importance of understanding how these factors interact to enhance educational outcomes. The research identifies several theoretical and methodological challenges in studying this relationship, recommending a comprehensive approach that considers multiple motivation constructs and behavioural mediators. By integrating these insights with motivational design principles, ACTA can develop tasks that are not only cognitively appropriate but also engaging and motivating for learners, ultimately leading to improved academic achievement. This approach ensures that educational interventions are both effective and tailored to the individual needs of students.

By leveraging insights from learning theories and instructional design research, ACTA can enhance the development of educational materials and tasks. This approach ensures that tasks are aligned with cognitive capabilities, engaging, and effective, ultimately leading to improved learning outcomes and task performance.

12.1.4 Human-Computer Interaction (HCI)

1. Usability Studies:

HCI research emphasizes the importance of usability and user experience in the design of technological interfaces. Usability studies provide insights into how users interact with technology, identifying potential issues and areas for improvement. By incorporating HCI principles, Applied Cognitive Task Analysis (ACTA) can ensure that tasks involving technology are intuitive, user-friendly, and efficient, thereby reducing cognitive load and enhancing performance.

- **Improving User Experience:**

The study by Paz (2023) examines the role of usability heuristics in improving user experience, particularly focusing on how these principles can be refined and adapted for specific contexts like mobile-based e-commerce applications. The research highlights the importance of applying usability heuristics such as consistency, feedback, and error prevention to enhance the intuitiveness and efficiency of technological interfaces. By integrating these principles, ACTA can design tasks that are easier to navigate and understand, thereby reducing cognitive load and improving overall performance.

- **Usability Testing**:

The study by Alqurni (2023) investigates various user-interface usability evaluation methods (UEM) and the usability features of e-learning applications and websites. It emphasizes the importance of usability approaches for the success of e-learning platforms like Moodle, Blackboard, and Google Classroom. This study shows that usability testing can identify specific areas where users encounter difficulties. By incorporating findings from usability testing into ACTA, tasks can be optimized to better align with users' cognitive processes, thereby enhancing usability and overall performance.

- **Reducing Cognitive Load:**

Research by Jingmiao (2024) explores the relationship between user interface usability and learning engagement in mobile learning applications, with a focus on the mediating impact of cognitive load. The study highlights that effective user interface design is a significant indicator of visual communication design, which can enhance student engagement. By applying these insights, ACTA can develop technology-based tasks that minimize unnecessary cognitive effort, leading to better task performance and user satisfaction.

2. Ergonomics:

Ergonomics focuses on designing tasks and environments that are physically and cognitively comfortable and sustainable. Integrating ergonomic principles into ACTA allows for a holistic approach to task design, addressing both physical and cognitive aspects to ensure efficiency and comfort.

- **Ergonomic Task Design:**

The study by Kalakoski et al. (2020) examines the impact of cognitively straining conditions, such as disruptions, interruptions, and information overload, on task performance and well-being at work. The research implements practices to manage cognitive strain in office work tasks and conditions, aiming to improve working conditions, workflow, well-being, and productivity. By incorporating ergonomic principles, ACTA can create tasks that are not only cognitively efficient but also physically sustainable, enhancing both performance and well-being.

- **Cognitive Ergonomics:**

The study by Kalakoski et al. (2020) examines the impact of cognitively straining conditions, such as disruptions, interruptions, and information overload, on task performance and well-being at work. The research implements practices to manage cognitive strain in office work tasks and conditions, aiming to improve working conditions, workflow, well-being, and productivity. This aligns with research in cognitive ergonomics, which focuses on designing systems that align with human cognitive capabilities. The study highlights that tasks should be designed to match users' mental models and cognitive strengths, reducing cognitive strain and enhancing performance. Integrating cognitive ergonomics into ACTA ensures that tasks are tailored to users' cognitive processes, improving overall efficiency and effectiveness.

- **Sustainable Task Design:**

The study by Faludi et al. (2020) investigates the implementation of sustainable design methods and tools, identifying needs and gaps in practice and research. The research emphasizes the importance of integrating sustainable design more tightly into mainstream product design and development, proposing a roadmap for collective progress in the sustainable design research community. This aligns with the concept of sustainable task design, which explores the long-term benefits of ergonomic task design. The research demonstrates that tasks designed with sustainability in mind can lead to improved productivity and well-being over time. By applying ergonomic principles, ACTA can create tasks that support long-term physical and cognitive health, ensuring sustained performance and reduced risk of burnout.

By incorporating HCI principles and ergonomic insights, ACTA can enhance task design to ensure that tasks involving technology are intuitive, user-friendly, and physically comfortable. This holistic approach reduces cognitive load, improves performance, and supports long-term well-being, making tasks more efficient and sustainable.

12.2 Cross-Sectoral Applications and Benefits

ACTA's interdisciplinary nature allows for its application across various sectors, each benefiting from its insights and methodologies.

12.2.1 Healthcare

1. Medical Training:

Applied Cognitive Task Analysis (ACTA) can be instrumental in designing training programs that enhance the cognitive skills required for complex medical procedures. By focusing on the cognitive demands of these tasks, ACTA ensures that healthcare professionals are well-prepared and efficient in their roles.

- **Enhancing Surgical Training**:

The study by Bhatia et al. (2022), which applies Cognitive Task Analysis (CTA) to break down the operative steps and cognitive processes involved in performing an open appendectomy, closely aligns with the principles of Applied Cognitive Task Analysis (ACTA). Both methodologies aim to capture the cognitive demands of complex tasks to enhance training and improve performance. In this case, ACTA could extend the findings of Bhatia et al. by using structured interviews to elicit both explicit and tacit knowledge from expert surgeons, focusing on the decision-making processes and critical cognitive cues that may not be immediately obvious but are crucial for effective surgical practice. By mapping out the operative steps and decision points, ACTA would help identify the cognitive skills necessary for successfully navigating surgical procedures, particularly in resource-limited environments such as low- and middle-income countries (LMICs). This approach would further enhance the development of teaching modules and training programs by incorporating cognitive strategies that address potential areas of error and reinforce key decision-making skills.

By focusing on both the procedural and cognitive elements of surgery, ACTA provides a comprehensive framework for improving skill acquisition and performance, ensuring that training programs are tailored to the specific cognitive challenges practitioners face during surgery. In essence, ACTA can help refine the insights gained through CTA, offering a more applied and practical approach to enhancing surgical training, particularly in settings where resources are limited and the cognitive demands of practitioners are especially high.

- **Simulation-Based Training:**

The study by Militello et al. (2023), which adapted Cognitive Task Analysis (CTA) for a large multi-site investigation with 102 anaesthesiologists working through high-fidelity simulated incidents, relates closely to the principles of Applied Cognitive Task Analysis (ACTA). Both CTA and ACTA are designed to capture the cognitive demands of high-stakes tasks, and ACTA's streamlined approach would provide practical solutions to the challenges identified in the study, such as standardizing interview techniques and managing large volumes of qualitative data.

In this study, the researchers implemented strategies to address standardization, effective training, and the analysis of complex qualitative data across diverse teams. ACTA's structured interview methods, including task diagrams and knowledge audits, could have facilitated a more efficient and standardized approach to data collection, especially in multi-site research. Additionally, ACTA's focus on capturing both explicit and tacit knowledge aligns with the study's findings that decision-making processes varied significantly among anaesthesiologists based on the simulated challenges. By using ACTA, the study could have enhanced the ability to extract these individual cognitive strategies more systematically, thereby refining training and scenario design.

Moreover, the research underscores the value of using simulations in CTA to replicate real-world scenarios and allow practitioners to refine their cognitive skills in controlled environments—an approach that ACTA excels in, particularly for high-stakes fields like healthcare, where decision-making, situational awareness, and adaptability are critical. ACTA can further support this by creating tailored training interventions that target the cognitive demands uncovered in the simulation, ultimately improving performance in real-world high-consequence situations.

- **Critical Decision-Making**:

The study by Fackler et al. (2009) explores the use of cognitive task analysis (CTA) techniques to understand key cognitive activities in critical care medicine, focusing on how work-hour limitations impact decision-making and workflow. By identifying cognitive activities such as pattern recognition, uncertainty management, and team coordination, the study demonstrates the value of CTA in analysing and improving decision-making processes in high-pressure environments. This research demonstrates that analysing cognitive tasks can enhance training programs for developing rapid decision-making and problem-solving skills, which is crucial for improving critical decision-making abilities.

By applying ACTA, the findings from Fackler et al. could be translated into more practical and applied training scenarios, allowing medical professionals to practice and refine their cognitive abilities in simulated environments. This would ultimately improve the decision-making process in critical care by aligning training with the cognitive demands of real-world medical scenarios, thereby reinforcing the importance of cognitive task analysis in developing rapid, effective decision-making skills in high-pressure fields like healthcare.

- **Improving Cognitive Skills:**

A systematic review by Edwards et al. (2021) evaluated the efficacy of Cognitive Task Analysis (CTA)-based training for surgical trainees. This review synthesized evidence from randomized controlled trials and observational studies, focusing on the impact of CTA on procedural knowledge and technical performance. The analysis included twelve studies, revealing that surgical trainees who underwent CTA-based training exhibited significantly enhanced procedural knowledge and superior technical performance compared to those who received conventional training. These findings suggest that CTA-based training effectively imparts complex cognitive processes to novices, aligning well with its successful applications in high-risk industries like the military and aviation. Relating this to Applied Cognitive Task Analysis (ACTA), the study demonstrates how ACTA can enhance simulation training by focusing on cognitive skill development. By targeting specific cognitive skills such as decision-making, problem-solving, and situational awareness, ACTA-informed simulations can improve cognitive preparedness and operational readiness, making them invaluable in high-stakes fields such as surgery.

- **Optimizing Training Effectiveness:**

The study by Ahmad et al. (2020) on the application of Cognitive Task Analysis (CTA) in orthopaedic and trauma surgery training highlights its potential to optimize training effectiveness, and its principles closely align with Applied Cognitive Task Analysis (ACTA). Faced with constraints such as reduced operating time and the need for senior-led care, orthopaedic trainees in the UK have seen a decline in hands-on training opportunities. To address this, the study systematically reviewed the benefits of CTA in training novice surgeons. Out of 33 initially identified studies, 7 met the inclusion criteria, encompassing 264 participants. The findings revealed that CTA provided both objective and subjective benefits, enhancing procedural knowledge and technical performance compared to traditional training methods. Participants expressed high satisfaction with CTA tools, noting their effectiveness as adjuncts to the apprenticeship model. The review concludes that CTA is a cost-effective, accessible method that allows for repeated practice, which is crucial in simulation training. By incorporating CTA into training programs, the effectiveness of training can be optimized, ensuring that trainees acquire the necessary cognitive and technical skills despite reduced hands-on opportunities. These findings align with the principles of ACTA, which emphasize understanding and improving cognitive processes in complex tasks. Integrating ACTA can further refine training programs by targeting specific cognitive skills, such as decision-making and problem-solving, leading to better preparedness and performance in real-world situations.

2. Patient Care:

Analysing cognitive tasks involved in patient care using ACTA can help create protocols that reduce cognitive load on healthcare providers. This reduction in cognitive load can improve the quality of care and patient outcomes.

- **Optimizing Workflow:**

The study by Swaby et al. (2022) explores how cognitive task analysis (CTA) methods can be used to understand expert clinical decision-making and optimize workflow in healthcare settings. By revealing the cognitive demands of clinical tasks, CTA can inform the development and refinement of complex interventions and clinical protocols, leading to streamlined processes and reduced cognitive load. By employing ACTA's structured interviews, healthcare interventions and protocols could be more effectively tailored to the real-world cognitive challenges faced by clinical practitioners. This approach ensures that workflows are optimized not only through procedural improvements but also by addressing the underlying cognitive demands that can lead to bottlenecks or errors. The use of ACTA in this context would enhance the design of clinical protocols by focusing on decision-making processes, ultimately leading to more efficient healthcare systems and improved patient outcomes. In essence, ACTA complements the findings of Swaby et al. (2022) by providing a practical framework for translating cognitive insights into actionable workflow improvements, particularly in high-demand healthcare environments where decision-making and efficiency are paramount.

- **Reducing Cognitive Load in Nursing:**

The study by Collins (2019) highlights the significant impact of cognitive overload on nursing performance, emphasizing how high volumes of complex information and frequent interruptions can lead to errors and decreased quality of care. Collins underscores the importance of reducing cognitive load by simplifying clinical workflows and standardizing communication platforms to enhance nurses' ability to focus and perform efficiently. This aligns closely with the principles of Applied Cognitive Task Analysis (ACTA), which aims to optimize task performance by analysing and restructuring cognitive demands. By using ACTA to identify and mitigate sources of cognitive strain, healthcare organizations can develop targeted strategies and protocols that support nurses in managing their tasks more effectively. This approach not only improves patient safety and care quality but also fosters a more satisfying work environment for nursing staff by reducing cognitive overload and enhancing overall performance.

- **Improving Clinical Decision Support Systems:**

Based on the study by Graham et al. (2023), cognitive task analysis (CTA) is presented as a valuable approach to enhance the design of decision support tools and healthcare technologies. By using CTA, researchers can gain a detailed understanding of the cognitive processes that healthcare providers use during tasks, which is crucial for developing effective clinical decision support systems (CDSS). The study emphasizes that aligning CDSS with these cognitive processes can reduce cognitive burden and improve decision-making. Integrating CTA findings into Applied Cognitive Task Analysis (ACTA) can lead to more effective design and implementation of CDSS, ultimately improving patient care and outcomes by providing timely and relevant support to healthcare professionals.

By leveraging ACTA in healthcare, training programs can be designed to enhance cognitive skills, and protocols can be developed to reduce cognitive load on healthcare providers. These approaches lead to better preparedness, efficiency, and quality of care, ultimately improving patient outcomes.

12.2.2. Education

- **Curriculum Development:**

Applied Cognitive Task Analysis (ACTA) can play a crucial role in informing the design of curricula that align with cognitive development stages. By understanding the cognitive processes involved in learning, ACTA ensures that educational materials are appropriately challenging and engaging for students.

- **Aligning with Cognitive Development:**

Research by Zhao, Zhao, & Li (2023) emphasizes on the alignment of teaching–learning–assessment of classroom project-based learning with curriculum standards. The research highlights the importance of matching the cognitive demands of instructional activities, assessments, and curriculum standards to enhance learning outcomes. By identifying discrepancies and deficiencies in the current instructional implementation, the study underscores the need for teacher training and instructional improvement. Integrating these insights into Applied Cognitive Task Analysis (ACTA) can assist educators in designing curricula that are developmentally appropriate and stimulating. ACTA can provide a framework for analysing and adjusting curriculum content to align better with students' cognitive abilities at different stages, promoting engagement and improving educational outcomes. This approach ensures that instructional activities and assessments are not only aligned with curriculum standards but also tailored to the cognitive development of learners, fostering a more effective learning environment.

- **Promoting Deep Learning**:

A study by Kovač et al. (2023) explores how curricula that emphasize deep learning and critical thinking can enhance cognitive development. The research highlights that curriculum designed with a focus on inquiry-based learning, problem-solving, and critical analysis lead to improved cognitive skills and deeper understanding. Incorporating these principles into ACTA can guide the creation of curricula that promote higher-order thinking and sustained cognitive engagement.

- **Personalized Learning Pathways:**

A study by Jiang et al. (2022) explores the use of dynamic personalized learning path planning algorithms to enhance learning achievements and efficiency by tailoring learning sequences to students' individual states and knowledge difficulties. The research demonstrates that by considering students' learning behaviour and test scores, personalized paths can be generated that significantly improve effective behaviour rates, course completion rates, and overall learning efficiency. This aligns with the principles of Applied Cognitive Task Analysis (ACTA), which can be used to identify students' cognitive needs and preferences. By integrating ACTA with personalized learning path algorithms, educators can develop curricula that cater to diverse learners, thereby enhancing engagement, knowledge mastery, and academic success.

- **Assessment Design:**

Understanding the cognitive demands of various tasks through ACTA can help in creating assessments that accurately measure students' knowledge and skills. This approach ensures that assessments are fair, reliable, and valid indicators of student performance.

- **Cognitive Validity in Assessments:**

Research by Field (2013) on Cambridge ESOL Listening tests explores cognitive validity, assessing how well these tests reflect real-world listening processes. The research argues that assessments should reflect the cognitive processes students use in real-world tasks. By incorporating Applied Cognitive Task Analysis (ACTA) into assessment design can significantly improve its effectiveness by focusing on critical cognitive skills such as critical thinking, problem-solving, and the practical application of knowledge. This approach leads to more accurate and meaningful evaluations of students' abilities, ensuring that assessments not only test theoretical knowledge but also reflect the complex cognitive processes required in real-world scenarios.

- **Formative Assessment Techniques:**

The study by Ozan and Kıncal (2018) on formative assessment practices reveals that such assessments can significantly enhance students' academic achievement and attitudes towards the subject, although no significant difference in self-regulation skills was noted. By comparing an experimental group with formative assessments to a control group, the research shows the positive impact of formative assessments on learning outcomes and student perceptions. Applying Applied Cognitive Task Analysis (ACTA) to the design of formative assessments can further improve their effectiveness by ensuring that these assessments are closely aligned with cognitive demands and provide timely, constructive feedback. This alignment helps create formative assessments that are more effective in addressing specific learning needs and improving student performance.

- **Reducing Test Anxiety**:

A study by Xu, Tian, & Yuan (2024) focuses on understanding how cognitive reappraisal can moderate the relationship between neuroticism and test anxiety, showing that individuals with higher cognitive reappraisal abilities experience lower levels of test anxiety despite higher neuroticism. This research indirectly supports the application of cognitive task analysis (ACTA) in reducing test anxiety. By applying ACTA principles, educators can design assessments that are better aligned with students' cognitive and emotional needs, thereby reducing test anxiety and improving performance. ACTA helps by providing insights into cognitive load and task complexity, enabling the creation of assessments that minimize unnecessary cognitive strain and emotional stress.

By leveraging ACTA in education, curriculum development can be tailored to align with cognitive development stages, and assessments can be designed to accurately measure students' knowledge and skills. These approaches ensure that educational materials are appropriately challenging, engaging, and supportive of student learning, leading to improved educational outcomes.

12.2.3. Industry

- **Training Programs:**

Applied Cognitive Task Analysis (ACTA) can be instrumental in designing effective training programs that enhance the cognitive skills needed for various industrial tasks. By focusing on the cognitive demands of these tasks, ACTA helps in creating training that leads to improved productivity and safety.

- **Enhancing Industrial Training**:

The study by Idris et al. (2022) on the E-Training System for Professional Development (TaSPoD) highlights the practical application of an e-training system in higher learning institutions to support educators in acquiring professional knowledge, skills, and values through structured training programs. By employing Applied Cognitive Task Analysis (ACTA), the research aimed to identify the most challenging cognitive elements within the training design. Through expert interviews, the study found difficulties in system design and functionality, as well as course content design, leading to three key recommendations: simplifying the graphical user interface, incorporating communication tools, and enabling customized applications for external documents. This approach underscores the importance of ACTA in designing effective training programs. Similarly, in the context of industrial training, using cognitive task analysis to dissect and address the cognitive components of tasks can lead to more targeted and effective training. This ensures that training programs are aligned with the cognitive demands of complex tasks, ultimately enhancing performance and safety in industrial environments. Integrating ACTA insights can optimize training systems, ensuring they are well-suited to prepare individuals for their specific roles.

- **Simulation-Based Training:**

The study by Idris et al. (2022) on the E-Training System for Professional Development (TaSPoD) demonstrates the application of e-training systems in higher education to enhance educators' professional competencies. Using Applied Cognitive Task Analysis (ACTA), the research identified challenges in system design and course content, leading to recommendations for improving user engagement through a simplified graphical user interface, communication tools, and customized applications for external documents. This approach aligns with research on simulation-based training in industrial settings, which emphasizes the use of cognitive task analysis to design simulations that replicate real-world scenarios. By providing a safe environment for practice, such simulations help workers develop and refine cognitive skills, leading to better preparedness and reduced risk of errors. Integrating simulation-based training into ACTA can thus enhance the effectiveness of both educational and industrial training programs, ensuring that participants are well-equipped to handle their roles.

- **Improving Safety Training:**

The study on the cognitive task analysis (CTA) based learning review of a serious accident at a British Sugar manufacturing plant by Pemberton and Mooney (2024) closely aligns with research on improving safety training through cognitive task analysis. This study demonstrated that CTA could provide a deeper understanding of the cognitive factors contributing to accidents and offer valuable content for scenario-based training exercises. These exercises effectively developed the tacit cognitive skills necessary for preventing similar incidents. By focusing on the cognitive aspects of safety procedures, the findings from this study suggest that training programs can be designed to enhance hazard recognition and decision-making skills. Similarly, ACTA focuses on capturing the cognitive demands and decision-making processes required in high-stakes environments. By emphasizing the cognitive aspects of safety procedures, ACTA could be used to design training programs that address hazard recognition and decision-making skills, just as Pemberton and Mooney's study did. ACTA's structured interviews and methods for eliciting expert knowledge would provide a practical framework for creating scenario-based training exercises that mirror real-world hazards, ensuring that workers develop the tacit skills necessary for preventing similar incidents. Thus, ACTA could enhance the safety training programs highlighted in the study by refining the cognitive content and helping trainees internalize critical decision-making strategies, further improving hazard recognition and operational safety.

- **Process Optimization:**

Analysing the cognitive aspects of industrial processes through ACTA can identify areas for improvement, leading to streamlined operations and reduced errors. This approach enhances efficiency and productivity while minimizing risks.

- **Streamlining Operations:**

Research by Arrabito and Banbury (2019) illustrates how CTA can uncover the cognitive processes involved in remote operations, which can inform the design of human-machine interfaces to improve collaboration and decision-making. By focusing on critical situation awareness elements required for effective decision-making in complex, time-sensitive UAS missions, the study aligns with the idea of using CTA to streamline operations. Understanding the cognitive demands of various tasks allows for the identification of process improvements that reduce cognitive load and enhance efficiency. The study by Arrabito and Banbury (2019) can be directly related to Applied Cognitive Task Analysis (ACTA), as both CTA and ACTA are designed to uncover and address the cognitive demands of complex tasks to improve decision-making and operational efficiency. In the context of remote operations for Unmanned Aircraft Systems (UAS), CTA focuses on understanding the cognitive processes involved, such as situation awareness and decision-making under pressure. ACTA shares this focus but offers a more streamlined approach, making it easier for practitioners to extract and apply cognitive insights in real-world settings. By identifying the critical situation awareness elements and cognitive load in UAS missions, Arrabito and Banbury's research aligns with ACTA's objective of improving human-machine interaction and collaboration. ACTA, with its structured interviews and focus on extracting both explicit and tacit knowledge, would similarly allow for a detailed understanding of how operators handle complex, time-sensitive tasks. This insight would then inform the design of more effective human-machine interfaces, training programs, and decision-support systems that are tailored to reduce cognitive load and improve efficiency, as highlighted in the study. Thus, ACTA can be seen as a practical extension of CTA in this context, providing a flexible, applied methodology to enhance UAS operations by optimizing cognitive demands, improving decision-making, and ensuring more effective collaboration between humans and machines.

- **Reducing Human Error:**

The 2013 report by the Office of Research and Development, U.S. Department of Transportation, can be directly related to Applied Cognitive Task Analysis (ACTA), as both CTA and ACTA share a focus on understanding cognitive demands to inform system design and improve human performance. In the context of railroad operations, ACTA would similarly play a key role by identifying the critical cognitive skills and challenges that workers face as they interact with evolving technologies such as positive train control and energy management systems. While the report highlights how CTA uncovers cognitive bottlenecks and anticipates performance issues to design better user interfaces and training programs, ACTA, as a streamlined and more applied version of CTA, would make this process more accessible to practitioners. ACTA's structured interviews and focus on both explicit and tacit knowledge would help extract crucial cognitive insights, enabling the development of interventions that address real-world operational challenges. By applying ACTA, human systems integration (HSI) efforts can be enhanced by ensuring that new technologies are not only optimized for operational efficiency but also aligned with the cognitive capacities of workers, reducing human error and enhancing safety in high-demand situations. Thus, ACTA would complement the work done by CTA in railroad operations, offering a more practical and flexible method to incorporate cognitive insights into the HSI process, ensuring that technological advancements are fully integrated into human performance requirements.

- **Optimizing Workflow:**

The study by Knisely et al. (2021) on Cognitive Task Analysis and Workload Classification (CTAWC) highlights an advanced methodology for decomposing cognitive tasks to identify and manage cognitive workload more effectively. By incorporating standard terminology from existing taxonomies and providing a structured framework for task decomposition, CTAWC offers a refined approach to analysing cognitive demands in automated systems. This enhanced method is particularly relevant for optimizing workflows, as it allows for precise identification of cognitive bottlenecks and sources of workload, which can then be addressed to improve task performance and efficiency. Integrating CTAWC into Applied Cognitive Task Analysis (ACTA) can significantly benefit workflow design by aligning processes with operators' cognitive capabilities, thereby enhancing productivity and reducing cognitive strain. This approach ensures that both the complexity of tasks and the cognitive resources required are well-managed, leading to more effective and user-friendly systems.

By leveraging ACTA in the industry, training programs can be designed to enhance cognitive skills, and processes can be optimized to reduce errors and improve efficiency. These approaches lead to better-prepared workers, safer environments, and streamlined operations, ultimately enhancing overall productivity and safety in industrial settings.

12.2.4 Military and Défense

- **Mission Planning:**

Applied Cognitive Task Analysis (ACTA) can significantly enhance mission planning protocols by considering the cognitive demands placed on military personnel. This approach improves decision-making processes and operational efficiency by addressing the cognitive complexities involved in planning and executing missions.

- **Cognitive Demands in Mission Planning:**

The study by Ste-Croix and Angel (2021) on cognitive task analysis (CTA) in dismounted infantry soldiers can be closely related to Applied Cognitive Task Analysis (ACTA), as both methodologies aim to dissect the cognitive demands involved in complex, high-stakes operations. In this study, CTA was used to analyse cognitive load and information flow in a battle position scenario, which aligns with ACTA's focus on uncovering the mental processes and decision-making strategies required in demanding environments like military operations. ACTA, with its structured interviews and ability to extract both explicit and tacit knowledge, could extend the findings of this study by providing additional insights into the cognitive strategies soldiers use during engagements. While the study used tools like the NASA Task Load Index to quantify cognitive load, ACTA would offer a more qualitative approach to understanding how soldiers manage situational awareness and high workload levels. These insights could be directly applied to enhance mission planning, training programs, and technological solutions like improved radios and UAVs by tailoring them to the actual cognitive challenges faced in real-world military scenarios. By integrating ACTA into military operations, the insights from studies like Ste-Croix and Angel (2021) can be further refined, optimizing workflows, reducing cognitive load, and enhancing decision-making processes in the field.

- **Decision-Making Support:**

The study by Ste-Croix and Angel (2021) examines the integration of Cognitive Task Analysis (CTA) into the design of interfaces for remote Unmanned Aircraft System (UAS) operations. The research highlights how CTA was employed to uncover the cognitive processes and information requirements of UAS operators and intelligence analysts working in geographically separated teams. By using a hybrid analysis technique, including mission function task analysis and goal-directed task analysis, the study identified key cognitive demands and situational awareness needs essential for effective decision-making in complex, time-sensitive missions. This application of CTA not only informs the design of human-machine interfaces but also supports the development of decision-support tools that enhance situational awareness and reduce cognitive overload. Integrating such findings into the ACTA framework ensures that operational tools and interfaces are better aligned with the cognitive capabilities of users, thereby improving decision-making accuracy and overall mission effectiveness.

- **Improving Operational Efficiency:**

The study by Harris-Thompson, Wiggins, and Ho (2006), which used Cognitive Task Analysis (CTA) to explore decision-making and situational awareness in Military Operations in Urban Terrain (MOUT), can be directly related to Applied Cognitive Task Analysis (ACTA). Both methodologies aim to capture the critical cognitive demands necessary for operational success, particularly in high-stakes environments like urban combat. In this study, CTA was used to identify the perceptual cues that house-clearing teams rely on, such as threat and environmental assessments, to make rapid and accurate decisions. ACTA would take a similar approach by delving into the cognitive processes that underlie these decisions, capturing both explicit knowledge (such as standard operating procedures) and tacit knowledge (like intuitive decision-making based on experience). By employing ACTA's structured interview methods, the study's findings could be further refined to identify the mental strategies that soldiers use in real-time, helping to enhance training simulations by making them more cognitively authentic. ACTA would allow for a deeper exploration of how teams coordinate actions and manage situational awareness in urban combat scenarios, providing practical insights to improve both training programs and operational efficiency. By aligning training with the cognitive demands of real-world operations, ACTA ensures that soldiers are better prepared for the complexities of urban warfare, reinforcing the value of incorporating both decision-making strategies and standard procedures into simulations.

- **Simulation Training:**

Integrating ACTA with simulation technologies can create realistic training scenarios that improve cognitive skills and preparedness for military personnel. Simulation training, informed by cognitive task analysis, helps in developing the skills necessary for complex and high-pressure environments.

- **Enhancing Realistic Training Scenarios:**

The study by Blacker et al. (2019) on cognitive training for military personnel highlights the untapped potential of such interventions to enhance operational readiness and performance by improving critical cognitive skills. This approach is closely related to the concept of enhancing realistic training scenarios through simulation-based training. The research demonstrates that by using cognitive task analysis (CTA) methods, training programs can be designed to replicate real-world scenarios, allowing military personnel to practice and refine their cognitive skills in a controlled environment. Specifically, the integration of Applied Cognitive Task Analysis (ACTA) into these simulations can tailor the training to address specific cognitive challenges faced in military operations. By focusing on targeted, outcome-based approaches and utilizing near transfer effects, ACTA-informed simulations can enhance training effectiveness and preparedness, ensuring that service members develop the necessary cognitive skills to perform optimally in various military-relevant scenarios.

- **Improving Cognitive Skills**:

The study by Blacker et al. (2019) explores the potential of cognitive training interventions to enhance the performance of military personnel, emphasizing the importance of improving cognitive skills for operational readiness and performance. The review highlights the application of cognitive training in various domains, including video games, working memory, and executive function training. This research underscores the need for targeted, outcome-based approaches to cognitive training that can be effectively applied in military settings. Connecting this with Applied Cognitive Task Analysis (ACTA), the study's focus on cognitive training aligns with ACTA's principles by demonstrating how understanding and targeting specific cognitive skills, such as decision-making and situational awareness, can enhance training effectiveness. ACTA provides a structured methodology for dissecting and analysing these cognitive processes, enabling the design of simulation-based training that mirrors real-world scenarios. By integrating insights from ACTA into cognitive training programs, military training can be tailored to address specific cognitive challenges, thereby improving cognitive preparedness and operational readiness. This approach ensures that training interventions are not only theoretically sound but also practically effective in enhancing the cognitive capabilities of military personnel.

- **Optimizing Training Effectiveness:**

The study by Ste-Croix and Angel (2021) on cognitive task analysis (CTA) for dismounted infantry soldiers is closely related to optimizing training effectiveness by leveraging detailed insights into cognitive demands to enhance training experiences. By employing a modified CTA approach, the researchers investigated the cognitive load and critical information requirements faced by soldiers in complex battle scenarios. This detailed understanding of perceptual and cognitive challenges is crucial for designing simulation-based training that mirrors real-world conditions. The study highlights that soldiers face substantial cognitive workload, particularly in maintaining situational awareness and during high-stress engagements. Integrating these insights into simulation training ensures that training scenarios accurately reflect the cognitive demands encountered in actual operations. This approach aligns with the principles of Applied Cognitive Task Analysis (ACTA), which emphasizes designing training and systems based on a thorough understanding of cognitive processes. By applying these principles, the study demonstrates how to create more relevant and challenging training environments, thereby enhancing military personnel's ability to manage complex and dynamic scenarios and improving overall mission performance.

By leveraging ACTA in military and defence settings, mission planning protocols can be designed to align with cognitive demands, and simulation training can be enhanced to improve cognitive skills and preparedness. These approaches lead to more effective decision-making, operational efficiency, and readiness for complex missions, ultimately enhancing military effectiveness and success.

12.2.5. Technology Development

- **Software Design:**

Applied Cognitive Task Analysis (ACTA) can play a pivotal role in informing the development of software that aligns with users' cognitive processes. By understanding how users interact with technology, ACTA helps create software that enhances usability and reduces cognitive load.

- **Improving Usability Through Cognitive Alignment:**

The study by Knisely et al. (2021) on Cognitive Task Analysis and Workload Classification (CTAWC) advances traditional Cognitive Task Analysis (CTA) by offering a more nuanced methodology for evaluating cognitive workload. CTAWC enhances the depth of analysis by integrating standardized terminology and frameworks, allowing for a precise classification of cognitive demands. This detailed approach builds on the principles of Applied Cognitive Task Analysis (ACTA), which emphasizes understanding and modelling cognitive processes to improve task performance and system design. By providing a structured model to evaluate task complexity and cognitive depth, CTAWC aligns with ACTA's goal of tailoring systems and interfaces to users' cognitive capabilities. This alignment ensures that systems and software are designed to better match users' cognitive models, thereby enhancing usability and reducing cognitive effort. The connection between CTAWC and ACTA underscores the importance of employing advanced CTA methodologies to refine user interactions and optimize system performance, ultimately leading to more effective and user-friendly designs.

- **Reducing Cognitive Load:**

The study by Papautsky et al. (2015) demonstrates the effective application of Cognitive Task Analysis (CTA) in designing unmanned aerial systems (UAS) control interfaces. By combining CTA with design thinking, the researchers developed interfaces for an autonomous helicopter that were specifically tailored to the cognitive demands and operational needs of UAS pilots. Through detailed interviews with subject-matter experts, CTA helped identify critical cognitive tasks and requirements, which informed the development of user-friendly human-system interface requirements. This integration of CTA ensured that the design minimized cognitive load by aligning interface elements with the pilots' cognitive capabilities and operational contexts. Applied Cognitive Task Analysis (ACTA) extends this approach by not only understanding cognitive tasks but also focusing on applying these insights to real-world training and system design. ACTA enhances the practical application of CTA by ensuring that the developed systems and training interventions address specific cognitive challenges and improve overall usability and performance in complex environments.

- **Enhancing User Experience**:

Papautsky et al. (2015) explore the integration of Cognitive Task Analysis (CTA) and design thinking to enhance the development of user interfaces for unmanned aerial systems (UAS). Their study highlights how CTA, through detailed interviews and analysis with subject-matter experts, informs the creation of human-system interface requirements and design concepts. This approach ensures that the systems meet the cognitive demands and operational needs of UAS operators. Applied Cognitive Task Analysis (ACTA), an extension of CTA, builds upon this foundation by focusing on applying these insights in real-world settings. While CTA provides a deep understanding of cognitive tasks and challenges, ACTA emphasizes the practical application of these insights to improve system performance and user experience. By integrating ACTA, the design process not only identifies cognitive requirements but also tailors training and system features to address specific cognitive challenges, further enhancing the effectiveness and usability of the system.

CONCLUSION

The interdisciplinary contributions to Applied Cognitive Task Analysis (ACTA) underscore its versatility and transformative potential across a wide array of fields. By integrating insights from cognitive science, educational psychology, human factors engineering, and social sciences, ACTA stands out as a robust framework for optimizing task performance and enhancing learning and training programs. The fusion of methodologies from these diverse disciplines enriches ACTA's approach, allowing it to address complex cognitive challenges with greater precision and efficacy.

In practical applications, ACTA's ability to dissect and analyse cognitive processes facilitates the design of training programs and systems that are tailored to specific cognitive demands. This not only enhances the effectiveness of training interventions but also ensures that they are aligned with the real-world needs of users, whether in military settings, healthcare, education, or technology development. The synergy between ACTA and fields such as systems engineering, instructional design, and human-computer interaction further broadens its impact, driving innovations that improve usability, reduce cognitive overload, and advance overall performance.

Moreover, the expanding applications of ACTA—from developing more intuitive user interfaces to optimizing high-stakes simulations—highlight its role as a pivotal tool in tackling cognitive challenges in diverse contexts. As research continues to evolve, the integration of ACTA with emerging technologies and methodologies promises to deepen our understanding of cognitive processes and refine our approaches to enhancing human performance. Thus, ACTA's interdisciplinary nature not only enriches its methodologies but also solidifies its value as a cornerstone in addressing and solving complex cognitive issues across multiple sectors.

REFERENCES

Ahmad, K., Bhattacharyya, R., & Gupte, C. (2020). Using cognitive task analysis to train orthopaedic surgeons - Is it time to think differently? A systematic review. *Annals of Medicine and Surgery (London)*, 59, 131–137. DOI: 10.1016/j.amsu.2020.09.031

Alqurni, J. S. (2023). Evaluating the user interface and usability approaches for e-learning systems. *International Journal of Information Technology and Web Engineering*, 18(1), 1–25. DOI: 10.4018/IJITWE.333638

Arrabito, G. R., Hou, M., & Banbury, S. (2019). *Cognitive task analysis for an unmanned aircraft system* (DRDC-RDDC-2019-D041). Defence Research and Development Canada.

Bhatia, M. B., Collings, A. T., & Gillio, A.. (2022). Enhancing surgical education for resource-limited settings: Open appendectomy cognitive task analysis. *Global Surgical Education : Journal of the Association for Surgical Education*, 1, 61. DOI: 10.1007/s44186-022-00064-7

Blacker, K. J., Hamilton, J., Roush, G., Pettijohn, K., & Biggs, A. T. (2019). Cognitive training for military application: A review of the literature and practical guide. *Journal of Cognitive Enhancement: Towards the Integration of Theory and Practice*, 3(1), 31–47. DOI: 10.1007/s41465-018-0076-1

Blitz, M. J., & Barfield, W. (2023). Memory enhancement and brain–computer interface devices: Technological possibilities and constitutional challenges. In V. Dubljević & A. Coin (Eds.), *Policy, identity, and neurotechnology* (Advances in Neuroethics, pp. 1-22). Springer, Cham. DOI: 10.1007/978-3-031-26801-4_12

Bui, P. U., Tong, D. H., & Lien, N. B. (2022). The effectiveness of experiential learning in teaching arithmetic and geometry in sixth grade. *Frontiers in Education*, 7. Advance online publication. DOI: 10.3389/feduc.2022.858631

Castro-Alonso, J. C., de Koning, B. B., & Fiorella, L. (2021). Five strategies for optimizing instructional materials: Instructor- and learner-managed cognitive load. *Educational Psychology Review*, 33, 1379–1407. DOI: 10.1007/s10648-021-09606-9

Collins, R. (2019). Clinician cognitive overload and its implications for nurse leaders. *Nurse Leader*, 18(1). Advance online publication. DOI: 10.1016/j.mnl.2019.11.007

Dessy, E., Van Puyvelde, M., & Mairesse, O. (2018). Cognitive performance enhancement: Do biofeedback and neurofeedback work? *Journal of Cognitive Enhancement: Towards the Integration of Theory and Practice*, 2, 12–42. DOI: 10.1007/s41465-017-0039-y

Edwards, T. C., Coombs, A. W., Szyszka, B., Logishetty, K., & Cobb, J. P. (2021). Cognitive task analysis-based training in surgery: A meta-analysis. *BJS Open*, 5(6), zrab122. Advance online publication. DOI: 10.1093/bjsopen/zrab122

Fackler, J. C., Watts, C., Grome, A., Miller, T., Crandall, B., & Pronovost, P. (2009). Critical care physician cognitive task analysis: An exploratory study. *Critical Care*, 13(R33). Advance online publication. DOI: 10.1186/cc7740

Faludi, J., Hoffenson, S., Kwok, S. Y., Saidani, M., Hallstedt, S. I., Telenko, C., & Martinez, V. (2020). A research roadmap for sustainable design methods and tools. *Sustainability*, 12(81), 8174. DOI: 10.3390/su12198174

Field, J. (2013). Examining listening. In Geranpayeh, A., & Taylor, L. (Eds.), *Examining Listening* (1st ed., pp. 77–158). Cambridge University Press.

Graham, L. A., Gray, C., Wagner, T. H., Illarmo, S., Hawn, M. T., Wren, S. M., Iannuzzi, J., & Harris, A. H. S. (2023). Applying cognitive task analysis to health services research. *Health Services Research*, 58(2), 415–422. DOI: 10.1111/1475-6773.14106

Harris-Thompson, D., Wiggins, S. L., & Ho, G. (2006). *Using cognitive task analysis to develop scenario-based training for house-clearing teams* (Final report, Contract # W7711-057971/001/TOR). Klein Associates Division, Applied Research Associates, Inc.

Idris, A. Y., Razak, R. A., & Rahman, S. S. A. (2022). Investigating knowledge task difficulties in an e-training system for professional development design. *Docens Series in Education*, 2, 51–65.

Jiang, B., Li, X., Yang, S., Kong, Y., Cheng, W., Hao, C., & Lin, Q. (2022). Data-driven personalized learning path planning based on cognitive diagnostic assessments in MOOCs. *Applied Sciences (Basel, Switzerland)*, 12(8), 3982. DOI: 10.3390/app12083982

Jingmiao, Y. (2024). Visual communication design for mobile learning apps: User interface usability and learning engagement. *Comunicar: Media Education Research Journal*, 78(32), 234–248. DOI: 10.58262/V32I78.19

Kalakoski, V., Selinheimo, S., Valtonen, T., Turunen, J., Käpykangas, S., Ylisassi, H., Toivio, P., Järnefelt, H., Hannonen, H., & Paajanen, T. (2020). Effects of a cognitive ergonomics workplace intervention (CogErg) on cognitive strain and well-being: A cluster-randomized controlled trial. A study protocol. *BMC Psychology*, 8(1), 1. DOI: 10.1186/s40359-019-0349-1

Klepsch, M., & Seufert, T. (2020). Understanding instructional design effects by differentiated measurement of intrinsic, extraneous, and germane cognitive load. *Instructional Science*, 48, 45–77. DOI: 10.1007/s11251-020-09502-9

Knisely, B., Joyner, J. S., & Vaughn-Cooke, M. (2021). Cognitive task analysis and workload classification. *MethodsX*, 8, 101235. DOI: 10.1016/j.mex.2021.101235

Kovač, V. B., Nome, D. Ø., Jensen, A. R., & Skreland, L. Lj. (2023). *The why, what and how of deep learning: Critical analysis and additional concerns*. Education Inquiry., DOI: 10.1080/20004508.2023.2194502

Laurence, P. G., & Macedo, E. C. (2023). Cognitive strategies in matrix-reasoning tasks: State of the art. *Psychonomic Bulletin & Review*, 30, 147–159. DOI: 10.3758/s13423-022-02160-7

Martin, F., Chen, Y., & Moore, R. L. (2020). Systematic review of adaptive learning research designs, context, strategies, and technologies from 2009 to 2018. *Educational Technology Research and Development*, 68, 1903–1929. DOI: 10.1007/s11423-020-09793-2

Mayer, R. E. (2014). Cognitive theory of multimedia learning. In Mayer, R. E. (Ed.), *The Cambridge handbook of multimedia learning* (2nd ed., pp. 43–71). Cambridge University Press., DOI: 10.1017/CBO9781139547369.005

McGee, H. M., & Crowley-Koch, B. (2019). Using behavioral systems analysis to improve large scale change initiatives in autism service organizations. *Perspectives on Behavior Science*, 42(4), 931–954. DOI: 10.1007/s40614-019-00231-0

Militello, , L. GSalwei, , M. EReale, , CSushereba, , CSlagle, , J. MGaba, , DWeinger, , M. BRask, , JFaiman, , JAndreae, , MBurden, , A. RAnders, , S. (2023). Adapting cognitive task analysis methods for use in a large sample simulation study of high-risk healthcare events. *Journal of Cognitive Engineering and Decision Making*, *17*(4), 315-331. DOI: 10.1177/15553434231192283

Ozan, C., & Kıncal, R. Y. (2018). The effects of formative assessment on academic achievement, attitudes toward the lesson, and self-regulation skills. *Educational Sciences: Theory & Practice*, 18(1), 85–118. DOI: 10.12738/estp.2018.1.0216

Papautsky, E. L., Dominguez, C., Strouse, R., & Moon, B. (2015). Integration of cognitive task analysis and design thinking for autonomous helicopter displays. *Journal of Cognitive Engineering and Decision Making*, 9(4), 283–294. DOI: 10.1177/1555343415602624

Paz, F., Huaracha, A., Paz-Sifuentes, F.-A., Moquillaza, A., Lecaros, A., Falconi, F., Aguirre, J., & Tapia, A. (2023). Comparing two sets of usability heuristics in a user experience evaluation: A case study in an e-commerce mobile application. In *Design, User Experience, and Usability* (pp. xx-xx). Springer. DOI: 10.1007/978-3-031-35702-2_5

Pemberton, I., & Mooney, D. (2024). How cognitive task analysis (CTA) can improve learning from serious accidents. In Golightly, D., Balfe, N., & Charles, R. (Eds.), *Contemporary Ergonomics and Human Factors 2024*. CIEHF.

Skulmowski, A., & Xu, K. M. (2022). Understanding cognitive load in digital and online learning: A new perspective on extraneous cognitive load. *Educational Psychology Review*, 34, 171–196. DOI: 10.1007/s10648-021-09624-7

Ste-Croix, C., & Angel, H. (2021). *Cognitive task analysis of a complex battle position scenario*. Human Systems Incorporated.

Swaby, L., Shu, P., Hind, D., & Sutherland, K. (2022). The use of cognitive task analysis in clinical and health services research — A systematic review. *Pilot and Feasibility Studies*, 8, 57. DOI: 10.1186/s40814-022-01002-6

Volpe, J. A. (2013). *Using cognitive task analysis to inform issues in human systems integration in railroad operations* (Final report). U.S. Department of Transportation, Research and Special Programs Administration, National Transportation Systems Center.

Vu, T., Magis-Weinberg, L., & Jansen, B. R. J. (2022). Motivation-achievement cycles in learning: A literature review and research agenda. *Educational Psychology Review*, 34, 39–71. DOI: 10.1007/s10648-021-09616-7

Zajda, J. (2021). Constructivist learning theory and creating effective learning environments. In *Globalisation and education reforms* (Vol. 25, pp. 43–60). Springer. DOI: 10.1007/978-3-030-71575-5_3

Zakeri, Z., Arif, A., Omurtag, A., Breedon, P., & Khalid, A. (2023). Multimodal assessment of cognitive workload using neural, subjective and behavioural measures in smart factory settings. *Sensors (Basel)*, 23(21), 8926. DOI: 10.3390/s23218926

Zhao, L., Zhao, B., & Li, C. (2023). Alignment analysis of teaching–learning-assessment within the classroom: How teachers implement project-based learning under the curriculum standards. *Disciplinary and Interdisciplinary Science Education Research*, 5, 13. DOI: 10.1186/s43031-023-00078-1

Section 5
Challenges and Ethical Considerations

Chapter 13
Challenges in Applying ACTA

ABSTRACT

This chapter examines the methodological and practical challenges of applying Applied Cognitive Task Analysis (ACTA) across complex domains. It highlights the difficulties in capturing and modeling cognitive processes in high-pressure environments, such as aviation and healthcare, where expert decision-making is critical. Additionally, it discusses the need for scalable and adaptable ACTA methodologies to accommodate varying expertise levels, as well as the integration of emerging technologies like AI and VR. The chapter also offers solutions, including hybrid methodologies and interdisciplinary collaboration, to enhance ACTA's effectiveness in real-world applications and ensure actionable insights.

INTRODUCTION

As Applied Cognitive Task Analysis (ACTA) continues to evolve as a critical methodology for understanding cognitive processes and enhancing performance in complex environments, it also faces several challenges and ethical considerations. These challenges span methodological, practical, and technological domains, highlighting the need for refined approaches to ensure accuracy, adaptability, and relevance in diverse applications. Furthermore, as ACTA integrates with emerging technologies and expands across new domains, ethical concerns regarding data privacy, equity, and the potential misuse of cognitive insights become increasingly significant. Addressing these challenges and ethical implications is essential for

DOI: 10.4018/979-8-3693-7733-8.ch013

sustaining the integrity and efficacy of ACTA methodologies while ensuring they are applied responsibly and equitably.

This section explores the multifaceted obstacles and ethical dilemmas encountered in the application of ACTA, emphasizing the importance of methodological rigor, practical feasibility, and technological advancements. It also delves into strategies for mitigating these challenges, offering solutions to enhance the scalability, applicability, and ethical grounding of ACTA across various sectors.

13.1 Addressing Methodological and Practical Challenges

While Applied Cognitive Task Analysis (ACTA) has proven to be a valuable tool for understanding cognitive processes and improving performance in complex environments, its implementation is not without significant challenges. These challenges are both methodological, involving the way cognitive processes are captured, modelled, and analysed, and practical, concerning the translation of these insights into usable interventions, training programs, or system designs. In high-stakes domains such as healthcare, aviation, and military operations, the dynamic nature of tasks, time constraints, and variations in expertise present further complications to effectively applying ACTA.

This section will explore the key methodological and practical hurdles that practitioners face when employing ACTA, focusing on the difficulties in capturing and modelling cognitive processes, managing cognitive load, addressing resource limitations, and ensuring that findings are actionable. By examining these challenges and drawing on recent research, we can gain a deeper understanding of how to refine ACTA methodologies to meet the demands of increasingly complex environments.

13.1.1 Capturing and Modelling Cognitive Processes

Applying Applied Cognitive Task Analysis (ACTA) presents several methodological and practical challenges, complicating its implementation across various domains. One of the primary challenges is the complexity of accurately capturing and modelling cognitive processes, particularly in high-stakes environments where tasks involve intricate decision-making and rapid information processing. The multifaceted nature of these tasks can lead to difficulties in designing appropriate data collection methods, such as interviews and observations, that effectively uncover the deep-seated cognitive structures underlying expert performance.

For example, recent studies, like those by Penney et al. (2022), and research into military and emergency response environments, underscore the difficulty of isolating and mapping cognitive processes in dynamic, high-pressure settings. These environments often require both intuitive and formal decision-making processes, depending on situational factors, further complicating the application of ACTA. Understanding these challenges is crucial for refining ACTA methodologies to enhance their accuracy and effectiveness in complex, high-stakes domains.

For instance, interviews designed to elicit expert knowledge might not always capture the tacit knowledge that experts draw upon in real-time, fluid situations. The study by Bielka et al. (2024) highlights this challenge by evaluating how anaesthesiologists manage difficult airway scenarios, such as "cannot intubate, can ventilate" (CI) and "cannot intubate, cannot ventilate" (CICV). In high-stakes medical environments, experts often rely on tacit knowledge, which is knowledge developed through experience and intuition, which is difficult to articulate during interviews. The study found that in initial simulations, 83.3% of participants deviated from the DAS guidelines, illustrating the limitations of explicit knowledge under pressure. However, after undergoing simulation-based training, there was a marked improvement in adherence to guidelines, decision-making, and response times both immediately after training and six months later. This suggests that while interviews may not fully capture the instinctual, experience-driven responses that guide real-time decisions, simulation training provides a way to uncover and refine these tacit cognitive processes. By engaging experts in realistic, dynamic scenarios, methods like simulation training, as demonstrated by Bielka et al., can help bridge the gap between explicit guidelines and the implicit knowledge required for expert performance in critical situations. Similarly, Penney et al. (2022) found that in emergency response situations, the use of standard interviews failed to fully capture the nuances of intuitive decision-making under pressure, which is often embedded in experience and context-specific cues. These findings suggest the need for more refined techniques, such as cognitive walkthroughs or simulated environments, to gain deeper insights into cognitive processes.

Another critical issue is the difficulty in isolating and mapping cognitive processes in environments where decision-making is influenced by rapidly changing situational factors. Penney et al. (2022) addressed this challenge in their study on threat assessment and decision-making across military, police, fire, and ambulance services. The study conducted a systematic review of English-language research published between 2000 and 2020, examining how personnel in these dynamic fields make critical decisions under pressure. It revealed that individuals across all four services use a blend of intuitive and formal decision-making processes, depending on situational complexity and individual factors. This duality of decision-making, where professionals switch between long-term planning and split-second choices, reflects the fluid nature of high-stakes environments like military operations or disaster response.

Penney et al. (2022) emphasized that isolating and capturing these cognitive processes is particularly challenging in real-world contexts, as cognitive task analysis (CTA) methods like Applied Cognitive Task Analysis (ACTA) are often tasked with mapping both long-term and immediate decision-making. The study illustrates that cognitive processes in such settings are not static; they constantly shift based on task complexity, time pressures, and emotional stressors. This complexity complicates the use of ACTA in these environments, as it requires simultaneously assessing long-term strategic decisions and the rapid, intuitive decisions made in moments of crisis. The findings by Penney et al. suggest that limiting decision-making to a single process or failing to train for this cognitive duality can restrict optimal performance, further underscoring the importance of adaptable decision-making models in emergency response and military contexts.

13.1.2 Translating Cognitive Models into Practical Applications

Another significant challenge is the practical application of ACTA findings. Translating detailed cognitive models into actionable interventions, training programs, or system designs requires interdisciplinary collaboration and a deep understanding of the target domain. For instance, recent research by Yang et al. (2021) addresses the pressing need for improved pilot training amidst the anticipated growth in the aviation industry and the current impact of Covid-19. It identifies key challenges, including the reliance on the personal experience of flight instructors for assessing pilot performance. The study proposes a conceptual framework that integrates machine learning (ML) techniques into simulator-based training. This framework aims to enhance the quality and efficiency of training by analysing large datasets from flight simulators to provide objective performance evaluations and personalized feedback. The use of ML can uncover latent patterns and optimize training outcomes by tailoring feedback to individual trainees, thus addressing the limitations of current training schemes. This study underscores the challenges of applying Applied Cognitive Task Analysis (ACTA) in the aviation industry, particularly in integrating cognitive insights into existing training frameworks. The reliance on traditional methods of performance assessment and the need for more objective evaluations highlight the difficulties in transferring ACTA findings into practical training programs. By incorporating ML techniques, the proposed framework aims to refine these methods, making it easier to align ACTA-derived cognitive insights with the evolving needs of pilot training. This integration demonstrates a step toward addressing the complexities of applying ACTA in dynamic environments like aviation, where accurate and effective training is crucial.

Similarly, in healthcare, integrating ACTA into clinical decision support systems often requires substantial adjustments to accommodate the fast-paced and dynamic nature of medical environments. Chen et al. (2023) reviews the current state of Clinical Decision Support Systems (CDSSs), which are designed to enhance patient care by providing evidence-based recommendations to clinicians. It examines the development, implementation, benefits, limitations, and future directions of CDSSs. The review highlights several challenges in adopting and optimizing these systems, including issues related to data privacy, system integration, and clinician acceptance. Despite their potential to improve clinical outcomes and efficiency, the effective implementation of CDSSs remains a complex task. The study's insights into the challenges of implementing CDSSs in healthcare are closely related to the application of Applied Cognitive Task Analysis (ACTA). Integrating ACTA into CDSSs involves significant adjustments to fit the dynamic and fast-paced nature of medical environments. ACTA can help in designing CDSSs that better align with clinicians' cognitive processes and decision-making needs by providing detailed analyses of task demands and information processing requirements. However, the challenges identified in the study, such as system integration and clinician acceptance, reflect the broader difficulties in applying ACTA to ensure that these systems effectively support clinical decision-making in real-world settings.

13.1.3 Scalability in Diverse Teams and Organizations

Scalability is another challenge when applying ACTA to large and diverse teams or organizations. Research by Ben Charif et al. (2022) emphasizes the need for improved methods and quality in scalability assessment tools to better support the expansion of healthcare innovations. The study's findings on scalability challenges are relevant to Applied Cognitive Task Analysis (ACTA), particularly when applied to large and diverse teams or organizations. Just as scalability tools struggle with variability and insufficient validation, ACTA faces difficulties in standardizing cognitive models across a broad spectrum of expertise levels. In healthcare settings, where professionals vary from novices to experts, replicating ACTA insights across different skill levels can be challenging. The study underscores the need for adaptable training frameworks that can account for this variability while preserving the accuracy of cognitive insights. Addressing these scalability issues is crucial for implementing effective ACTA-based training interventions in complex and heterogeneous environments like healthcare.

Furthermore, the variability in expertise levels presents unique challenges in standardizing ACTA applications. Each level of expertise might require different cognitive models or training frameworks, which means that ACTA tools and methods must be highly adaptable. For example, novices may benefit from more structured and detailed cognitive task analyses that break down tasks into more manageable parts, while experts may require less detailed analyses that focus on complex decision-making processes.

The study by Taylor et al. (2023) underscores the importance of CTA in understanding the cognitive processes in high-performance sport coaching, particularly in adapting to varying levels of athlete expertise. The analysis revealed that coaches use knowledge flexibly, projecting the needs of their athletes while handling day-to-day challenges. This directly relates to the challenges in ACTA, where tools must be tailored for different expertise levels. Novice coaches benefit from structured guidance, while experts require more nuanced cognitive strategies for decision-making in dynamic environments.

13.1.4 Integration with Emerging Technologies

Moreover, the integration of ACTA with emerging technologies, such as artificial intelligence (AI) and virtual reality (VR), introduces new complexities. While these technologies hold great potential for enhancing cognitive task analysis, they also require a deep understanding of both the technological tools and the cognitive processes they aim to support. Recent studies, such as those by Wu et al. (2024), underscore the challenge of ensuring that AI-driven tools accurately reflect the nuances of human cognitive processes, rather than oversimplifying them or introducing biases that could distort the analysis. The study by Wu et al. provides a comprehensive taxonomy of AI-enhanced VR applications in medical settings, illustrating the need for precise and nuanced integration of these technologies to avoid potential distortions and to ensure they truly enhance cognitive task analysis.

Additionally, the use of Virtual Reality (VR) in training presents unique challenges in maintaining the fidelity of cognitive simulations, which is crucial for ensuring that skills developed in virtual environments effectively translate to real-world scenarios. The study by Wu et al. (2024) offers a comprehensive examination of AI-enhanced VR applications in medical settings, introducing a systematic taxonomy that classifies techniques and applications based on different phases of medical diagnosis and treatment. This detailed analysis underscores the importance of accurately reflecting real-world conditions within VR systems to ensure that the cognitive simulations maintain high fidelity. Such fidelity is essential for making sure that the skills and knowledge acquired in VR training are applicable and effective in actual clinical settings. Therefore, the study highlights the need for rigorous development and evaluation of VR systems to support accurate and effective cognitive task analysis.

13.1.5 Ensuring Actionable Findings

One of the critical challenges in applying Applied Cognitive Task Analysis (ACTA) is ensuring that its findings are not only accurate but also actionable in real-world environments. Industries such as healthcare, aviation, and military operations often operate under significant time and resource constraints, which can hinder the full implementation of ACTA recommendations. These environments demand not only rapid decision-making but also the flexibility to adapt insights into everyday workflows, making it difficult to translate cognitive models into practical, operational interventions.

This challenge is compounded by the dynamic nature of many industries. As operational contexts and cognitive demands evolve, ACTA-based interventions must continuously adapt to remain relevant. Research by Petersson et al. (2022) highlights this issue in the healthcare sector, where strategic capacity and external conditions like regulatory requirements limit the ability to implement cognitive insights fully. Similar challenges exist in other high-stakes environments, where the mismatch between theoretical recommendations and the fast-paced reality of the workplace can lead to the underutilization of ACTA findings.

Further complicating this issue is the need for ongoing validation of ACTA recommendations. Once an intervention is designed based on cognitive task analysis, it must be monitored and adapted over time to ensure its effectiveness. The static nature of traditional cognitive models poses a significant challenge when applied in fluid environments, such as emergency response or critical care, where conditions are continually shifting. Without iterative validation and adaptation, ACTA findings risk becoming obsolete or misaligned with the evolving cognitive demands of the tasks being performed. As Riley et al. (2024) emphasize, ongoing external validation is crucial for maintaining the predictive accuracy and practical utility of models, particularly when applied across different target populations and evolving settings.

In addition to these challenges, ensuring that ACTA findings are practical and usable often requires significant interdisciplinary collaboration. Cognitive scientists, instructional designers, domain experts, and organizational leaders must work together to bridge the gap between theoretical insights and real-world application. However, as Li et al. (2024) highlight in their analysis of interdisciplinary collaboration during COVID-19 research, fostering such collaboration can be challenging, with notable gaps in integrating certain fields. In fast-paced industries, similar challenges arise, making it difficult to promote the cross-functional collaboration necessary for effectively implementing ACTA in training programs and system designs. This lack of collaboration can prevent the full potential of ACTA from being realized, especially in environments where cognitive overload and decision-making under pressure are routine.

Moreover, the introduction of emerging technologies such as AI and VR can further complicate the actionability of ACTA insights. While these technologies have the potential to enhance cognitive analysis and training, they also introduce new layers of complexity, such as the risk of oversimplifying human cognitive processes or introducing biases into AI-driven systems. For instance, as Hung et al. (2020) demonstrate, VR-based cognitive training offers immersive and interactive experiences that can better engage individuals and support real-life skill generalization. However, despite these benefits, the implementation of VR still requires careful consideration to avoid external distractions and ensure consistent effectiveness across cognitive domains. These technological complexities require ACTA practitioners to continuously adapt their methodologies to ensure the accuracy and applicability of cognitive models in technologically advanced environments.

In summary, while ACTA provides valuable cognitive insights, the challenge lies in ensuring that these findings are both practical and sustainable within the constraints of real-world environments. The dynamic nature of industries such as healthcare, aviation, and defence, combined with time and resource limitations, presents significant obstacles to translating ACTA recommendations into actionable outcomes. Overcoming these challenges requires a focus on continuous validation, adaptability, and cross-disciplinary collaboration to ensure that ACTA remains a valuable tool for improving cognitive performance and decision-making across various fields.

13.2 Solutions for Common Issues Encountered in ACTA Projects

13.2.1 Hybrid Methodologies for Enhanced Task Analysis

To address the methodological and practical challenges in implementing Applied Cognitive Task Analysis (ACTA), several solutions have been proposed and refined through recent research and practice. One effective approach is the development of hybrid methodologies that integrate ACTA with other task analysis techniques. For instance, Papautsky et al. (2020) combines cognitive task analysis with participatory design to create task flows that better reflect real-world conditions, such as uncertainty, time pressure, and high-risk environments. This hybrid method addresses some of the limitations of traditional task flows by accounting for both cognitive and behavioural aspects of work, making it more adaptable and flexible in complex work domains. This approach aligns with the concept of integrating ACTA with other methodologies to address the challenges in capturing the dynamic and complex nature of tasks in real-world settings. By combining ACTA with techniques like participatory design, the study provides a more comprehensive and practical framework for developing effective task flows, especially in high-stakes environments like the one involving a submarine watch team's use of an updated imaging system.

Similarly, a study from Gore et al. (2018) explores the integration of ACTA with ethnographic methods to enhance the depth of task analysis in complex settings. This approach allows for a more nuanced understanding of the cognitive and social aspects of professional tasks, as it combines the detailed, task-focused insights from ACTA with the contextual richness provided by ethnographic observation. By capturing both the explicit and tacit knowledge of experts, this hybrid methodology offers a more holistic view of task performance and cognitive processes.

The study by Papautsky et al. (2020) and Gore et al. (2018) underscore the importance of refining task analysis techniques to address real-world complexities. For instance, the research on engineering professionals in the petroleum additives industry demonstrates how ACTA can be adapted to document and accelerate cognitive expertise by incorporating socio-cognitive insights and context-specific knowledge. This approach highlights undocumented expertise and provides a basis for further development and application of ACTA in various domains. By integrating ACTA with participatory design and ethnographic methods, researchers can overcome some of the limitations of traditional task analysis. These hybrid approaches not only enhance the validity and applicability of ACTA but also support the creation of more accurate and actionable task analyses that are better suited to the dynamic and multifaceted nature of real-world tasks.

13.2.2 Leveraging AI and Machine Learning for Data Analysis

Advancements in technology have also played a significant role in overcoming some of the traditional limitations of ACTA. The integration of artificial intelligence (AI) and machine learning into ACTA processes offers new possibilities for automating certain aspects of data collection and analysis. For instance, AI-driven tools can now assist in the identification of cognitive patterns and decision-making processes, making it easier to process large datasets and draw actionable insights more efficiently. A relevant study by Almubark et al. (2020) exemplifies these advancements by investigating and comparing the classification performance of machine learning algorithms using behavioural data from neuropsychological tests and cognitive tasks. The study utilized a multilayer perceptron (MLP) network and several traditional machine learning algorithms to analyse data from individuals with mild cognitive impairment (MCI), mild Alzheimer's disease (AD), and control groups. The findings demonstrated that MLP outperformed traditional algorithms in classifying cognitive data, achieving approximately 90% sensitivity and specificity when combining neuropsychological and cognitive task data.

This study highlights how machine learning can enhance the analysis of cognitive data by providing more accurate classifications and insights into cognitive impairments. The use of deep learning models like MLP illustrates the potential for AI to improve the sensitivity and specificity of diagnostic tools, which aligns with the broader goal of integrating AI into ACTA processes. By automating and refining data analysis, AI technologies can help address some of the traditional limitations of ACTA, such as the need for extensive manual data processing and the challenge of capturing complex cognitive patterns.

13.2.3 Fostering Interdisciplinary Collaboration

Another critical solution is fostering closer interdisciplinary collaboration between cognitive scientists, domain experts, and instructional designers. This collaboration ensures that the cognitive insights derived from ACTA are effectively translated into practical applications, such as training programs or system designs. A study by Tjiam et al. (2012) illustrates this approach by focusing on the integration of cognitive task analysis with instructional design in the development of simulator-based training programs. The study emphasizes that while many studies of simulator-based surgical skills training have concentrated on psychomotor skill acquisition, surgical procedures require both psychomotor and cognitive skills. To address this, the authors advocate for the collaboration of educational experts and physicians in designing effective training programs.

Tjiam et al. (2012) describe a theory-driven approach to designing simulator-based skill training that combines cognitive task analysis with the four-component instructional design model. This integration helps to ensure that the training program not only addresses the psychomotor aspects but also incorporates the cognitive processes necessary for expert performance. The study highlights how such an approach was applied to develop simulator-based training for the nephrostomy procedure, demonstrating the practical benefits of combining cognitive insights with instructional design principles.

This example underscores the importance of interdisciplinary collaboration in creating training programs that are both effective and grounded in a thorough understanding of cognitive processes. By working together, cognitive scientists, domain experts, and instructional designers can develop training solutions that are more comprehensive and aligned with real-world requirements. This collaborative approach ensures that cognitive insights from ACTA are effectively incorporated into practical applications, ultimately enhancing the effectiveness of training programs and system designs.

In addition, as manufacturing evolves towards smart and semi-automated systems, it is crucial to consider human factors to maximize the performance potential of experts and avoid undesirable side effects. A recent study by Morgenstern et al. (2024) provides valuable insights into this aspect by proposing a cognition-based framework that integrates several psychological concepts to understand cognitive processes in manufacturing environments. Morgenstern et al. (2024) emphasize the importance of describing human sub-tasks during machine operation to gain a deeper understanding of cognitive processes involved in managing complex production systems. Their research employs an observational eye tracking study combined with retrospective think-aloud interviews to analyse cognitive processes such as process monitoring, anomaly detection, diagnosis, problem-solving, and resolution. This approach highlights the role of situation awareness, individual expertise, and cognitive team processes in effective anomaly resolution.

This study is particularly relevant to the field of Applied Cognitive Task Analysis (ACTA), which focuses on understanding the cognitive processes underlying complex tasks. By integrating cognitive insights with system design, Morgenstern et al. (2024) demonstrate how ACTA can be applied to describe and support cognitive (team) processes in smart manufacturing. The framework proposed in the study reflects an interdisciplinary approach that aligns with the principles of ACTA, emphasizing the need for collaboration between cognitive scientists, domain experts, and system designers to develop effective and user-centred solutions.

The findings from Morgenstern et al. (2024) underscore the importance of incorporating cognitive insights into the design of production systems to enhance their effectiveness and adaptability. By using ACTA methods to analyse and describe cognitive processes, the study provides a foundation for developing more human-centred and efficient manufacturing technologies. Morgenstern et al. (2024) illustrate how ACTA can be applied to address the cognitive challenges in smart manufacturing environments. Their research highlights the value of interdisciplinary collaboration in translating cognitive insights into practical applications, ultimately contributing to the development of systems that better support human operators in complex production settings.

13.2.4 Continuous Feedback Loops and Iterative Refinement

A significant solution for improving the application of ACTA is the implementation of continuous feedback loops during task analysis and training development. While initial cognitive insights are valuable, they often require ongoing refinement, especially in dynamic environments where the cognitive demands of tasks may change over time. Research by IJzerman et al. (2023) emphasizes the importance of iterative validation to ensure that interventions remain relevant and effective. The study outlines a six-phase iterative approach for developing, testing, and refining an implementation strategy for a complex eHealth intervention, which aligns directly with the concept of iterative validation. By alternating between theory and practice and continuously refining the strategy based on real-world feedback, the approach demonstrates the importance of regularly assessing and updating interventions to address context-specific challenges and improve sustainability. The process of piloting the intervention (phase 4), identifying challenges, and then connecting them back to broader theory (phase 5) reflects the iterative validation method, ensuring that the intervention evolves to meet the practical needs of users while staying grounded in theory. This iterative refinement allows for continuous improvement and adjustment, ensuring that the intervention remains effective in its intended clinical context, which is crucial for long-term success.

In environments such as healthcare or aviation, where decision-making processes evolve alongside technological advancements, regular feedback can help ensure that ACTA-based interventions remain aligned with the latest operational needs and cognitive challenges. For instance, in a pilot training context, feedback from trainees and instructors could be used to adjust training modules and refine ACTA findings. This iterative approach allows for the dynamic updating of task analyses, ensuring that training interventions are not only reactive but also proactive in anticipating future changes in task requirements.

13.2.5 Adapting ACTA to Remote and Virtual Work Environments

With the increasing prevalence of remote work and virtual collaboration, especially in the wake of the COVID-19 pandemic, it is essential to explore how ACTA can be adapted for these emerging work contexts. In distributed teams, particularly in industries like software development, consulting, and education, understanding the cognitive demands of remote decision-making and collaboration presents unique challenges.

Recent research by Morrison-Smith and Ruiz (2020) highlights the cognitive complexities faced by virtual teams, including maintaining situational awareness, managing communication breakdowns, and addressing the cognitive load associated with using virtual tools. These challenges are compounded by factors like geographical distance, asynchronous communication, and diverse team configurations, all of which can lead to misalignments in decision-making and reduced efficiency.

Integrating ACTA with methods designed to capture remote work dynamics, such as real-time virtual observations and screen-recorded task analyses, could provide valuable insights. By analysing how virtual teams manage tasks and make decisions in distributed environments, ACTA can identify critical cognitive demands and inform the development of more effective virtual collaboration tools. Additionally, ACTA can guide the design of more intuitive virtual platforms that minimize cognitive load, enhance user efficiency, and facilitate smoother communication across remote teams. These tools can help mitigate the cognitive challenges associated with remote work, making collaboration more seamless and effective.

13.2.6 Leveraging Virtual Reality (VR) and Augmented Reality (AR) in ACTA

Emerging technologies like Virtual Reality (VR) and Augmented Reality (AR) can further augment the effectiveness of ACTA by providing immersive environments for simulating complex, real-world tasks. VR and AR can be used to recreate high-stakes scenarios, allowing cognitive scientists to observe and capture task performance in controlled, yet highly realistic settings. This approach can be especially valuable in industries such as healthcare, military training, and aviation, where direct observation in real-world environments may be costly or impractical.

A study by Mergen et al. (2024) demonstrates how VR simulations in healthcare training help uncover cognitive bottlenecks and decision-making blind spots that are often overlooked in traditional observational methods. By using VR to simulate high-pressure surgical procedures or emergency response scenarios, ACTA practitioners can gain deeper insights into how individuals respond to stress, cognitive overload, and time-sensitive decisions. This immersive environment allows for a more accurate analysis of decision-making processes, revealing critical cognitive steps and potential points of failure. These insights can then be used to tailor more effective training programs and interventions, helping learners develop the necessary skills to perform under pressure. Moreover, the repeatability of VR simulations provides ample opportunities for practitioners to refine their cognitive strategies and improve their performance in realistic, yet controlled, environments.

13.2.7 Scalability of ACTA Across Diverse Domains

The scalability of ACTA remains a challenge, particularly when applying its methods across diverse sectors or large, multi-site organizations. However, recent research suggests that leveraging automated analysis tools and cloud-based platforms can help scale ACTA more efficiently. Ben Charif et al. (2022) addressed the scalability of healthcare innovations and discuss the potential for automated platforms to support large-scale task analysis efforts. By adopting cloud-based solutions, ACTA findings can be more easily shared, updated, and refined across geographically dispersed teams and sites, reducing the administrative burden while ensuring consistency in data collection and analysis.

Additionally, standardizing the data collection process and developing sector-specific templates for cognitive task analysis can make ACTA more accessible and scalable across industries with varying operational needs. For instance, creating industry-tailored versions of ACTA for fields like finance, manufacturing, and education could help extend the methodology's reach and usability, allowing more sectors to benefit from the cognitive insights it provides.

CONCLUSION

In conclusion, Chapter 13 has explored the multifaceted challenges that emerge when applying Applied Cognitive Task Analysis (ACTA) in diverse and high-stakes environments. These challenges span methodological, practical, and technological domains, underscoring the complexity of accurately capturing and modelling cognitive processes, translating cognitive models into actionable interventions, and scaling ACTA across varied expertise levels and organizational contexts.

One of the primary difficulties lies in capturing and modelling cognitive processes in dynamic, high-pressure settings like healthcare and aviation, where tasks demand both intuitive and formal decision-making. The intricate nature of these environments complicates the design of data collection methods, making it challenging to uncover the deep cognitive structures underlying expert performance.

The practical application of ACTA findings is another significant challenge, as translating these insights into usable training programs, system designs, or interventions requires interdisciplinary collaboration and a deep understanding of the target domain. The chapter highlighted the complexities involved in scaling ACTA across diverse teams and industries with varying expertise levels, particularly in ensuring that cognitive models remain relevant and actionable across different operational contexts.

Additionally, the integration of emerging technologies such as AI, VR, and machine learning offers both opportunities and new complexities for ACTA. While these technologies enhance cognitive analysis by automating data collection and improving precision, they also introduce challenges related to maintaining the fidelity of cognitive simulations and ensuring that AI-driven tools accurately reflect human cognitive processes.

To address these challenges, several solutions have emerged, such as the development of hybrid methodologies that combine ACTA with other task analysis techniques like participatory design and ethnographic methods. This allows ACTA to capture both cognitive and behavioural aspects of tasks, making it more adaptable to real-world conditions. Leveraging AI and machine learning can further automate and enhance cognitive task analysis by uncovering patterns and optimizing performance evaluation processes.

Moreover, the implementation of continuous feedback loops and iterative refinement ensures that ACTA-based interventions remain relevant and effective over time, especially in dynamic environments where task demands evolve rapidly. The growing prevalence of remote and virtual work environments also necessitates adapting ACTA to capture the cognitive challenges unique to distributed teams and virtual collaboration.

Lastly, scalability remains a key challenge, but recent advancements in automated analysis tools and cloud-based platforms offer promising solutions for scaling ACTA across large, multi-site organizations. By standardizing data collection processes and developing sector-specific templates, ACTA can become more accessible and scalable across industries with varying needs.

In summary, while ACTA faces considerable challenges in its application, the ongoing development of hybrid methodologies, technological advancements, and interdisciplinary collaboration provides a pathway for overcoming these obstacles. These efforts will ensure that ACTA continues to evolve and remains a valuable tool for optimizing cognitive performance and decision-making in complex, dynamic environments across diverse sectors.

REFERENCES

Almubark, I., Chang, L. C., Shattuck, K. F., Nguyen, T., Turner, R. S., & Jiang, X. (2020). A 5-min cognitive task with deep learning accurately detects early Alzheimer's disease. *Frontiers in Aging Neuroscience*, 12, 603179. DOI: 10.3389/fnagi.2020.603179

Ben Charif, A., Zomahoun, H. T. V., Gogovor, A., Samri, M. A., Massougbodji, J., Wolfenden, L., Ploeg, J., Zwarenstein, M., Milat, A. J., Rheault, N., Ousseine, Y. M., Salerno, J., Markle-Reid, M., & Légaré, F. (2022). Tools for assessing the scalability of innovations in health: A systematic review. *Health Research Policy and Systems*, 20(1), 34. DOI: 10.1186/s12961-022-00830-5

Bielka, K., Kuchyn, I., & Fomina, H.. (2024). Difficult airway simulation-based training for anaesthesiologists: Efficacy and skills retention within six months. *BMC Anesthesiology*, 24, 44. DOI: 10.1186/s12871-024-02423-x

Chen, Z., Liang, N., Zhang, H., Li, H., Yang, Y., Zong, X., Chen, Y., Wang, Y., & Shi, N. (2023). Harnessing the power of clinical decision support systems: Challenges and opportunities. *Open Heart*, 10(1), e002432. DOI: 10.1136/openhrt-2023-002432

Gore, J., Banks, A. P., & McDowall, A. (2018). Developing cognitive task analysis and the importance of socio-cognitive competence/insight for professional practice. *Cognition Technology and Work*, 20(4), 555–563. DOI: 10.1007/s10111-018-0502-2

Hung, S. C.-w., Ho, A. Y.-n., Lai, I. H.-w., Lee, C. S.-w., Pong, A. S.-k., & Lai, F. H.-y. (2020). Meta-analysis on the effectiveness of virtual reality cognitive training (VRCT) and computer-based cognitive training (CBCT) for individuals with mild cognitive impairment (MCI). *Electronics (Basel)*, 9(12), 2185. DOI: 10.3390/electronics9122185

IJzerman, R. V., van der Vaart, R., & Breeman, L. D. (2023). An iterative approach to developing a multifaceted implementation strategy for a complex eHealth intervention within clinical practice. *BMC Health Services Research*, 23, 1455. DOI: 10.1186/s12913-023-10439-

Li, Y., Liu, S., Zeng, A., Wu, J., Zhang, J., Zhang, W., & Li, S. (2024). Interdisciplinary dynamics in COVID-19 research: Examining the role of computer science and collaboration patterns. *Systems*, 12(4), 113. DOI: 10.3390/systems12040113

Mergen, M., Graf, N., & Meyerheim, M. (2024). Reviewing the current state of virtual reality integration in medical education: A scoping review. *BMC Medical Education*, 24, 788. DOI: 10.1186/s12909-024-05777-5

Morgenstern, T., Klichowicz, A., Bengler, P., Todtermuschke, M., & Bocklisch, F. (2024). Towards a cognition-based framework describing interdisciplinary expert team processes for cognitive robotics in Industry 5.0 technologies. *Applied Sciences (Basel, Switzerland)*, 14(10), 4121. DOI: 10.3390/app14104121

Morrison-Smith, S., & Ruiz, J. (2020). Challenges and barriers in virtual teams: A literature review. *SN Applied Sciences*, 2, 1096. DOI: 10.1007/s42452-020-2801-5

Penney, G., Launder, D., Cuthbertson, J., & Thompson, M. B. (2022). Threat assessment, sense making, and critical decision-making in police, military, ambulance, and fire services. *Cognition Technology and Work*, 24(3), 423–439. DOI: 10.1007/s10111-022-00694-3

Petersson, L., Larsson, I., Nygren, J. M., Nilsen, P., Neher, M., Reed, J. E., Tyskbo, D., & Svedberg, P. (2022). Challenges to implementing artificial intelligence in healthcare: A qualitative interview study with healthcare leaders in Sweden. *BMC Health Services Research*, 22(1), 850. DOI: 10.1186/s12913-022-08215-8

Riley, R. D., Archer, L., Snell, K. I. E., Ensor, J., Dhiman, P., & Martin, G. P.. (2024). Evaluation of clinical prediction models (part 2): How to undertake an external validation study. *BMJ (Clinical Research Ed.)*, 384, e074820. DOI: 10.1136/bmj-2023-074820

Taylor, J., Ashford, M., & Jefferson, M. (2023). High-performance coach cognition in the wild: Using applied cognitive task analysis for practical insights—Cognitive challenges and curriculum knowledge. *Frontiers in Psychology*, 14, 1154168. Advance online publication. DOI: 10.3389/fpsyg.2023.1154168

Wu, Y., Hu, K., Chen, D. Z., & Wu, J. (2024). AI-enhanced virtual reality in medicine: A comprehensive survey (Version 3). *arXiv*. https://doi.org//arXiv.2402.03093DOI: 10.48550

Yang, S., Yu, K., Lammers, T., & Chen, F. (2021). Artificial intelligence in pilot training and education – Towards a machine learning aided instructor assistant for flight simulators. In Stephanidis, C., Antona, M., & Ntoa, S. (Eds.), *HCI International 2021 - Posters* (Vol. 1420, pp. 689–694). Springer., DOI: 10.1007/978-3-030-78642-7_78

Chapter 14
Ethical Considerations in ACTA

ABSTRACT

This chapter addresses the ethical considerations essential to Applied Cognitive Task Analysis (ACTA). Key topics include the importance of obtaining informed consent, minimizing harm during data collection, and maintaining confidentiality and data integrity, especially in sensitive environments like healthcare and military settings. It also examines the risks of dual-use dilemmas, where cognitive insights could be misused, and stresses the need for responsible dissemination of findings. By adopting ethical practices, including robust informed consent, distress protocols, and cybersecurity measures, ACTA researchers can ensure their work adheres to high ethical standards while fostering trust and safeguarding participant welfare.

INTRODUCTION

Applied Cognitive Task Analysis (ACTA) is a vital methodology used to understand the cognitive demands associated with complex tasks, often in high-stakes environments such as healthcare, military operations, aviation, and education. While the insights gained from ACTA can significantly improve performance, training, and decision-making, the methodology also brings forward several ethical challenges. These challenges span the entirety of the research process, from participant selection and data collection to analysis, dissemination, and the application of findings.

As ACTA focuses on uncovering deep cognitive processes, it often involves the collection of sensitive personal data from participants. Ensuring that these processes are conducted with full respect for participant autonomy, privacy, and well-being is critical to maintaining both the integrity of the research and the trust of the individuals involved. Moreover, in environments where ACTA findings can shape policy, operational strategies, or technological applications, researchers must

DOI: 10.4018/979-8-3693-7733-8.ch014

carefully consider the potential for unintended consequences and the dual-use dilemmas that can arise. The need for ethical foresight is especially pronounced as emerging technologies such as artificial intelligence (AI), virtual reality (VR), and augmented reality (AR) become more integrated into ACTA studies.

This chapter will address the key ethical considerations that researchers and practitioners must navigate when employing ACTA. These include ensuring informed consent, protecting confidentiality, managing dual-use dilemmas, and promoting ethical dissemination practices. It will also explore the challenges of applying ACTA in sensitive environments, such as healthcare and military operations, where the stakes are particularly high, and the ethical complexities are more pronounced.

Furthermore, this chapter will discuss the importance of maintaining transparency, avoiding bias, and fostering interdisciplinary collaboration to ensure that ACTA findings are not only accurate but also ethically sound and beneficial to all stakeholders. As ACTA continues to evolve alongside advancements in technology and methodology, understanding and addressing these ethical concerns will remain crucial to its responsible and effective application. Through this exploration of ethical considerations, we aim to provide a comprehensive guide for researchers to navigate the complexities of conducting ACTA in an ethically responsible manner.

14.1 Ensuring Ethical Practices in Data Collection and Analysis

Applied Cognitive Task Analysis (ACTA) is a powerful tool for understanding the cognitive processes underlying complex tasks. However, its effectiveness depends not only on the accuracy of the analysis but also on the ethical practices employed during data collection and analysis. Ensuring ethical practices in ACTA is critical to maintaining the integrity of the research and protecting the rights of participants.

One key ethical consideration in data collection is obtaining informed consent from all participants. Informed consent involves fully informing participants about the nature of the study, the methods used, and the potential risks and benefits associated with their involvement. Recent research emphasizes the importance of transparency and clear communication in obtaining informed consent, particularly in complex cognitive task analyses where the procedures may not be immediately understood by participants (Xu et al., 2020). Xu et al. (2020) also emphasize the importance of robust relationships between researchers and participants, which can lead to more informed and voluntary participation. Additionally, it underscores the role of Human Research Ethics Committees (HRECs) in guiding the consent process and ensuring that ethical standards are met, which is crucial in maintaining the integrity of research in fields like ACTA. In addition, Blease et al. (2018) also highlight the importance of transparency and comprehensive disclosure in the informed consent process, which is critical in both psychotherapy and research settings to ensure ethical practice. This can be applied to ACTA by emphasizing that participants should be provided with all relevant information, including potential risks, benefits, and the context of the research, to make informed decisions about their participation. Blease et al. (2017) support the idea that ethical informed consent goes beyond mere procedural compliance; it requires a deep commitment to transparency, evidence-based disclosure, and consideration of the individual characteristics and needs of participants. This can be applied to ACTA research to reinforce the importance of robust informed consent practices that fully respect the rights and autonomy of participants.

In addition to informed consent, researchers must also consider the potential for harm, including psychological stress or discomfort, that may arise during the data collection process. For example, in high-stakes environments where participants may be exposed to scenarios simulating real-world stressors, it is crucial to establish protocols that minimize the risk of distress and provide support resources if needed. Researchers should implement distress protocols to effectively manage and mitigate potential harm throughout the research process. These protocols are vital in ensuring that participants do not suffer any harm, risk, or disadvantage as a result of their involvement in a study. Harm can take various forms—emotional, physical, social, or psychological—and may occur at any stage, from data collection through to dissemination (UK Statistics Authority, 2022). To address these concerns, UK Statistics Authority (2022) outlines that distress protocols should be created prior to data collection and tailored to the specific risks associated with each study. Typically, two separate distress protocol documents are required: one for participants and one for researchers. The participant distress protocol should outline possible risks and the steps to be taken if a participant experiences distress. This includes immediate actions such as providing access to counselling services or the option to withdraw from the study without penalty. The researcher distress protocol, on the other hand, should ensure that researchers are equipped to handle distressing situations, both for themselves and participants, and that they are supported throughout the research process. By proactively establishing these protocols, researchers uphold ethical standards and ensure the well-being of all individuals involved in the study. This approach is especially critical in studies involving high-risk scenarios, where the potential for distress is significant.

When analysing data, maintaining objectivity and avoiding bias is essential. This includes being vigilant against confirmation bias, where researchers might unintentionally focus on data that supports their hypotheses while disregarding conflicting evidence. To combat this, recent studies suggest incorporating techniques such as triangulation, where data from multiple sources or methods are cross-validated to ensure a more balanced and accurate analysis (Noble & Heale, 2019). Triangulation is a well-established technique used to enhance the credibility and validity of research findings by integrating various data sources, methods, or perspectives. It helps researchers overcome biases that might arise from relying on a single method or viewpoint. By combining different approaches, triangulation offers a more comprehensive and nuanced understanding of the research problem (Noble & Heale, 2019).

Triangulation is a valuable technique for enhancing the credibility and validity of research findings by incorporating multiple data sources, methods, or perspectives. Data triangulation involves collecting data from different times, locations, or participant groups to ensure that findings are not confined to a single context or group. This approach helps confirm that the results are generalizable and not affected by the unique circumstances of any one setting. Investigator triangulation includes utilizing multiple researchers in the analysis process, which reduces individual biases and improves the reliability of the results. By having various researchers contribute to data interpretation, the risk of personal bias is minimized, leading to a more balanced and accurate understanding. Theory triangulation entails applying multiple theoretical frameworks to interpret the data. This method allows researchers to address potential limitations of any single perspective and gain a more comprehensive understanding of the phenomenon being studied. Finally, methodological triangulation involves employing various data collection methods, such as combining qualitative and quantitative approaches. This technique facilitates cross-verification of findings and strengthens the robustness of the research conclusions (Noble & Heale, 2019). By integrating these triangulation techniques, researchers can achieve a more balanced and accurate analysis. Triangulation allows for cross-validation of data and interpretations, reducing the risk of bias and increasing the credibility of the research findings. This approach aligns with recent recommendations for improving objectivity and ensuring that research conclusions are well-supported by diverse sources of evidence.

Furthermore, the use of artificial intelligence (AI) in research, including cognitive task analysis (ACTA), introduces complex ethical challenges that must be rigorously addressed to ensure the integrity and fairness of the findings. According to Resnik and Hosseini (2024), researchers must identify, describe, and control AI-related biases and random errors to prevent skewed results and ensure accuracy. Transparency about the use of AI, including explaining its limitations in accessible language, is crucial for maintaining the credibility of research (Resnik & Hosseini, 2024). Engaging with stakeholders and impacted communities is essential to address concerns about bias and ensure that AI applications are equitable and relevant. Furthermore, clear documentation of synthetic data and ethical education on responsible AI use are important for avoiding misleading conclusions and promoting ethical practices in ACTA (Resnik & Hosseini, 2024). This comprehensive approach helps safeguard against potential ethical issues and supports the responsible application of AI in research.

14.2 Maintaining Confidentiality and Integrity in Sensitive Environments

Maintaining confidentiality and integrity is of utmost importance in sensitive environments where ACTA is applied, such as in military, healthcare, or corporate settings. In these environments, the data collected can be highly sensitive, involving proprietary information, personal data, or insights into critical operations. Protecting this data is essential not only for ethical reasons but also to maintain trust between researchers and participants.

One of the primary challenges in maintaining confidentiality is the potential for data breaches, especially when digital tools and online platforms are used for data collection and storage. Ensuring robust cybersecurity measures is critical in protecting sensitive information from unauthorized access. A study by Cremer et al. (2022) highlights the critical issue of cybersecurity in the context of escalating cybercrime, which cost the global economy nearly USD 1 trillion in 2020 and saw average cyber insurance claims rise substantially. The study underscores the growing necessity for robust cybersecurity measures, including encryption, secure servers, and regular audits, to protect sensitive information from unauthorized access. It reveals a significant gap in open databases and cybersecurity resources, which exacerbates the risk of data breaches and undermines efforts to manage and mitigate cyber risks effectively. This emphasizes the importance of implementing stringent cybersecurity protocols to ensure the confidentiality and integrity of research data.

In addition to implementing technical safeguards, ethical considerations in data management also necessitate establishing clear protocols for data access and sharing. According to research by Pimenta Rodrigues et al. (2024) data breaches expose critical personal, health, and financial information, as evidenced by incidents such as the breach of 7.4 billion records by Le Figaro. The study, which analysed 428 global data breaches from 2018 to 2019, underscores the importance of stringent security measures and effective data protection regulations. It highlights the need for protocols that restrict data access to authorized personnel only and ensure that data sharing practices are meticulously anonymized to prevent re-identification. This is especially crucial in collaborative environments, such as defence or industrial projects, where data is shared across different teams or organizations. By adhering to these ethical standards, researchers and organizations can mitigate the risks associated with data breaches and protect sensitive information from unauthorized exposure.

Maintaining the integrity of data is crucial in scientific research, where the stakes are often high and the pressure to present favourable outcomes can be significant. The study on research integrity by Ciubotariu and Bosch (2022) highlights the importance of responsible science communication as a critical component of robust scientific practice. It underscores the need for adherence to norms such as Rigor,

Reproducibility, and Responsibility (the 3R's), and emphasizes the role of education and institutions in promoting these principles. This includes training in good scientific practice and responsible communication to prevent the manipulation or misrepresentation of data. The study outlines a framework for responsible science communication and stresses that researchers must report their findings truthfully and completely, even when the results are unfavourable. Such ethical guidelines are essential for upholding the highest standards of scientific integrity and ensuring that research findings remain accurate and reliable.

Finally, in environments where research findings have significant implications for policy decisions or operational strategies, it is essential to maintain a clear distinction between the roles of researchers and decision-makers. Gleditsch (2023) emphasizes that research often struggles to provide actionable policy implications due to limitations such as failing to discuss objectives, targeting factors without evaluating intervention effectiveness, and not addressing uncertainties or potential unintended effects. This highlights the need for researchers to avoid conflicts of interest and present their findings objectively. Instead of offering policy prescriptions as an afterthought, research should inform debates, draw on a cumulative body of evidence, and provide new data to enhance problem description and analysis. Maintaining objectivity and avoiding external pressures are crucial to ensuring that research contributions remain credible and effectively support policy development.

14.3 Managing Dual-Use Dilemmas in ACTA

Dual-use dilemmas are prevalent in various fields of research, including Applied Cognitive Task Analysis (ACTA). These dilemmas arise when the findings of research, while intended for beneficial purposes, also have the potential to be used in harmful ways. In the context of ACTA, dual-use concerns are particularly relevant in fields like military and healthcare, where the same cognitive insights and training methodologies designed to enhance efficiency and effectiveness could be repurposed for detrimental applications. For example, research in the military domain aimed at improving decision-making skills and training efficiency through cognitive task analysis can also be used to enhance the execution of lethal operations or develop more effective warfare strategies. Similarly, in healthcare, ACTA studies designed to optimize clinical decision-making and patient outcomes might inadvertently contribute to the development of techniques that could be used for psychological manipulation or invasive surveillance.

A study by Kavouras and Charitidis (2019) examining dual-use research in biotechnology highlights the broader implications of these dilemmas. The study discusses the increasing ethical debates surrounding dual-use issues, especially in the wake of significant events like the anthrax attacks in 2001 and the creation of the mammalian transmissible H5N1 influenza virus in 2012. These events underscored the potential for scientific advancements to be misused, prompting renewed discussions among scientists, policymakers, and the public about the need for robust ethical guidelines and governance frameworks.

This study is particularly relevant to ACTA because it underscores the necessity for researchers to be vigilant about the potential misuse of their findings. Just as biotechnological research must navigate the balance between openness and confidentiality, ACTA practitioners must consider how their work could be repurposed in unintended and potentially harmful ways. The study also highlights the importance of implementing ethical oversight throughout the research process. It suggests that establishing clear ethical guidelines and engaging in continuous ethical reflection are crucial for mitigating the risks associated with dual-use research.

To address these concerns, it is recommended to involve policymakers, and other stakeholders in the research process to develop safeguards that prevent misuse while allowing for beneficial advancements. This collaborative approach is also applicable to ACTA projects, where cross-disciplinary engagement can help ensure that cognitive insights are used responsibly and ethically. By adopting similar strategies, ACTA researchers can better navigate the dual-use dilemmas inherent in their work, balancing the pursuit of knowledge with the need to protect against its potential misuse.

In addition, the study by Selgelid (2009) explores the complexities of the dual-use dilemma in life science research, where scientific advancements intended for beneficial purposes can also be misused for harmful applications, such as the creation of biological weapons. This dilemma poses significant ethical challenges, as the dual-use nature of research brings into question the balance between scientific openness and security concerns.

Selgelid (2009) critically examines several controversial publications that have been central to debates about dual-use life science research and highlights the need for more ethical input into these discussions. The study notes that the majority of debates have primarily involved science and security experts, rather than ethicists, underscoring a gap in the governance of dual-use research. This lack of ethical engagement has led to calls for integrating ethical foresight into the research process—a strategy that involves assessing potential future impacts of research and implementing measures to prevent its misuse.

By advocating for increased ethical oversight, Selgelid (2009) emphasizes the importance of anticipating and addressing dual-use dilemmas proactively. The study suggests that involving ethicists, alongside scientists and policymakers, in the governance of dual-use research is crucial for developing robust safeguards that prevent harmful applications while still promoting scientific progress. This approach aligns with the concept of ethical foresight, aiming to ensure that research is conducted responsibly and ethically, with careful consideration of its potential dual-use implications.

14.4 Ethical Considerations in Participant Selection and Representation

Participant selection and representation are critical ethical considerations in ACTA studies, particularly when aiming to understand and improve the cognitive processes of domain experts. While ACTA typically involves expert participants, ensuring a diverse range of experts is still essential. Diversity among experts can enhance the validity and generalizability of the findings, leading to more robust and inclusive solutions in system design and training.

A recent study by Chatters et al. (2024) underscores the importance of diversity in clinical trials and highlights barriers to achieving representative participant pools. Although the study focuses on clinical trials, its recommendations are relevant to ACTA research. Chatters et al. (2024) found that logistical challenges, resource limitations, and lack of awareness contribute to underrepresentation in research. Similarly, in ACTA studies, ensuring that expert participants represent a range of experiences and backgrounds can mitigate biases and improve the applicability of the research findings.

Chatters et al. (2024) recommend strategies such as increased outreach, clear communication about the benefits of participation, and securing additional resources. Applying these strategies to ACTA can help ensure that the cognitive models developed are comprehensive and applicable across different contexts. By addressing these concerns, ACTA researchers can enhance the inclusivity and effectiveness of their interventions and solutions.

Moreover, ethical considerations in research extend significantly to the process of informed consent, which is crucial for safeguarding participants' rights and ensuring they make autonomous decisions about their involvement (Arellano et al., 2023). This is particularly relevant in Applied Cognitive Task Analysis (ACTA) studies, where expert participants provide valuable insights into cognitive processes. In ACTA studies, applying these principles is vital to ensure that expert participants are fully informed about the research, including how their cognitive processes will be analysed and how the results will be utilized. The study highlights that challenges in obtaining informed consent can arise in various contexts, such as cluster-randomized or decentralized trials. Although ACTA may not directly encounter these specific challenges, ensuring that informed consent is appropriately adapted to the study's context remains essential. This includes addressing potential privacy concerns and providing transparent information about the study's scope and application. By adhering to these informed consent principles, researchers in ACTA can uphold ethical standards and ensure that expert participants are fully aware of their involvement and its implications.

14.5 Ethical Issues in the Dissemination of ACTA Findings

The dissemination of ACTA findings, particularly in sensitive or high-stakes environments, raises several ethical issues. Researchers must carefully consider how their findings are communicated to avoid misinterpretation, sensationalism, or misuse. The responsible reporting of results is essential to ensure that the information is accurately conveyed and that it contributes positively to the field.

Understanding and practicing good research ethics is essential for ensuring the integrity and societal value of scientific work. Cuschieri (2022) highlights that adherence to ethical principles, such as avoiding deception, harm, and conflicts of interest, is crucial to prevent research misconduct, including data fabrication and plagiarism. This ethical foundation directly relates to the responsible dissemination of findings. Researchers must carefully consider the impact of their results on participants and the broader community to avoid potential harm, such as stigma or discrimination. By contextualizing findings and transparently communicating their limitations, researchers can prevent misuse and ensure that their work is understood within the appropriate context, thereby upholding both ethical standards and the credibility of their research.

Another crucial ethical responsibility in research is the effective dissemination of findings, particularly when they have significant implications for public policy or industry practices. Taylor (2019) emphasizes that researchers must not only publish their results but also ensure that these findings are communicated clearly and accessibly to participants and relevant communities. This transparency is essential not only as a matter of respect for those who contributed to the research but also to foster broader societal benefit. Proper dissemination helps to prevent misinterpretation, reduces the risk of harm, and ensures that research outcomes are used to inform and improve decision-making processes. Taylor argues that while practical challenges exist, such as the complexity of results and logistical issues, the commitment to making findings understandable and relevant to non-expert audiences is a fundamental aspect of ethical research practice. This approach aligns with the broader goal of enhancing public understanding and effectively influencing policy and practice through well-communicated research insights. Researchers should strive to communicate their findings in a way that is accessible to non-expert audiences, ensuring that the insights gained from ACTA are used to inform and improve decision-making processes across various domains.

CONCLUSION

In conclusion, the ethical considerations in Applied Cognitive Task Analysis (ACTA) are multifaceted and crucial to ensuring the integrity and social value of research outcomes. This chapter has addressed several key areas where ethical practices must be carefully applied: data collection and analysis, maintaining confidentiality and integrity, managing dual-use dilemmas, participant selection and representation, and the dissemination of findings.

Ensuring ethical practices in data collection and analysis is fundamental. Informed consent, transparency, and minimizing harm are essential components that uphold the integrity of ACTA research. The emphasis on robust informed consent procedures and distress protocols reflects a commitment to protecting participant rights and well-being. By incorporating techniques such as triangulation and adhering to strict objectivity, researchers can mitigate biases and enhance the reliability of their findings, aligning with recommendations from current research (Noble & Heale, 2019).

Maintaining confidentiality and data integrity in sensitive environments is another critical consideration. With the growing risks of data breaches and cyber threats, implementing strong cybersecurity measures and clear data access protocols is vital. The ethical management of data ensures that sensitive information is protected and trust between researchers and participants is upheld (Cremer et al., 2022; Pimenta Rodrigues et al., 2024). Furthermore, the integrity of research findings must be preserved by adhering to rigorous scientific standards and responsible communication practices (Ciubotariu & Bosch, 2022; Gleditsch, 2023).

Managing dual-use dilemmas requires vigilance and proactive ethical oversight. The potential for research findings to be misused in harmful ways necessitates the development of safeguards and continuous ethical reflection throughout the research process (Kavouras & Charitidis, 2019; Selgelid, 2009). Engaging with stakeholders and implementing ethical guidelines can help navigate these challenges and ensure that cognitive insights are used responsibly.

Participant selection and representation are critical to the validity and inclusivity of ACTA studies. Ensuring a diverse range of participants enhances the applicability and robustness of findings, while addressing barriers to representation improves research quality (Chatters et al., 2024). Ethical practices in informed consent further safeguard participant autonomy and clarify the implications of their involvement (Arellano et al., 2023).

Finally, the ethical dissemination of ACTA findings is essential to prevent misinterpretation and misuse. Clear, transparent communication that respects participant contributions and conveys research limitations is fundamental to maintaining research integrity and societal benefit (Cuschieri, 2022; Taylor, 2019). By prioritizing ethical standards throughout the research process, ACTA practitioners can contribute valuable insights while upholding the principles of responsible and impactful research.

Overall, integrating these ethical considerations into ACTA research not only enhances the credibility and utility of the findings but also ensures that the research process respects and protects the rights and well-being of all involved. As ACTA continues to evolve and address complex cognitive processes, maintaining a strong ethical foundation will remain paramount to its success and societal impact.

REFERENCES

Arellano, L., Alcubilla, P., & Leguízamo, L. (2023). *Ethical considerations in informed consent*. IntechOpen. DOI: 10.5772/intechopen.1001319

Blease, C., Kelley, J. M., & Trachsel, M. (2018). Informed consent in psychotherapy: Implications of evidence-based practice. *Journal of Contemporary Psychotherapy*, 48(2), 69–78. DOI: 10.1007/s10879-017-9372-9

Chatters, R., Dimairo, M., & Cooper, C.. (2024). Exploring the barriers to, and importance of, participant diversity in early-phase clinical trials: An interview-based qualitative study of professionals and patient and public representatives. *BMJ Open*, 14, e075547. DOI: 10.1136/bmjopen-2023-075547

Ciubotariu, I. I., & Bosch, G. (2022). Improving research integrity: A framework for responsible science communication. *BMC Research Notes*, 15(1), 177. DOI: 10.1186/s13104-022-06065-5

Cremer, F., Sheehan, B., & Fortmann, M.. (2022). Cyber risk and cybersecurity: A systematic review of data availability. *The Geneva Papers on Risk and Insurance. Issues and Practice*, 47(4), 698–736. DOI: 10.1057/s41288-022-00266-6

Cuschieri, S. (2022). Understanding research ethics. In *A roadmap to successful scientific publishing* (pp. 13–25). Springer. DOI: 10.1007/978-3-030-99295-8_2

Gleditsch, K. S. (2023). *This research has important policy implications....* Peace Economics, Peace Science and Public Policy. DOI: 10.1515/peps-2023-0002

Kavouras, P., & Charitidis, C. A. (2019). Dual use in modern research. In Iphofen, R. (Ed.), *Handbook of research ethics and scientific integrity* (pp. 1–17). Springer. DOI: 10.1007/978-3-319-76040-7_7-1

Noble, H., & Heale, R. (2019). Triangulation in research, with examples. *Evidence-Based Nursing*, 22(3), 89–91. DOI: 10.1136/ebnurs-2019-103145

Pimenta Rodrigues, , G. AMarques Serrano, , A. LLopes Espiñeira Lemos, , A. NCanedo, , E. DMendonça, , F. L. L. dde Oliveira Albuquerque, , RSandoval Orozco, , A. LGarcía Villalba, , L. J. (2024). Understanding data breach from a global perspective: Incident visualization and data protection law review. *Data, 9*(2), 27. DOI: 10.3390/data9020027

Resnik, D. B., & Hosseini, M. (2024). The ethics of using artificial intelligence in scientific research: New guidance needed for a new tool. *AI and Ethics*. Advance online publication. DOI: 10.1007/s43681-024-00493-8

Selgelid, M. J. (2009). Governance of dual-use research: An ethical dilemma. *Bulletin of the World Health Organization*, 87(9), 720–723. DOI: 10.2471/BLT.08.051383

Taylor, J. (2019). Reporting research findings to participants is an ethical imperative. *BMJ (Clinical Research Ed.)*, 367, l6324. DOI: 10.1136/bmj.l6324

UK Statistics Authority. (2022, October 21). *Ethical considerations associated with qualitative research methods*. Office for Statistics Regulation. https://uksa.statisticsauthority.gov.uk/publication/ethical-considerations-associated-with-qualitative-research-methods/pages/4/

Xu, A., Baysari, M. T., Stocker, S. L., Leow, L. J., Day, R. O., & Carland, J. E. (2020). Researchers' views on, and experiences with, the requirement to obtain informed consent in research involving human participants: A qualitative study. *BMC Medical Ethics*, 21(1), 93. DOI: 10.1186/s12910-020-00538-7

Section 6
Practical Implications and Case Studies

Chapter 15
Implementing ACTA Findings

ABSTRACT

This chapter explores the practical application of Applied Cognitive Task Analysis (ACTA) findings across various sectors. It outlines strategies for translating cognitive insights into real-world outcomes, including tailoring findings to stakeholder needs, developing training programs, designing user-centered interfaces, and integrating ACTA findings into policies. Emphasizing continuous evaluation and iteration, the chapter showcases success stories from industries like healthcare, military, and education. By applying ACTA, organizations can improve performance, safety, and efficiency while fostering a culture of continuous improvement and innovation.

INTRODUCTION

Applied Cognitive Task Analysis (ACTA) has emerged as a powerful tool for uncovering the cognitive processes that underpin complex tasks across various fields. Its strength lies in its ability to dissect expert decision-making and thought patterns, providing deep insights that can be leveraged to improve performance, safety, and efficiency. By mapping these cognitive structures, ACTA offers a clear pathway for understanding how experts approach tasks in high-stakes, dynamic environments, making it an invaluable resource across domains such as healthcare, aviation, education, and defence.

This chapter focuses on strategies for effectively translating ACTA findings into real-world applications. By drawing on recent research and success stories from different industries, we explore how these cognitive insights can be turned into actionable solutions. The discussion will highlight how organizations have used ACTA to optimize training programs, refine operational procedures, and enhance

DOI: 10.4018/979-8-3693-7733-8.ch015

system designs, ultimately showing how the practical application of ACTA can drive significant improvements in both individual and team performance.

15.1. Strategies for Translating ACTA Insights into Real-World Applications

Successfully implementing Applied Cognitive Task Analysis (ACTA) findings in real-world settings requires a structured and systematic approach that ensures cognitive insights are effectively integrated into practical applications. This involves addressing the needs of various stakeholders, creating tailored training and tools, developing comprehensive policies, and maintaining an iterative process for continuous improvement. Here are several key strategies to achieve these objectives:

15.1.1. Tailoring Insights to Stakeholder Needs

A critical step in translating ACTA findings into actionable outcomes is tailoring insights to the specific needs and contexts of different stakeholders, including managers, practitioners, and end-users. Stakeholders often have diverse needs and priorities, and it is essential to align the findings of cognitive task analyses with their operational goals and contextual realities.

The study by Graham et al. (2023) on adapting Cognitive Task Analysis (CTA) for health services research highlights the significance of understanding cognitive processes in clinical environments to develop effective decision-support tools and healthcare technology. This approach aligns with the importance of stakeholder engagement throughout the ACTA process. By involving stakeholders such as healthcare providers in the CTA process, researchers can ensure that the resulting frameworks and tools are more relevant and actionable. For example, involving managers and practitioners early on can help identify key performance indicators and practical challenges, leading to solutions that align with organizational objectives and are feasible in practice. The study's emphasis on a systematic and iterative approach to identifying cognitive processes supports the idea that engaging stakeholders throughout the analysis and development stages enhances the relevance and effectiveness of the recommendations derived from these methods.

Continuing from the discussion on the importance of tailoring insights to stakeholder needs, the study by Knisely et al. (2021) introduces the Cognitive Task Analysis and Workload Classification (CTAWC) framework, which enhances the traditional Cognitive Task Analysis (CTA) methodology by focusing specifically on the evaluation of cognitive workload and task complexity in automated systems. This study provides a compelling example of how insights from cognitive analyses can be effectively aligned with the needs of different stakeholders, particularly in environments where automation plays a significant role.

CTAWC improves upon conventional CTA by integrating standardized terminology and classifications to describe expected cognitive workload during task performance. This approach allows for a more detailed and nuanced understanding of the cognitive demands placed on human operators in supervisory roles, where monitoring and intervention in automated processes are critical. The enhanced depth provided by CTAWC in decomposing cognitive tasks helps identify specific sources of cognitive workload, making it easier for stakeholders to address these demands through targeted training and system design.

By offering a framework to evaluate cognitive depth and task complexity systematically, the CTAWC model provides stakeholders with a robust tool for empirical studies and practical applications. Managers can use this framework to develop more effective training programs and operational protocols, ensuring that operators are well-prepared to handle the cognitive demands of their roles. Practitioners and end-users benefit from the insights gained through CTAWC by having clearer guidelines and support mechanisms tailored to their specific tasks and responsibilities.

In essence, the work of Knisely et al. (2021) demonstrates the importance of tailoring cognitive analysis findings to the specific needs and contexts of stakeholders. By focusing on the practical applications of cognitive workload assessments and providing a standardized model for task analysis, the CTAWC framework ensures that the outcomes of cognitive task analyses are not only theoretically sound but also practically relevant and actionable in real-world settings. This approach aligns with the broader goals of ACTA, emphasizing the need to customize insights and recommendations to meet the diverse needs of different stakeholder groups effectively.

15.1.2. Developing Training and Educational Programs

Developing targeted training and educational programs is one of the most effective ways to implement ACTA findings. These programs should focus on enhancing the specific cognitive skills and strategies identified as critical during the ACTA process. By doing so, organizations can ensure that employees are well-equipped to perform their tasks efficiently and effectively.

For example, a study by Swaby et al. (2022) illustrated the use of Cognitive Task Analysis (CTA) methods to improve clinical decision-making in healthcare settings. The study highlighted how understanding expert decision-making processes through CTA could inform the creation of refined training protocols that enhance the cognitive abilities of healthcare professionals. By focusing on the tacit knowledge and decision-making strategies used by experienced clinicians, the study was able to develop training programs that were better aligned with the realities of clinical practice. This approach not only helped in improving the skills of less experienced clinicians but also ensured that training was directly relevant to the challenges they would face in real-world scenarios.

In a similar vein, organizations across various sectors can leverage ACTA findings to develop training programs tailored to their specific needs. By identifying the key cognitive skills and strategies that contribute to successful task performance, these programs can provide employees with targeted practice and feedback. This focus on practical, scenario-based learning helps to reinforce the knowledge and skills necessary for effective decision-making and problem-solving, ultimately leading to enhanced performance and productivity.

Additionally, training programs informed by ACTA findings can be designed to be iterative, allowing for continuous refinement based on ongoing feedback and new insights. This dynamic approach ensures that training remains relevant and effective, adapting to the evolving needs of the organization and its workforce. By systematically developing and implementing such targeted educational initiatives, organizations can build a strong foundation of cognitive skills that are crucial for achieving their strategic goals and maintaining competitive advantage.

15.1.3. Designing User-Centred Interfaces and Tools

ACTA insights are particularly valuable when it comes to designing user-centred interfaces and tools that align with the cognitive processes of users. By understanding the cognitive demands of tasks, designers can create interfaces and tools that support users' natural workflows, reducing cognitive load and enhancing performance.

The study by Arrabito and Banbury (2019) demonstrates the effectiveness of using CTA to inform the design of human-machine interfaces for unmanned aircraft systems (UAS). In their research, CTA was employed to analyse the cognitive processes involved in operating UAS, particularly in complex, time-sensitive missions. This analysis revealed critical elements of situation awareness required by operators to make effective decisions. The findings led to the development of interfaces that improved collaboration and decision-making by integrating UAS operators and intelligence analysts within the same operational space. The hybrid analysis technique used, which combined mission function task analysis, goal-directed task analysis, and CTA, highlighted how understanding the cognitive demands of tasks can directly inform the design of tools that support effective decision-making in high-pressure environments.

Similarly, the research by Gallegos Ayala et al. (2023) on brain-computer interfaces (BCIs) underscores the importance of aligning interface design with cognitive processes. Their study introduced a novel algorithm for extracting frontal theta oscillations from electroencephalographic (EEG) recordings to assess mental workload across different cognitive tasks. The algorithm demonstrated significant improvements in classifying mental workload conditions compared to previous methods. By providing real-time and continuous assessments of mental workload, this research supports the development of BCIs that adapt to users' cognitive states, thereby optimizing human-computer interaction and enhancing user experience.

The work of Arrabito and Banbury (2019) and Gallegos Ayala et al. (2023) illustrate how understanding cognitive processes and mental workload can lead to more effective and user-friendly designs. For instance, by identifying the specific cognitive requirements and challenges faced by users, designers can create interfaces that minimize unnecessary cognitive load, streamline workflows, and facilitate better decision-making. In practice, designing user-centred interfaces involves aligning interface elements with the cognitive processes and tasks that users engage in. This approach helps in creating tools that are tailored to users' needs, making them more efficient and reducing the likelihood of errors.

15.1.4. Integrating ACTA Findings into Policy and Procedure Development

ACTA findings can also play a crucial role in informing policy and procedure development. By understanding the cognitive demands of tasks, organizations can develop policies and procedures that are more aligned with the actual workflows and cognitive processes of their employees.

A report by the U.S. Department of Transportation (2013) highlighted how cognitive task analysis can be used to enhance Human Systems Integration (HSI) in railroad operations. The study demonstrated that cognitive task analysis methods are instrumental in identifying and addressing the cognitive demands associated with new technologies and operational changes. By analysing how workers interact with complex systems, the report provided actionable insights for improving system design and operational procedures. This approach led to the development of policies and procedures that are more aligned with the cognitive realities of railroad tasks, resulting in enhanced efficiency, safety, and overall performance.

For example, the report discussed how cognitive task analysis was employed to understand the cognitive processes involved in operating locomotives and managing railroad systems. This analysis revealed critical areas where cognitive overload could occur, such as during the monitoring of automated systems or the coordination of tasks under time constraints. Based on these insights, the report recommended specific changes to system interfaces and operational protocols to better support users' cognitive needs. This led to more intuitive interfaces, improved training programs, and optimized decision-making processes.

Similarly, other industries can benefit from integrating ACTA (Applied Cognitive Task Analysis) findings into their policy development processes. By applying ACTA methods, organizations can gain a deep understanding of the cognitive demands associated with various tasks and workflows. This knowledge allows for the design of policies and procedures that accurately reflect the cognitive realities of their employees' work environments.

In conclusion, the integration of ACTA findings into policy and procedure development offers significant benefits across various industries. By aligning policies with the cognitive demands of tasks and workflows, organizations can enhance performance, improve safety, and achieve better outcomes in their operations.

15.1.5. Continuous Evaluation and Iteration

Implementing ACTA findings into real-world applications is not a one-time process; it requires continuous evaluation and iteration. This ongoing process ensures that the interventions remain relevant and effective as tasks and technologies evolve.

A study by Sutton et al. (2020) underscores the evolving nature of Computerized Clinical Decision Support Systems (CDSS) and highlights the need for ongoing assessment to ensure these systems effectively support clinical decision-making. Despite the advancements in CDSS, there remain uncertainties regarding their impact on providers, patient outcomes, and overall costs. The study reviews the current state of CDSS, detailing both successful applications and notable setbacks, thus emphasizing the importance of iterative refinement. Evidence-based recommendations provided in the study stress the necessity for regular updates, user feedback, and performance monitoring to mitigate risks and enhance system efficacy. Similarly, in the context of ACTA, continuous evaluation and iteration are crucial for adapting cognitive task analysis findings to changing contexts and technological advancements. Establishing robust mechanisms for feedback from users and stakeholders, monitoring performance outcomes, and making iterative design improvements ensure that ACTA-based interventions remain relevant and effective. This approach not only aligns interventions with actual needs but also helps in optimizing task performance and cognitive workload management.

Organizations should establish mechanisms for regular feedback from users and stakeholders, as well as ongoing monitoring of performance outcomes. For example, Saqr et al. (2019) explored the use of social network analysis and machine learning models to monitor student engagement in collaborative online discussions. Their findings underscore the importance of continuous evaluation in understanding how users interact with systems and identifying areas for improvement. By adopting a similar approach, organizations can continuously refine and adapt their implementations, ensuring they remain aligned with evolving needs and technologies.

In summary, effectively translating ACTA insights into real-world applications requires a comprehensive and iterative approach that involves tailoring findings to stakeholder needs, developing targeted training and tools, integrating findings into policies and procedures, and continuously evaluating and iterating on solutions. By adopting these strategies, organizations can unlock the full potential of ACTA, enhancing performance, safety, and efficiency across various sectors.

15.2. Success Stories and Lessons Learned from Various Sectors

The successful application of Applied Cognitive Task Analysis (ACTA) findings across different sectors offers valuable lessons and highlights best practices for translating cognitive insights into practical outcomes. These success stories demonstrate how understanding cognitive processes can significantly enhance performance, safety, and efficiency across diverse fields.

15.2.1. Healthcare: Enhancing Clinical Decision-Making

In the healthcare sector, Applied Cognitive Task Analysis (ACTA) has been pivotal in refining clinical decision-making processes. The study by Swaby et al. (2022) exemplifies how cognitive task analysis can optimize workflows by informing the development of clinical protocols and interventions. By meticulously analysing the cognitive steps involved in expert decision-making, healthcare organizations can identify critical points where errors are most likely to occur. This detailed understanding of expert cognition allows for the development of targeted training programs and decision support tools specifically designed to address these vulnerabilities. While Swaby et al. utilizes CTA methods to gain insights into the nuanced, tacit knowledge that experienced clinicians use in high-stress, high-complexity environments, these findings can be effectively adapted and applied within an ACTA framework. By documenting and transferring this expert knowledge, healthcare organizations can not only reduce errors but also enhance the overall quality of care. The practical application of these insights, as facilitated by ACTA, ensures that interventions are both effective and feasible in real-world settings, aligning clinical protocols with the actual practices of expert clinicians.

Cognitive Task Analysis (CTA) has been instrumental in enhancing Electronic Health Record (EHR) interfaces by reducing the cognitive load on clinicians, thereby improving decision-making accuracy and efficiency. A study by Pfaff et al. (2020) utilized a CTA method specifically designed to uncover the cognitive demands experienced by clinicians when using EHR systems. The study identified 145 unique cognitive demands across various roles and settings, revealing significant issues such as the EHR's inability to help clinicians maintain a comprehensive view of patient information, support deep reasoning about patient states, and facilitate collaborative work. These findings emphasize the need for EHR designs that better align with the cognitive processes of clinicians, thereby reducing mental strain and enhancing overall usability. By addressing these cognitive challenges through an Applied Cognitive Task Analysis (ACTA) approach, EHR interfaces can be refined to support more efficient and accurate clinical decision-making. This approach ensures that the design and implementation of EHR systems are grounded in a deep understanding of clinicians' cognitive workflows, ultimately leading to improved healthcare outcomes.

In conclusion, the successful application of Applied Cognitive Task Analysis (ACTA) across diverse sectors demonstrates its critical role in translating cognitive insights into practical improvements. By thoroughly analysing cognitive processes, ACTA helps optimize workflows, enhance safety, and boost efficiency, as exemplified by studies like those of Swaby et al. (2022) and Pfaff et al. (2020). These examples highlight how ACTA can refine clinical protocols and EHR systems by aligning designs with the cognitive demands of users, ultimately improving decision-making and care quality. The continued integration and refinement of ACTA methodologies not only drive advancements within specific fields but also provide a valuable framework for other industries aiming to leverage cognitive insights for better performance and outcomes.

15.2.2. Military and Défense: Improving Mission Planning and Simulation Training

In military and defence settings, Applied Cognitive Task Analysis (ACTA) has proven indispensable for enhancing mission planning and simulation training, especially in preparing personnel for complex and high-risk scenarios. Research by Ste-Croix and Angel (2021) provides a comprehensive examination of how ACTA can be applied to understand the cognitive demands placed on dismounted infantry soldiers. In this study, a cognitive task analysis was conducted using a Complex Battle Position Scenario developed by subject matter experts (SMEs). This approach aimed to investigate, identify, and characterize the critical information requirements and cognitive demands experienced by soldiers in simulated combat environments.

The study involved twelve SMEs, including Sergeants, Warrant Officers, and Officers, who were guided through the scenario to identify key tasks and cognitive load associated with each. The research utilized tools such as the NASA Task Load Index and various questionnaires to capture the cognitive burden during different tasks, revealing that the highest workload levels were associated with engagements and maintaining situational awareness. This insight is crucial for developing simulation training programs that effectively replicate the cognitive challenges faced in real combat situations.

The study also explored the potential for technological solutions to alleviate cognitive burden, highlighting the need for improved radios, GPS, and unmanned air vehicles (UAVs). The SMEs indicated that these technologies could enhance situational awareness and reduce workload, aligning with the findings from Ste-Croix and Angel (2021) that underscore the importance of integrating ACTA insights into the design of training tools and technologies.

Incorporating these findings into military training and mission planning ensures that training scenarios are more reflective of real-world cognitive demands, thereby improving preparedness and performance. By using ACTA to analyse and address cognitive challenges, military organizations can create more effective training programs and enhance the overall readiness of their personnel for complex and high-risk environments.

In addition, the study by Cannon-Bowers et al. (2013) explores the use of a comprehensive cognitive task analysis technique to improve the efficacy and effectiveness of medical training within the Department of Défense. By employing a scenario-based approach, the technique aims to generate detailed training requirements, performance metrics, and simulator requirements for medical tasks such as cricothyroidotomy and haemorrhage control. This method involves experts performing the tasks multiple times, with each iteration focusing on different aspects of the training development process.

This approach aligns with the broader application of ACTA in military settings, where understanding cognitive demands and improving simulation training are critical. By generating actionable data to inform simulation-based training system design, the technique helps ensure that training scenarios are relevant and effective, addressing the specific needs and challenges encountered in real combat situations. Thus, the study highlights the importance of utilizing advanced cognitive task analysis methods to enhance training programs, making them more effective in preparing personnel for complex and high-risk scenarios.

In summary, both the study and the broader application of ACTA in military and defence settings emphasize the role of detailed cognitive analysis in developing effective training programs and simulation tools. This ensures that personnel are better prepared for the cognitive challenges they will face in real-world operations.

15.2.3. Industrial Applications: Supporting Human-Machine Interaction

In industrial settings, particularly where human-machine interaction is critical, Applied Cognitive Task Analysis (ACTA) has become essential for enhancing safety and efficiency. A significant study conducted by Roth et al. (2020) focuses on the application of Cognitive Task Analysis (CTA) within the context of the U.S. Nuclear Regulatory Commission's (NRC) guidelines for evaluating control room designs and technologies. This study underscores the evolving nature of work activities in nuclear power plants, which are increasingly cognitive due to automation and supervisory control demands.

The study highlights how traditional Task Analysis (TA) methods, which have been integral to Human Factors Engineering (HFE) in safety reviews, are being complemented by more advanced cognitive methodologies. CTA is particularly valuable in this context because it delves into the cognitive challenges associated with supervisory control tasks that are not easily observable but are crucial for effective task performance. As automation increases, the complexity of human-system interactions requires a deeper understanding of cognitive processes, including decision-making, information processing, and perceptual skills.

Roth et al. (2020) advocate for the inclusion of CTA in the review of licensee applications for new nuclear power plant designs and modifications. This approach ensures that control room designs and technologies are evaluated not only for their physical and operational aspects but also for their impact on cognitive workload and human performance. By providing detailed summaries and use cases for various CTA methods, the study offers valuable guidance for improving system designs to better support human operators in high-stress, high-complexity environments.

In the context of Roth et al.'s (2020) study on CTA and its application in nuclear power plants, Applied Cognitive Task Analysis (ACTA) further enhances the approach by focusing on practical implementations that directly support human-machine interactions. While CTA provides valuable insights into cognitive processes, ACTA translates these insights into actionable design improvements for complex systems. For instance, in environments like nuclear control rooms where cognitive demands are high, ACTA can be used to develop systems that reduce cognitive load and enhance usability. By applying ACTA principles, designers can create interfaces and operational procedures that align with the cognitive capabilities of operators, improving their efficiency and decision-making. This includes designing control panels that present information in a way that minimizes confusion and supports quick, accurate responses. The integration of ACTA ensures that cognitive challenges identified through CTA are addressed in practical, user-centred ways, thereby enhancing overall system performance and safety in high-stress, complex environments like those described by Roth et al. (2020). This alignment of cognitive insights with practical design solutions underscores the critical role of ACTA in optimizing human-machine interactions across industrial settings.

The study by Taylor et al. (2023) explores how Applied Cognitive Task Analysis (ACTA) can be employed to understand the cognitive challenges and knowledge use among high-performance sport coaches, revealing significant insights into their adaptive strategies and decision-making processes. While the study on high-performance sport coaching explores ACTA to understand the cognitive demands and knowledge use among coaches, its principles can be relevant to industrial applications, especially those involving human-machine interaction. The research highlights how ACTA can uncover cognitive challenges and the flexible use of knowledge in complex, performance-critical environments. Similarly, in industrial settings where human-machine interaction is crucial, ACTA serves a comparable role by identifying cognitive demands and optimizing system designs to support operators. Just as the study reveals how coaches adapt their strategies to meet athletes' needs, ACTA helps in designing user interfaces and workflows that align with operators' cognitive capabilities, enhancing safety and efficiency. By applying ACTA methodologies to understand cognitive processes and challenges, both in sports and industrial settings, we gain insights that can drive improvements in performance and system design, ultimately supporting more effective human-machine collaboration in high-stakes environments.

Overall, the application of ACTA in industrial settings illustrates its pivotal role in optimizing human-machine interaction. By focusing on cognitive demands and designing systems that support human operators, industries can achieve higher productivity and safety levels, reflecting the ongoing importance of cognitive analysis in technology integration and workflow optimization.

15.2.4. Education: Enhancing Learning Outcomes

In the education sector, Applied Cognitive Task Analysis (ACTA) has proven essential for developing pedagogical practices that significantly improve learning outcomes. ACTA provides educators with a structured approach to understanding how learners interact with complex tasks, allowing for the design of instructional materials that are aligned with cognitive processes. Research by Alias (2024) on microlearning strategies demonstrates how ACTA can be used to design educational experiences that break down complex cognitive tasks into smaller, more manageable components. This segmentation helps create targeted instructional materials that address the specific strengths and weaknesses of learners. By tailoring the educational content to fit these cognitive processes, educators can enhance engagement, motivation, and ultimately improve knowledge retention and application.

Further supporting this approach, recent research highlights the benefits of incorporating cognitive task analysis into curriculum design. For instance, Abd Wahid et al. (2023) explored the application of Applied Cognitive Task Analysis (ACTA) to address challenges in designing professional ethics instructional frameworks for technical educators. This study utilized ACTA to identify and analyse the cognitive difficulties faced by educators when developing and delivering ethics education.

The research found that tasks such as assessing learners' profiles, setting learning objectives, and selecting appropriate instructional methods were particularly challenging for educators. By employing a structured approach to cognitive task analysis, Abd Wahid et al. (2023) were able to identify specific cognitive hurdles and propose strategies to mitigate them. These strategies included conducting preliminary assessments of students' profiles, enhancing educators' knowledge of teaching theories, and adopting interactive pedagogical approaches.

The findings from Abd Wahid et al. (2023) underscore how ACTA can be effectively integrated into curriculum design to address cognitive complexities and improve instructional practices. By focusing on the cognitive demands of teaching professional ethics, the study illustrates the value of ACTA in creating more targeted and effective educational frameworks. This research contributes to the growing body of evidence that supports the use of cognitive task analysis to enhance curriculum design, ensuring that educational materials and methods are better aligned with the cognitive processes of both educators and learners.

In the realm of enhancing educational outcomes through technology, the study by Idris et al. (2022) provides significant insights into the application of Applied Cognitive Task Analysis (ACTA) in the design of e-training systems. This research focused on validating the design of the E-Training System for Professional Development (TaSPoD), employing ACTA to identify and address cognitive challenges encountered by users. The study reveals that experts faced difficulties with two key elements: system design and functionality, and course content design.

By utilizing ACTA, Idris et al. (2022) were able to elicit detailed feedback on the cognitive demands associated with the e-training system. This feedback led to the formulation of targeted recommendations to enhance user engagement and effectiveness. The proposed improvements included the development of a more intuitive graphical user interface with simple navigation features, the integration of communication tools like chat programs and video conferencing, and the customization of the application for handling external documents. These enhancements aim to address cognitive overload and make the training environment more supportive and efficient for users.

This study underscores the importance of incorporating ACTA into the design and development of educational technology. By focusing on cognitive challenges and user needs, ACTA helps in creating more effective training systems that cater to the cognitive processes of learners. The improvements suggested by Idris et al. (2022) not only enhance the functionality and usability of e-training systems but also contribute to better learning outcomes by providing a more engaging and supportive learning environment.

In conclusion, Applied Cognitive Task Analysis (ACTA) has proven instrumental in enhancing educational outcomes through its focus on understanding and addressing cognitive demands in both traditional and technological contexts. Research by Alias (2024) demonstrates how ACTA can be effectively employed to develop a microlearning framework by breaking down complex tasks into manageable components, thereby improving engagement, motivation, and knowledge retention. Similarly, Abd Wahid et al. (2023) highlight the use of ACTA in addressing cognitive challenges associated with designing professional ethics instructional frameworks for technical educators, showing how targeted strategies can improve curriculum development and teaching practices. Furthermore, Idris et al. (2022) illustrate the application of ACTA in the design of e-training systems, identifying and addressing cognitive difficulties to enhance user interaction and training effectiveness. Collectively, these studies underscore ACTA's significant role in refining educational practices and tools, ensuring that instructional designs are better aligned with cognitive processes and ultimately fostering more effective learning environments.

CONCLUSION

Implementing ACTA findings into real-world applications requires a multifaceted approach that takes into account the diverse needs of stakeholders, develops targeted interventions, and continuously evaluates and refines solutions. The success stories and lessons learned from various sectors—including industrial settings, education, and healthcare—demonstrate the transformative potential of ACTA in enhancing performance, safety, and efficiency.

In industrial contexts, ACTA helps in designing systems that better align with human cognitive processes, leading to improved safety protocols and more effective human-machine interactions. In educational settings, ACTA has proven effective in developing curricula that address cognitive challenges and support more effective teaching and learning strategies. Similarly, in healthcare, ACTA contributes to refining medical training and improving clinical practices by highlighting cognitive demands and optimizing training approaches.

The application of ACTA extends beyond merely identifying problems; it involves developing practical, evidence-based interventions that address these issues. Successful implementation requires collaboration among researchers, practitioners, and stakeholders to ensure that solutions are relevant and impactful. Moreover, the iterative nature of applying ACTA means that interventions must be continuously assessed and adjusted based on feedback and evolving needs.

By leveraging insights gained from ACTA, organizations can unlock the full potential of their workforce, enhance operational efficiency, and achieve strategic objectives. The ongoing evolution of ACTA methodologies—driven by continuous research and adaptation—will further enhance its applicability and impact across various domains. Future research should focus on expanding ACTA's applications, refining its methods, and exploring new areas where its principles can drive meaningful improvements.

In summary, the effective implementation of ACTA not only addresses immediate cognitive challenges but also fosters a culture of continuous improvement and innovation. Embracing ACTA's transformative potential will help organizations navigate complex environments and achieve excellence in their respective fields.

REFERENCES

Abd Wahid, S. N., Abdul Razak, R., & Halili, S. H. (2023). Investigating knowledge task difficulties in designing professional ethics instructional framework for technical educators. *Jurnal Penyelidikan Pendidikan*, 41, 45–57.

Alias, N. F. (2024). *Developing pedagogical practices of microlearning educational experience in higher education* (Doctoral dissertation). University of Malaya.

Arrabito, G. R., Hou, M., & Banbury, S. (2019). *Cognitive task analysis for an unmanned aircraft system* (DRDC-RDDC-2019-D041). Defence Research and Development Canada.

Cannon-Bowers, J., Bowers, C., Stout, R., Ricci, K., & Hildabrand, A. (2013). Using cognitive task analysis to develop simulation-based training for medical tasks. *Military Medicine*, 178(suppl_10), 15–21. DOI: 10.7205/MILMED-D-13-00211

Gallegos Ayala, G. I., Haslacher, D., Krol, L. R., Soekadar, S. R., & Zander, T. O. (2023). Assessment of mental workload across cognitive tasks using a passive brain-computer interface based on mean negative theta-band amplitudes. *Frontiers in Neuroergonomics*, 4, 1233722. DOI: 10.3389/fnrgo.2023.1233722

Graham, L. A., Gray, C., Wagner, T. H., Illarmo, S., Hawn, M. T., Wren, S. M., Iannuzzi, J., & Harris, A. H. S. (2023). Applying cognitive task analysis to health services research. *Health Services Research*, 58(2), 415–422. DOI: 10.1111/1475-6773.14106

Idris, A. Y., Abdul Razak, R., & Abdul-Rahman, S. S. (2022). Investigating knowledge task difficulties in an e-training system for professional development design. *Docens Series in Education*, 2, 51–65.

Knisely, B. M., Joyner, J. S., & Vaughn-Cooke, M. (2021). Cognitive task analysis and workload classification. *MethodsX*, 8, 101235. DOI: 10.1016/j.mex.2021.101235

Pfaff, M. S., Eris, O., Weir, C., Anganes, A., Crotty, T., Rahman, M. T., Ward, M., & Nebeker, J. (2020). Analysis of the cognitive demands of electronic health record use. *Journal of Biomedical Informatics*, 113, 103633. DOI: 10.1016/j.jbi.2020.103633

Roth, E., O'Hara, J., Dickerson, K., Hughes, N., Fleger, S., & Ki, D. (2020). *Cognitive task analysis: Technical basis and guidance development*. Office of Nuclear Regulatory Research, United States Nuclear Regulatory Commission.

Saqr, M., Nouri, J., & Santolini, M. (2019, March). *Towards group-aware learning analytics: Using social network analysis and machine learning to monitor and predict performance in collaborative learning*. Paper presented at the 13th Annual International Technology, Education and Development Conference. DOI: 10.21125/inted.2019.1881

Ste-Croix, C., & Angel, H. (2021). *Cognitive task analysis of a complex battle position scenario*. Human Systems Incorporated.

Sutton, R. T., Pincock, D., Baumgart, D. C., Sadowski, D. C., Fedorak, R. N., & Kroeker, K. I. (2020). An overview of clinical decision support systems: Benefits, risks, and strategies for success. *NPJ Digital Medicine*, 3, 17. DOI: 10.1038/s41746-020-0221-y

Swaby, L., Shu, P., Hind, D., & Sutherland, K. (2022). The use of cognitive task analysis in clinical and health services research: A systematic review. *Pilot and Feasibility Studies*, 8(1), 57. DOI: 10.1186/s40814-022-01002-6

Taylor, J., Ashford, M., & Jefferson, M. (2023). High performance coach cognition in the wild: Using applied cognitive task analysis for practical insights—Cognitive challenges and curriculum knowledge. *Frontiers in Psychology*, 14, 1154168. DOI: 10.3389/fpsyg.2023.1154168

Chapter 16
Evaluating the Impact of ACTA

ABSTRACT

This chapter examines methods for evaluating the effectiveness of Applied Cognitive Task Analysis (ACTA)-based interventions, focusing on both short-term and long-term impacts. Key assessment strategies include pre- and post-intervention evaluations, longitudinal studies, behavioral observations, self-reported measures, and performance metrics. These methods provide comprehensive insights into how ACTA improves decision-making, training efficiency, adaptability, and organizational learning. By assessing cognitive processes and task performance over time, the chapter highlights the value of ACTA in optimizing training, enhancing operational efficiency, and supporting sustained organizational development across various fields.

INTRODUCTION

Applied Cognitive Task Analysis (ACTA) is a method designed to understand the cognitive processes involved in complex tasks, with the aim of improving training, decision-making, and performance across various professional settings. Evaluating the impact of ACTA-based interventions is essential to determine their effectiveness and validate their continued application in different fields. By measuring how well these interventions enhance task performance, reduce errors, or streamline processes, organizations can justify their investment in ACTA.

This chapter focuses on the methods used to assess the effectiveness and impact of ACTA-based interventions. It explores key evaluation techniques, including performance metrics, feedback mechanisms, and long-term outcome tracking, to provide a comprehensive view of how ACTA interventions contribute to organizational success. Additionally, we will examine the broader benefits these interventions

DOI: 10.4018/979-8-3693-7733-8.ch016

offer, such as improved efficiency, safety, and decision-making, and how they can lead to sustained growth and innovation in various industries.

16.1 Methods for Assessing the Effectiveness and Impact of ACTA-Based Interventions

Evaluating the effectiveness of Applied Cognitive Task Analysis (ACTA)-based interventions involves using a comprehensive and multi-dimensional approach that incorporates both quantitative and qualitative methods. These methods provide a holistic view of the intervention's impact, capturing not only measurable changes in performance but also subjective experiences and behavioural modifications. The following are key methods used to assess the impact of ACTA-based interventions:

16.1.1 Pre- and Post-Intervention Assessments

Pre- and post-intervention assessments are one of the most common and straightforward methods for evaluating the effectiveness of ACTA interventions. These assessments involve measuring the knowledge, skills, and performance levels of participants before and after the intervention to evaluate changes over time.

- Implementation and Analysis: Typically, participants undergo a baseline assessment prior to the intervention, followed by a similar assessment after completing the intervention. This approach allows researchers to directly compare the pre-intervention and post-intervention data, identifying any improvements or declines in performance.
- Recent Research: For instance, a study by Edwards et al. (2021) systematically reviewed the efficacy of Cognitive Task Analysis (CTA)-based training in surgical education. The study included randomized controlled trials and observational studies that measured the impact of CTA-based interventions on the procedural knowledge and technical performance of novices, such as surgical trainees and medical students. The results demonstrated significant improvements in both procedural knowledge and technical performance for those who underwent CTA-based training compared to those who received conventional training methods. Specifically, the standardized mean difference (SMD) for procedural knowledge was 1.36, and for technical performance, it was 2.06, both showing statistically significant enhancements (Edwards et al., 2021). This evidence of improved outcomes through CTA-based training supports the applicability of similar methods within ACTA interventions. By demonstrating significant improvements in pre- and post-assessments, this

study reinforces the value of using such assessments to evaluate the impact of ACTA-based training programs.

By applying pre- and post-intervention assessments, researchers and educators can obtain a clear and quantifiable understanding of the impact of training programs like ACTA. This method not only validates the training's effectiveness but also provides insights into areas that may require further improvement or adaptation.

16.1.2 Longitudinal Studies

Longitudinal studies are critical for assessing the long-term impact and sustainability of ACTA-based interventions. Unlike pre- and post-intervention assessments, which provide a snapshot of immediate changes, longitudinal studies track the performance and outcomes of individuals or groups over an extended period.

- Implementation and Analysis: By collecting data at multiple time points, longitudinal studies can identify trends and long-term outcomes, such as retention of skills, adaptation to new challenges, and sustained behavioural changes. This method is particularly valuable in understanding how ACTA-based training affects cognitive processes and task performance in the long run.
- Recent Research: Swaby et al. (2022) conducted a study that examined the use of Cognitive Task Analysis (CTA) methods to understand expert clinical decision-making across various settings. Although the study primarily focused on documenting and analysing expert knowledge and decision-making processes, its findings have implications for longitudinal research. The study reviewed 81 articles on CTA applications, highlighting how these methods help elucidate expert decision-making under different conditions. The insights gained from such CTA studies can be instrumental in designing longitudinal studies that track the long-term effects of ACTA-based interventions. The study by Swaby et al. (2022) emphasizes the increasing use of CTA methods to improve understanding of complex decision-making in clinical environments. This background provides a foundation for applying similar methodologies in longitudinal studies to evaluate how ACTA-based training impacts performance and decision-making over time. By integrating CTA insights into longitudinal research, we can better understand how training interventions contribute to sustained improvements and adaptability in real-world settings.

By incorporating longitudinal studies, researchers can gain a deeper and more comprehensive understanding of the long-term effects of ACTA-based interventions. This approach extends beyond immediate performance changes, allowing for the assessment of skill retention, adaptability to new challenges, and sustained behavioural improvements over time. Longitudinal research, informed by CTA methods as highlighted by Swaby et al. (2022), provides valuable insights into the enduring impact and efficacy of training programs. It helps identify how interventions influence cognitive processes and task performance in the long run, ensuring that the benefits of ACTA-based training are not only realized in the short term but also sustained and meaningful over time.

16.1.3 Behavioural Observations

Behavioural observations involve the real-time monitoring and recording of participants' actions during task performance. This method provides direct insights into how ACTA training influences behaviour and task execution, allowing for a nuanced understanding of the intervention's impact on practical skills.

- Implementation and Analysis: Observational data can be collected in naturalistic settings where participants perform their regular duties or in controlled environments designed to simulate real-world conditions. Researchers typically use a structured coding scheme to document specific behaviours, such as decision-making processes, error rates, and adaptive strategies. This systematic approach allows for the detailed analysis of how training impacts task performance and cognitive processes.
- Recent Research: In a study by Arrabito and Banbury (2019), cognitive task analysis (CTA) methods were used to inform the design of human-machine interfaces for Unmanned Aircraft Systems (UAS). The study involved conducting CTA to understand the cognitive processes of UAS operators and intelligence analysts. Behavioural observations were integral to this research, as they provided real-time insights into how operators and analysts interact with the system, make decisions, and handle complex, time-sensitive situations. The findings highlighted the importance of integrating behavioural observations with CTA to refine interface design and improve operational efficiency. By capturing how operators' actions align with cognitive processes, the study demonstrated how observational data can enhance the application of CTA in practical settings. In addition, the study's focus on critical situation awareness elements further underscores the value of behavioural observations. Observing how operators manage critical situations and interact with the system offers a deeper understanding of the cognitive demands and

practical challenges of the task. This real-time data is crucial for developing effective training programs and designing interfaces that support improved decision-making and task performance.

The study's findings underscore the importance of integrating behavioural observations with CTA to refine interface design and enhance operational efficiency. By capturing operators' actions and aligning them with the cognitive processes identified through CTA, the research illustrated how observational data can inform the practical application of ACTA. This integration of CTA and behavioural observations helps to bridge the gap between theoretical understanding and real-world implementation, showcasing how ACTA can be effectively used to improve system design and training in complex environments. This approach ensures that the cognitive demands and practical challenges faced by users are addressed, leading to more effective and user-centred solutions.

16.1.4 Self-Reported Measures

Self-reported measures, such as surveys, questionnaires, and interviews, are used to gather subjective data from participants regarding their experiences, perceptions, and attitudes toward the ACTA interventions. These measures are important for capturing the psychological and emotional aspects of training, which can influence overall effectiveness.

- Implementation and Analysis: Self-reported measures often involve collecting data through various tools that ask participants to provide feedback on their experiences with the training. This can include perceptions of the training's relevance, perceived improvements in cognitive skills, and the overall impact on their professional performance. Analysing this data helps to understand how participants view the intervention and how it affects their engagement and learning outcomes.
- Recent Research: A study by Gore and McDowall (2018) focused on the application of Applied Cognitive Task Analysis (ACTA) in engineering professionals. This longitudinal project involved training engineers to use ACTA techniques to document cognitive processes within their field. Self-reported measures were employed to assess the effectiveness of the ACTA methods from the participants' perspectives. Engineers provided feedback on their experiences with the training and the perceived impact on their cognitive expertise and practical skills. The study revealed that self-reported measures were instrumental in understanding how ACTA techniques impacted participants' cognitive processes and professional practice. Engineers reported high face

validity for the ACTA methods, indicating that these techniques effectively captured detailed cognitive and perceptual skills that were previously undocumented. This subjective feedback was crucial for refining ACTA techniques and validating their effectiveness in a professional context. The study demonstrated how integrating self-reported measures with ACTA can enhance the evaluation and refinement of cognitive training interventions.

Self-reported measures are a valuable tool for gaining insights into participants' subjective experiences with ACTA interventions. By capturing personal feedback on the relevance, effectiveness, and impact of the training, these measures provide a deeper understanding of how cognitive training affects individuals beyond objective performance metrics. The study by Gore and McDowall (2018) exemplifies how self-reported data can complement cognitive analysis techniques by highlighting participants' perceptions of the training's effectiveness and its influence on their cognitive skills and professional practice. This integration of subjective feedback with ACTA helps to refine training methods, ensuring they are both effective and well-received by participants. Overall, self-reported measures enhance the evaluation of ACTA interventions, offering a comprehensive view of their impact and guiding future improvements in cognitive training programs.

16.1.5 Performance Metrics

Performance metrics are crucial for objectively assessing the effectiveness of ACTA interventions, providing quantifiable evidence of their impact on task performance and efficiency. These metrics often include task completion time, error rates, accuracy, and productivity levels.

- Implementation and Analysis: Performance metrics are typically collected through a combination of direct observations, surveys, and specialized assessment tools. These metrics allow researchers to measure specific outcomes and evaluate how ACTA interventions influence various aspects of job performance.
- Recent Research: In the study by Ste-Croix and Angel (2021), cognitive task analysis was used to evaluate the information requirements and cognitive load experienced by dismounted infantry soldiers during complex scenarios. Performance metrics were captured using tools such as the NASA Task Load Index, which provided valuable data on cognitive demands and task performance. The study revealed that roles like Company 2IC and Platoon Commander experienced higher cognitive workloads, and identified opportunities for technological improvements to alleviate cognitive burden. This

application of performance metrics underscores their importance in assessing the effectiveness of ACTA interventions and guiding enhancements to operational efficiency.

The use of performance metrics provides a rigorous and objective framework for evaluating the effectiveness of ACTA interventions. By measuring specific outcomes such as task completion time, error rates, and accuracy, researchers can obtain detailed insights into how ACTA influences job performance and operational efficiency. The study by Ste-Croix and Angel (2021) exemplifies the application of performance metrics in high-stakes environments, highlighting how these metrics can reveal critical aspects of cognitive load and task demands. The findings underscore the value of performance metrics in identifying areas for improvement and guiding the development of interventions that enhance both individual and organizational effectiveness. As ACTA continues to evolve, integrating robust performance metrics will be essential for validating its impact and refining its application in diverse settings.

16.1.6 Review of ACTA-Based Interventions: Evaluation Framework and Impact

Evaluating the effectiveness and impact of Applied Cognitive Task Analysis (ACTA)-based interventions requires a comprehensive, multi-dimensional approach that integrates both quantitative and qualitative methods. This holistic evaluation captures not only measurable changes in performance but also subjective experiences and behavioural adjustments.

Key methods for assessing ACTA interventions include pre- and post-intervention assessments, longitudinal studies, behavioural observations, self-reported measures, and performance metrics. Pre- and post-intervention assessments provide a direct comparison of participants' knowledge, skills, and performance before and after the intervention, offering clear evidence of its impact. Longitudinal studies are crucial for understanding the long-term effects of ACTA-based interventions. By tracking participants' performance and outcomes over time, these studies reveal how well skills and knowledge are retained and how individuals adapt to new challenges. In addition, behavioural observations give real-time insights into how ACTA training influences task execution and decision-making. This method helps in understanding practical applications and refining training programs based on observed behaviours. Self-reported measures capture participants' subjective experiences, providing valuable feedback on the relevance and effectiveness of the training. This helps to assess how well the intervention meets participants' needs and its impact on their cognitive processes. Performance metrics offer objective data on task performance, including aspects such as completion time, error rates, and accuracy. These metrics provide quantifiable evidence of how ACTA interventions affect job performance and operational efficiency.

By integrating these diverse methods, researchers and practitioners can obtain a comprehensive understanding of ACTA-based interventions. This approach ensures that the effectiveness of training programs is thoroughly evaluated, supporting the continued refinement and application of ACTA across various fields and industries. Such an evaluation framework not only highlights the benefits of ACTA interventions but also guides ongoing improvements, contributing to enhanced performance and organizational development.

16.2 Long-Term Outcomes and Benefits for Organizations and Industries

The long-term outcomes and benefits of ACTA-based interventions extend well beyond immediate performance improvements. These benefits encompass enhanced organizational capabilities, reduced training costs, and increased adaptability to complex and dynamic environments. Understanding these long-term effects helps organizations justify and optimize their investment in ACTA-based training.

16.2.1 Enhanced Decision-Making and Problem-Solving Skills

ACTA interventions are designed to improve individuals' cognitive abilities, which in turn enhances decision-making and problem-solving skills. This is particularly valuable in high-stakes environments such as healthcare, military operations, and emergency services, where effective decision-making can significantly impact outcomes.

- Research and Evidence: ACTA has been shown to improve decision-making skills in various professional settings. For example, Militello et al. (2023) adapted CTA methods for a large-scale, multi-site study involving 102 anaesthesiologists working through high-fidelity simulated incidents. This research utilized cognitive interviews to reveal how decision processes varied under different scenarios. The study faced challenges in standardizing interview techniques and analysing extensive qualitative data but successfully provided valuable insights into decision-making. The findings highlighted individual differences in how anaesthesiologists approached complex cases, demonstrating the efficacy of ACTA in enhancing decision-making and problem-solving skills in high-stakes environments. This application of ACTA not only addresses immediate decision-making challenges but also contributes to long-term improvements in cognitive processes crucial for effective problem-solving in critical situations. Similarly, Karam Ahmad et al. (2020) explored the application of CTA in orthopaedic and trauma training. Their systematic review indicated that CTA tools offer significant benefits in training novice surgeons, including improved decision-making and problem-solving skills. The study highlighted that CTA provides both objective and subjective advantages over traditional methods, enhancing trainees' ability to manage complex surgical procedures effectively. By incorporating CTA into orthopaedic training, the study demonstrated how these methods support better decision-making and problem-solving in high-pressure environments, further validating the role of ACTA in improving cognitive skills in critical fields.

16.2.2 Improved Training Efficiency

ACTA-based training focuses on the critical cognitive elements of tasks, leading to a more efficient training process. By targeting the core cognitive processes required for task performance, ACTA can reduce the time and resources needed for training while maximizing learning outcomes.

- Research and Evidence: Simon and Richards (2022) demonstrate the efficacy of ACTA in enhancing training efficiency through their work with elite coaches preparing athletes for the Tokyo 2020 Olympic and Paralympic Games. Their study utilized ACTA to develop a Shared Mental Model (SMM) that aligned coaching and sport science experts to improve performance. By breaking down the complex task of the 200m race start into discrete components, ACTA provided a detailed understanding of the cognitive and skill requirements, which had previously been overlooked. This granular analysis enabled the refinement of training processes, making them more focused and efficient.

The study highlighted how ACTA's approach facilitated more efficient training by enabling coaches to address specific cognitive demands and skill components, thus optimizing the use of time and resources. The application of ACTA led to significant real-world impacts, including the achievement of an Olympic medal, multiple Paralympic medals, and World Records. This demonstrates how targeting critical cognitive elements through ACTA can lead to substantial improvements in training efficiency and performance outcomes. Furthermore, the insights gained from ACTA enabled the development of tailored and effective training plans, enhancing overall training efficiency and ensuring clearer alignment of roles and responsibilities within the coaching team.

16.2.3 Increased Adaptability to Changing Environments

In today's rapidly evolving work environments, adaptability is crucial for maintaining a competitive edge. ACTA-based interventions help individuals and organizations adapt to new technologies, processes, and regulations by providing a deeper understanding of cognitive processes and task demands.

- Research and Evidence: The study by Stanik et al. (2022) on preclinical patient care demonstrates how ACTA can facilitate adaptability in high-stress and complex environments, such as emergency medical services. The project aimed to develop a combined medical device that would reduce the mental and physical load on emergency teams who often need to carry and manage multiple heavy and complex medical devices. By applying a customized, fully digital version of the ACTA method, the study systematically elicited requirements for the user interface design of the new combined device. Through the creation of a detailed hierarchical task diagram, interviews with subject matter experts, and collaborative design workshops, the study identified critical cognitive and physical demands faced by emergency teams. This comprehensive approach enabled the development of innovative solutions that could better meet the needs of users in dynamic and unpredictable environments. The adaptability demonstrated by this study in developing and refining a new device to streamline emergency response tasks highlights the value of ACTA in helping organizations and individuals respond effectively to technological and procedural changes in their work environments. This case illustrates how ACTA-based methodologies can support the adaptation to new tools and workflows, improving efficiency and effectiveness under rapidly changing conditions. By focusing on the cognitive aspects of task performance and integrating insights from diverse stakeholders, ACTA fosters an environment where adaptability is not only possible but optimized, thus maintaining a high level of performance despite evolving challenges.

16.2.4 Sustained Organizational Learning and Development

ACTA promotes a culture of continuous learning and development within organizations. By regularly assessing cognitive tasks and updating training programs based on these analyses, organizations can ensure their workforce remains competent and current with the latest skills and knowledge.

- Research and Evidence: Taylor et al. (2023) examined the cognitive challenges faced by high-performance endurance sport coaches using Applied

Cognitive Task Analysis (ACTA). Despite extensive empirical work in sport coaching, significant gaps remain in understanding coaching expertise at various performance levels. This study addressed these gaps by applying ACTA to investigate the cognitive skills and knowledge used by seven high-performance endurance sport coaches from a national governing body. The findings revealed prominent and ongoing challenges in day-to-day coaching practice, requiring substantial adaptive skills. Notably, coaches were found to use their knowledge flexibly and conditionally to meet the demands of their roles, with a novel insight being the use of curriculum knowledge to project the future needs of athletes. These results suggest valuable opportunities for utilizing ACTA to explore cognitive challenges in sport coaching and enhance development practices. By identifying the cognitive demands and adaptive skills required for high-performance coaching, organizations can refine their training programs to better support coaches. This approach fosters sustained organizational learning by ensuring that coaching practices evolve in response to identified challenges, ultimately contributing to improved performance outcomes. In parallel, Marshall and Finlayson (2022) focused on the cognitive skills required of surgical nurses managing deteriorating patients and identified common errors made by less experienced nurses. Their study used ACTA to reveal critical cognitive demands such as early recognition of physiological deterioration and decision-making skills. The findings emphasized the importance of continuously updating training programs to address these cognitive demands, promoting sustained learning and development within healthcare settings. Together, these studies underscore the effectiveness of ACTA in fostering a culture of continuous improvement and adaptability within organizations. By regularly assessing and addressing cognitive challenges, organizations can enhance their training programs, support ongoing professional development, and maintain high standards of performance and care.

16.2.5 Review of ACTA-Based Interventions: Long-Term Outcomes and Strategic Benefits

The long-term outcomes of ACTA-based interventions extend far beyond immediate performance gains, offering significant benefits to organizations across various sectors. By enhancing decision-making and problem-solving skills, ACTA equips individuals to perform more effectively under pressure, particularly in high-stakes environments like healthcare and emergency response. Studies, such as Militello et al. (2023) and Ahmad et al. (2020), demonstrate how ACTA improves cognitive processes in complex tasks, resulting in better decision-making and problem-solving outcomes.

In addition to improving individual performance, ACTA enhances training efficiency by focusing on critical cognitive elements, reducing time and resources while maximizing learning outcomes. Simon and Richards (2022) highlighted this in their work with elite athletes, where ACTA helped streamline training processes, leading to significant real-world successes.

ACTA also fosters adaptability in rapidly changing environments. By providing deeper insights into cognitive demands, organizations can better navigate transitions and integrate new technologies, as shown in Stanik et al. (2022), where ACTA contributed to the development of innovative medical devices for emergency teams.

Finally, ACTA supports sustained organizational learning by promoting continuous assessment and refinement of cognitive skills, as seen in studies like Taylor et al. (2023) and Marshall and Finlayson (2022). This continuous improvement process ensures that organizations remain competitive, maintain high-performance standards, and keep their workforce current with evolving industry practices.

In conclusion, the strategic value of ACTA lies in its comprehensive approach to addressing cognitive challenges, enhancing training methodologies, and supporting both individual and organizational growth. By improving decision-making and problem-solving abilities, streamlining training processes, and promoting adaptability and continuous learning, ACTA-based interventions offer significant long-term benefits for organizations across various industries. This holistic approach ensures that organizations not only achieve but also sustain excellence in today's complex and rapidly changing professional environments.

CONCLUSION

In conclusion, the evaluation of Applied Cognitive Task Analysis (ACTA)-based interventions highlights their profound and lasting impact across various industries. A comprehensive, multi-dimensional assessment framework combining both quantitative and qualitative methods is essential to capture the full range of benefits ACTA offers. Pre- and post-intervention assessments, longitudinal studies, behavioural observations, self-reported measures, and performance metrics collectively provide a thorough understanding of ACTA's effectiveness, both in immediate performance improvements and long-term outcomes.

ACTA significantly enhances decision-making and problem-solving skills, particularly in high-pressure environments where precise cognitive processes are crucial. By focusing on critical cognitive elements, ACTA interventions also increase training efficiency, reducing time and resources while maximizing learning outcomes. Furthermore, ACTA fosters adaptability, helping individuals and organizations navigate rapidly changing environments and integrate new technologies more effectively.

Sustained organizational learning and development are key long-term benefits of ACTA-based interventions, promoting continuous assessment and refinement of cognitive skills. This ensures that organizations remain agile, competitive, and capable of meeting the challenges posed by evolving industry demands. Ultimately, the strategic value of ACTA lies in its holistic approach to addressing cognitive challenges, improving training methodologies, and driving both individual and organizational growth, making it a crucial tool for achieving and sustaining excellence in complex professional environments.

REFERENCES

Ahmad, K., Bhattacharyya, R., & Gupte, C. (2020). Using cognitive task analysis to train orthopaedic surgeons: Is it time to think differently? A systematic review. *Annals of Medicine and Surgery (London)*, 59, 131–137. DOI: 10.1016/j.amsu.2020.09.031

Arrabito, G. R., Hou, M., & Banbury, S. (2019). *Cognitive task analysis for an unmanned aircraft system (DRDC-RDDC-2019-D041)*. Defence Research and Development Canada.

Edwards, T. C., Coombs, A. W., Szyszka, B., Logishetty, K., & Cobb, J. P. (2021). Cognitive task analysis-based training in surgery: A meta-analysis. *BJS Open*, 5(6), zrab122. Advance online publication. DOI: 10.1093/bjsopen/zrab122

Gore, J., Banks, A. P., & McDowall, A. (2018). Developing cognitive task analysis and the importance of socio-cognitive competence/insight for professional practice. *Cognition Technology and Work*, 20(4), 555–563. DOI: 10.1007/s10111-018-0502-2

Marshall, D., & Finlayson, M. (2022). Applied cognitive task analysis methodology: Fundamental cognitive skills surgical nurses require to manage patient deterioration. *Nursing Praxis in Aotearoa New Zealand*, 38(1), 25–37. DOI: 10.36951/27034542.2022.04

Militello, , L. GSalwei, , M. EReale, , CSushereba, , CSlagle, , J. MGaba, , DWeinger, , M. BRask, , JFaiman, , JAndreae, , MBurden, , A. RAnders, , S. (2023). Adapting cognitive task analysis methods for use in a large sample simulation study of high-risk healthcare events. *Journal of Cognitive Engineering and Decision Making, 17*(4), 315–331. DOI: 10.1177/15553434231192283

Simon, S., & Richards, P. (2022). (Manuscript submitted for publication). Investigating cognitive observational difficulties and expert skills in elite Olympic and Paralympic coaches using an applied cognitive task analysis. *Journal of Expertise*.

Stanik, C., Puhlfürß, T., Mahler, A., Sasu, P. B., Reip, W., & Maalej, W. (2022). Lessons learned from customizing and applying ACTA to design a novel device for emergency medical care. *arXiv*. https://arxiv.org/abs/2108.05622

Ste-Croix, C., & Angel, H. (2021). *Cognitive task analysis of a complex battle position scenario*. Human Systems Incorporated.

Swaby, L., Shu, P., & Hind, D.. (2022). The use of cognitive task analysis in clinical and health services research: A systematic review. *Pilot and Feasibility Studies*, 8, 57. DOI: 10.1186/s40814-022-01002-6

Taylor, J., Ashford, M., & Jefferson, M. (2023). High performance coach cognition in the wild: Using applied cognitive task analysis for practical insights—Cognitive challenges and curriculum knowledge. *Frontiers in Psychology*, 14. Advance online publication. DOI: 10.3389/fpsyg.2023.1154168

Compilation of References

Abdalla, M. M., Oliveira, L. G. L., Azevedo, C. E. F., & Gonzalez, R. K. (2018). Quality in Qualitative Organizational Research: Types of triangulation as a methodological alternative. *Administração: Ensino e Pesquisa*, 19(1), 66–98. DOI: 10.13058/raep.2018.v19n1.578

Adarkwah, M. A. (2021). The power of assessment feedback in teaching and learning: A narrative review and synthesis of the literature. *SN Social Sciences*, 1, 75. DOI: 10.1007/s43545-021-00086-w

Ahmad, K., Bhattacharyya, R., & Gupte, C. (2020). Using cognitive task analysis to train orthopaedic surgeons - Is it time to think differently? A systematic review. *Annals of Medicine and Surgery (London)*, 59, 131–137. DOI: 10.1016/j.amsu.2020.09.031

Ahmad, M. I., Keller, I., & Robb, D. A. (2023). A framework to estimate cognitive load using physiological data. *Personal and Ubiquitous Computing*, 27(4), 2027–2041. DOI: 10.1007/s00779-020-01455-7

Aitchanov, B., Nussipbekov, A., and Zhaparov, M. (2012). Microlearning of web fundamentals based on mobile learning. *IJCSI International Journal of Computer Science Issues, 9*, (6:3), 148150.

Aksu, Ş. H., Çakıt, E., & Dağdeviren, M. (2024). Mental workload assessment using machine learning techniques based on EEG and eye tracking data. *Applied Sciences (Basel, Switzerland)*, 14(6), 2282. DOI: 10.3390/app14062282

Al Mamun, M. A., & Lawrie, G. (2023). Student-content interactions: Exploring behavioural engagement with self-regulated inquiry-based online learning modules. *Smart Learning Environments*, 10(1), 1. DOI: 10.1186/s40561-022-00221-x

Alam, T., & Benaida, M. (2022). Smart curriculum mapping and its role in outcome-based education. *Informatica (Vilnius)*, 46(4). Advance online publication. DOI: 10.31449/inf.v46i4.3717

Al-Areibi, I., Dickson, B. A., & Kotsopoulos, D. (2022). An analysis of synchronous and asynchronous online undergraduate motivation during the COVID pandemic. *International Journal of E-Learning & Distance Education*, 37(1). Advance online publication. DOI: 10.55667/ijede.2022.v37.i1.1213

Alasmari, T. M. (2020). Can mobile learning technology close the gap caused by gender segregation in the Saudi educational institutions? *Journal of Information Technology Education*, 19, 655–670. DOI: 10.28945/4634

Alias, N. F. (2024). *Developing pedagogical practices of microlearning educational experience in higher education* (Doctoral dissertation). University of Malaya.

Almubark, I., Chang, L. C., Shattuck, K. F., Nguyen, T., Turner, R. S., & Jiang, X. (2020). A 5-min cognitive task with deep learning accurately detects early Alzheimer's disease. *Frontiers in Aging Neuroscience*, 12, 603179. DOI: 10.3389/fnagi.2020.603179

Alqurashi, E. (2017). Microlearning: A pedagogical approach for technology integration. *The Turkish Online Journal of Educational Technology*, 16, 942–947.

Alqurni, J. S. (2023). Evaluating the user interface and usability approaches for e-learning systems. *International Journal of Information Technology and Web Engineering*, 18(1), 1–25. DOI: 10.4018/IJITWE.333638

Al-Zahrani, A. M. (2024). Enhancing postgraduate students' learning outcomes through flipped mobile-based microlearning. *Research in Learning Technology*, 32. Advance online publication. DOI: 10.25304/rlt.v32.3110

Amanda, F. F., Sumitro, S., Lestari, S. R., & Ibrohim, I. (2023). Enhancing critical thinking and problem-solving skills by complexity science-problem based learning model. *Multidisciplinary Journal of Educational Research*. Advance online publication. DOI: 10.17583/remie.9409

Arellano, L., Alcubilla, P., & Leguízamo, L. (2023). *Ethical considerations in informed consent*. IntechOpen. DOI: 10.5772/intechopen.1001319

Armah, J. K., Bervell, B., & Bonsu, N. O. (2023). Modelling the role of learner presence within the community of inquiry framework to determine online course satisfaction in distance education. *Heliyon*, 9(5). Advance online publication. DOI: 10.1016/j.heliyon.2023.e15803

Arrabito, G. R., Hou, M., & Banbury, S. (2019). *Cognitive task analysis for an unmanned aircraft system* (DRDC-RDDC-2019-D041). Defence Research and Development Canada.

Arrabito, G. R., Hou, M., & Banbury, S. (2019). *Cognitive task analysis for an unmanned aircraft system (DRDC-RDDC-2019-D041)*. Defence Research and Development Canada.

Asma, H., & Dallel, S. (2020). Cognitive load theory and its relation to instructional design: Perspectives of some Algerian university teachers of English. *Arab World English Journal*, 11(4), 110–127. DOI: 10.24093/awej/vol11no4.8

Ataizi, M. (2012). Situated Cognition. In Seel, N. M. (Ed.), *Encyclopedia of the Sciences of Learning*. Springer. DOI: 10.1007/978-1-4419-1428-6_16

Bahati, B., Fors, U., Hansen, P., Nouri, J., & Mukama, E. (2019). Measuring learner satisfaction with formative e-assessment strategies. *International Journal of Emerging Technologies in Learning*, 14(7), 61–79. DOI: 10.3991/ijet.v14i07.9120

Barthakur, A., Dawson, S., & Kovanovic, V. (2023). Advancing learner profiles with learning analytics: A scoping review of current trends and challenges. In *Proceedings of the 13th International Learning Analytics and Knowledge Conference (LAK23)*. DOI: 10.1145/3576050.3576083

Bellhäuser, H., Dignath, C., & Theobald, M. (2023). Daily automated feedback enhances self-regulated learning: A longitudinal randomized field experiment. *Frontiers in Psychology*, 14, 1125873. DOI: 10.3389/fpsyg.2023.1125873

Ben Charif, A., Zomahoun, H. T. V., Gogovor, A., Samri, M. A., Massougbodji, J., Wolfenden, L., Ploeg, J., Zwarenstein, M., Milat, A. J., Rheault, N., Ousseine, Y. M., Salerno, J., Markle-Reid, M., & Légaré, F. (2022). Tools for assessing the scalability of innovations in health: A systematic review. *Health Research Policy and Systems*, 20(1), 34. DOI: 10.1186/s12961-022-00830-5

Berlin, C., & Praetorius, G. (2023). Applied Cognitive Task Analysis (ACTA) of marine piloting in a Swedish Context. *Human Factors in Transportation*, 95, 709–718. DOI: 10.54941/ahfe1003856

Bhatia, M. B., Collings, A. T., & Gillio, A.. (2022). Enhancing surgical education for resource-limited settings: Open appendectomy cognitive task analysis. *Global Surgical Education : Journal of the Association for Surgical Education*, 1, 61. DOI: 10.1007/s44186-022-00064-7

Bielka, K., Kuchyn, I., & Fomina, H.. (2024). Difficult airway simulation-based training for anaesthesiologists: Efficacy and skills retention within six months. *BMC Anesthesiology*, 24, 44. DOI: 10.1186/s12871-024-02423-x

Biggs, J. (2014). Constructive alignment in university teaching. *HERDSA Review of Higher Education*, 1, 5–22.

Blacker, K. J., Hamilton, J., Roush, G., Pettijohn, K., & Biggs, A. T. (2019). Cognitive training for military application: A review of the literature and practical guide. *Journal of Cognitive Enhancement: Towards the Integration of Theory and Practice*, 3(1), 31–47. DOI: 10.1007/s41465-018-0076-1

Black, P., & Wiliam, D. (2009). Developing the theory of formative assessment. *Educational Assessment, Evaluation and Accountability*, 21(1), 5–31. DOI: 10.1007/s11092-008-9068-5

Blease, C., Kelley, J. M., & Trachsel, M. (2018). Informed consent in psychotherapy: Implications of evidence-based practice. *Journal of Contemporary Psychotherapy*, 48(2), 69–78. DOI: 10.1007/s10879-017-9372-9

Bleich, C. (2018). [*Types of Microlearning for Your Employees.* EdgePoint Learning. https://www.edgepointlearning.com/blog/types-of-microlearning/]. *Top (Madrid)*, 11.

Blitz, M. J., & Barfield, W. (2023). Memory enhancement and brain–computer interface devices: Technological possibilities and constitutional challenges. In V. Dubljević & A. Coin (Eds.), *Policy, identity, and neurotechnology* (Advances in Neuroethics, pp. 1-22). Springer, Cham. DOI: 10.1007/978-3-031-26801-4_12

Bloom, B. S. (1956) Taxonomy of Educational Objectives. In B. S., Bloom, M. D., Engelhart, E. J., Furst, W. H., Hill, & D. R. Krathwohl. (Eds.). *Handbook: The Cognitive Domain.* David McKay Publication.

Blyznyuk, T., & Kachak, T. (2024). Benefits of interactive learning for students' critical thinking skills improvement. *Journal of Vasyl Stefanyk Precarpathian National University*, 11(1), 94–102. DOI: 10.15330/jpnu.11.1.94-102

Bosch, C., & Laubscher, D. J. (2022). Promoting self-directed learning as learning presence through cooperative blended learning. *International Journal of Learning. Teaching and Educational Research*, 21(9), 18–33. DOI: 10.26803/ijlter.21.9.2

Bui, P. U., Tong, D. H., & Lien, N. B. (2022). The effectiveness of experiential learning in teaching arithmetic and geometry in sixth grade. *Frontiers in Education*, 7. Advance online publication. DOI: 10.3389/feduc.2022.858631

Callisen, L. (2019). *Why Micro Learning is the Future of Training in the Workplace.* eLearning Industry. https://elearningindustry.com/micro-learning-future-of-trainingworkplace

Cannon-Bowers, J., Bowers, C., Stout, R., Ricci, K., & Hildabrand, A. (2013). Using cognitive task analysis to develop simulation-based training for medical tasks. *Military Medicine*, 178(Suppl_10), 15–21. DOI: 10.7205/MILMED-D-13-00211

Cardino Jr, J. M., & Ortega-Dela Cruz, R. A. (2020). Understanding of learning styles and teaching strategies towards improving the teaching and learning of mathematics. *LUMAT: International Journal on Math. Science and Technology Education*, 8(1), 19–43. DOI: 10.31129/LUMAT.8.1.1348

Castro-Alonso, J. C., de Koning, B. B., & Fiorella, L. (2021). Five strategies for optimizing instructional materials: Instructor- and learner-managed cognitive load. *Educational Psychology Review*, 33, 1379–1407. DOI: 10.1007/s10648-021-09606-9

Cevallos Menéndez, I. Y., Cobeña Napa, M. A., Mendoza Moreira, M. L., & Vélez Zambrano, G. G. (2019). The importance of formative assessment in the learning teaching process. *International Journal of Social Science and Humanity*, 3(2), 238–249. DOI: 10.29332/ijssh.v3n2.322

Chakladar, D., & Roy, P. P. (2023). Cognitive workload estimation using physiological measures: A review. *Cognitive Neurodynamics*. Advance online publication. DOI: 10.1007/s11571-023-10051-3

Chatters, R., Dimairo, M., & Cooper, C.. (2024). Exploring the barriers to, and importance of, participant diversity in early-phase clinical trials: An interview-based qualitative study of professionals and patient and public representatives. *BMJ Open*, 14, e075547. DOI: 10.1136/bmjopen-2023-075547

Chen, Z., Liang, N., Zhang, H., Li, H., Yang, Y., Zong, X., Chen, Y., Wang, Y., & Shi, N. (2023). Harnessing the power of clinical decision support systems: Challenges and opportunities. *Open Heart*, 10(1), e002432. DOI: 10.1136/openhrt-2023-002432

Choi, J. I., & Hannafin, M. (1995). Situated cognition and learning environments: Roles, structures, and implications for design. *Educational Technology Research and Development*, 43(2), 53–69. DOI: 10.1007/BF02300472

Choudhary, H., & Pandita, D. (2023). Maximizing learning outcomes in the digital age: The role of microlearning for Gen Z. *Development and Learning in Organizations*, Vol. ahead-of-print No. ahead-of-print. DOI: 10.1108/DLO-02-2023-0038

Christou, P. A. (2023). The use of artificial intelligence (AI) in qualitative research for theory development. *The Qualitative Report*, 28(9), 2739–2755. DOI: 10.46743/2160-3715/2023.6536

Cicekci, M. A., & Sadik, F. (2019). Teachers' and students' opinions about students' attention problems during the lesson. *Journal of Education and Learning*, 8(6), 15–27. DOI: 10.5539/jel.v8n6p15

Ciubotariu, I. I., & Bosch, G. (2022). Improving research integrity: A framework for responsible science communication. *BMC Research Notes*, 15(1), 177. DOI: 10.1186/s13104-022-06065-5

Clark, R. E., Feldon, D., van Merriënboer, J. J. G., Yates, K., & Early, S. (2008). Cognitive task analysis. In Spector, J. M., Merrill, M. D., van Merriënboer, J. J. G., & Driscoll, M. P. (Eds.), *Handbook of research on educational communications and technology* (3rd ed., pp. 577–593). Macmillan/Gale.

Collins, A., & Halverson, R. (2018). *Rethinking education in the age of technology: The digital revolution and schooling in America*. Teachers College Press.

Collins, R. (2019). Clinician cognitive overload and its implications for nurse leaders. *Nurse Leader*, 18(1). Advance online publication. DOI: 10.1016/j.mnl.2019.11.007

Contrino, M. F., Reyes-Millán, M., Vazquez-Villegas, P., & Membrillo-Hernández, J. (2024). Using an adaptive learning tool to improve student performance and satisfaction in online and face-to-face education for a more personalized approach. *Smart Learning Environments*, 11(1). Advance online publication. DOI: 10.1186/s40561-024-00292-y

Cremer, F., Sheehan, B., & Fortmann, M.. (2022). Cyber risk and cybersecurity: A systematic review of data availability. *The Geneva Papers on Risk and Insurance. Issues and Practice*, 47(4), 698–736. DOI: 10.1057/s41288-022-00266-6

Creswell, J. W., & Clark, V. L. P. (2017). *Designing and conducting mixed methods research* (3rd ed.). SAGE Publications.

Cuschieri, S. (2022). Understanding research ethics. In *A roadmap to successful scientific publishing* (pp. 13–25). Springer. DOI: 10.1007/978-3-030-99295-8_2

Davis, F. D. (1986). *A technology acceptance model for empirically testing new end-user information systems: Theory and results* [Doctoral dissertation, Massachusetts Institute of Technology].

de Vries, P., Van den Bogaard, M., & Overschie, M. (2018, September). Microlearning to support authentic learning in continuing education and for engineering students. In *SEFI2018 - Creativity, Innovation and Entrepreneurship for Engineering Education Excellence* (Conference proceedings).

Dessy, E., Van Puyvelde, M., & Mairesse, O. (2018). Cognitive performance enhancement: Do biofeedback and neurofeedback work? *Journal of Cognitive Enhancement: Towards the Integration of Theory and Practice*, 2, 12–42. DOI: 10.1007/s41465-017-0039-y

Dewey, J. (1938). *Experience and education*. Macmillan.

Dhakal, A., & Bobrin, B. D. (2023). *Cognitive Deficits*. Europe PMC.

Diaz Redondo, R. P., Rodríguez, M., Escobar, J., & Vilas, A. F. (2021). Integrating micro-learning content in traditional e-learning platforms. *Multimedia Tools and Applications*, 80, 3121–3151. DOI: 10.1007/s11042-020-09523-z

Du, X., Zhang, L., & Hung, J. L. (2022). Understand group interaction and cognitive state in online collaborative problem solving: Leveraging brain-to-brain synchrony data. *International Journal of Educational Technology in Higher Education*, 19, 52. DOI: 10.1186/s41239-022-00356-4

Edwards, T. C., Coombs, A. W., Szyszka, B., Logishetty, K., & Cobb, J. P. (2021). Cognitive task analysis-based training in surgery: A meta-analysis. *BJS Open*, 5(6), zrab122. Advance online publication. DOI: 10.1093/bjsopen/zrab122

El Yazidi, R. (2023, March-April). Strategies for promoting critical thinking in the classroom. International. *Journal of English Literature and Social Sciences*, 8(2). Advance online publication. DOI: 10.22161/ijels

Eslit, E. R. (2023). *Unveiling the cognitive depths: Delving into language learning and acquisition amidst the post-pandemic era* (ED628255). ERIC. https://files.eric.ed.gov/fulltext/ED628255.pdf

Ettarh, R., Al-hussaini, H., Hassan, S., & Kilarkaje, N. (2018). Declining attention span in medical students: Fact or fiction? *The FASEB Journal*, 32(S1). Advance online publication. DOI: 10.1096/fasebj.2018.32.1_supplement.506.6

Fackler, J. C., Watts, C., Grome, A., Miller, T., Crandall, B., & Pronovost, P. (2009). Critical care physician cognitive task analysis: An exploratory study. *Critical Care*, 13(R33). Advance online publication. DOI: 10.1186/cc7740

Fairhurst, N., Koul, R., & Sheffield, R. (2023). Students' perceptions of their STEM learning environment. *Learning Environments Research*, 26(3), 977–998. DOI: 10.1007/s10984-023-09463-z

Faludi, J., Hoffenson, S., Kwok, S. Y., Saidani, M., Hallstedt, S. I., Telenko, C., & Martinez, V. (2020). A research roadmap for sustainable design methods and tools. *Sustainability*, 12(81), 8174. DOI: 10.3390/su12198174

Fengyu, Z. (2023). The impact of vocabulary learning methods on students' vocabulary application skills. *English Language Teaching and Linguistics Studies*, 5(4), 206. DOI: 10.22158/eltls.v5n4p206

Fidan, M. (2023). The effects of microlearning-supported flipped classroom on pre-service teachers' learning performance, motivation and engagement. *Education and Information Technologies*, 28, 12687–12714. DOI: 10.1007/s10639-023-11639-2

Field, J. (2013). Examining listening. In Geranpayeh, A., & Taylor, L. (Eds.), *Examining Listening* (1st ed., pp. 77–158). Cambridge University Press.

Fiock, H. (2020). Designing a community of inquiry in online courses. *International Review of Research in Open and Distance Learning*, 21(1), 135–153. DOI: 10.19173/irrodl.v20i5.3985

Fitria, T. N. (2022). Microlearning in teaching and learning process: A review. *Cendekia: Jurnal Ilmu Sosial Bahasa dan Pendidikan, 2(*4), 114-135. DOI: 10.55606/cendikia.v2i4.473

Fitzgerald, J. H., & Tisdell, C. C. (2019, June). The impact of educational micro-content on the student learning experience. In *Proceedings of the 2019 International Conference on Mathematics, Science and Technology Teaching and Learning* (pp. 17-22). Association for Computing Machinery. DOI: 10.1145/3348400.3348412

Flavell, J. H. (1979). Metacognition and cognitive monitoring: A new area of cognitive–developmental inquiry. *The American Psychologist*, 34(10), 906–911. DOI: 10.1037/0003-066X.34.10.906

Fu, S., Gu, H., & Yang, B. (2020). The affordances of AI-enabled automatic scoring applications on learners' continuous learning intention: An empirical study in China. *British Journal of Educational Technology*, 51(5). Advance online publication. DOI: 10.1111/bjet.12995

Gacusan, J. C., Dangis, S. J., & Afalla, B. (2023). Tailoring education for alpha learners: Harnessing learning styles for maximum learning outcomes. *NURTURE*, 17(3), 302–313. DOI: 10.55951/nurture.v17i3.335

Gallegos Ayala, G. I., Haslacher, D., Krol, L. R., Soekadar, S. R., & Zander, T. O. (2023). Assessment of mental workload across cognitive tasks using a passive brain-computer interface based on mean negative theta-band amplitudes. *Frontiers in Neuroergonomics*, 4, 1233722. DOI: 10.3389/fnrgo.2023.1233722

Gardner, H. (2006). *Multiple intelligences: New horizons* (Completely rev. and updated.). BasicBooks.

Garrison, D. R., Anderson, T., & Archer, W. (2000). Critical inquiry in a text-based environment: Computer conferencing in higher education model. *The Internet and Higher Education*, 2(2-3), 87–105. DOI: 10.1016/S1096-7516(00)00016-6

Germuth, A. A. (2018, December). Professional Development that Changes Teaching and Improves Learning. *Journal of Interdisciplinary Teacher Leadership*, 2(1). Advance online publication. DOI: 10.46767/kfp.2016-0025

Ghafar, Z. N., Abdullkarim, S. T., & Mahmud, I. (2023). Microlearning as a learning tool for teaching and learning in acquiring language: Applications, advantages, and influences on the language. *Canadian Journal of Educational and Social Studies*, 3(2), 45–62. DOI: 10.53103/cjess.v3i2.127

Gherghel, C., Yasuda, S., & Kita, Y. (2023). Interaction during online classes fosters engagement with learning and self-directed study both in the first and second years of the COVID-19 pandemic. *Computers & Education*, 200, 104795. DOI: 10.1016/j.compedu.2023.104795

Ghimire, N. B. (2020). Teachers' professional development in community campuses of TU: Expectation and responsibility. *Siddhajyoti Interdisciplinary Journal*, 1, 63–73. DOI: 10.3126/sij.v1i0.34921

Gill, P., & Baillie, J. (2018). Interviews and focus groups in qualitative research: An update for the digital age. *British Dental Journal*. Advance online publication. DOI: 10.1038/sj.bdj.2018.815

Gill, P., Stewart, K., Treasure, E., & Chadwick, B. (2008). Methods of data collection in qualitative research: Interviews and focus groups. *British Dental Journal*, 204(6), 291–295. DOI: 10.1038/bdj.2008.192

Giurgiu, L. (2017). Microlearning an evolving trend. *Science Bulletin*, 22(1), 18–23. DOI: 10.151/bsaft-2017-003

Gleditsch, K. S. (2023). *This research has important policy implications....* Peace Economics, Peace Science and Public Policy. DOI: 10.1515/peps-2023-0002

Gong, B., Collins, C., & Amrein-Beardsley, A. (2022). An International Professional Development Collaboration: Supporting Teachers in the Kingdom of Saudi Arabia through an Immersion Program in the United States. [GESPR]. *Gulf Education and Social Policy Review*, 2(2), 115–143. DOI: 10.18502/gespr.v2i1.10045

Gore, J. (2013). *Transforming knowledge: Capturing engineers' cognitive expertise. Paper presented at the 11th International Conference on Naturalistic Decision Making*, Marseilles, France.

Gore, J., Banks, A., & Mcdowall, A. (2015). Advancing ACTA: Developing Socio-Cognitive Competence/Insight. *Proceedings of the 12th International Conference on Naturalistic Decision Making*, 1–4. Retrieved from http://www2.mitre.org/public/ndm/papers/GoreBanksMcDowall41315.pdf

Gore, J., Banks, A. P., & McDowall, A. (2018). Developing cognitive task analysis and the importance of socio-cognitive competence/insight for professional practice. *Cognition Technology and Work*, 20(4), 555–563. DOI: 10.1007/s10111-018-0502-2

Göschlberger, B., & Bruck, P. A. (2017, December). Gamification in mobile and workplace integrated microlearning. In *Proceedings of the 19th International Conference*. DOI: 10.1145/3151759.3151795

Gottler, A. (2023). Collaboration between instructional designers and subject matter experts in digital transformation projects. *Studies in Technology Enhanced Learning*, 3(2). Advance online publication. DOI: 10.21428/8c225f6e.93df9a6e

Graham, L. A., Gray, C., Wagner, T. H., Illarmo, S., Hawn, M. T., Wren, S. M., Iannuzzi, J., & Harris, A. H. S. (2023). Applying cognitive task analysis to health services research. *Health Services Research*, 58(2), 415–422. DOI: 10.1111/1475-6773.14106

Gunawardena, M., Bishop, P., & Aviruppola, K. (2024). Personalized learning: The simple, the complicated, the complex and the chaotic. *Teaching and Teacher Education, 139*(December 2023), 104429. DOI: 10.1016/j.tate.2023.104429

Haleem, A., Javaid, M., Qadri, M. A., & Suman, R. (2022). Understanding the role of digital technologies in education: A review. *Sustainable Operations and Computers*, 3, 275–285. DOI: 10.1016/j.susoc.2022.05.004

Hamdoun, M. W. (2021, February 5). *Constructive Alignment Approach: Enhancing Learning and Teaching* [Paper presentation]. The Nile TESOL 2021 International Virtual Conference. Sphinx Publishing Company.

Han, F. (2021). The relations between teaching strategies, students' engagement in learning, and teachers' self-concept. *Sustainability*, 13(9), 5020. DOI: 10.3390/su13095020

Hari, J. (2022). *Stolen Focus: Why You Can't Pay Attention – and How to Think Deeply Again*. Crown.

Harris-Thompson, D., Wiggins, S. L., & Ho, G. (2006). *Using cognitive task analysis to develop scenario-based training for house-clearing teams* (Final report, Contract # W7711-057971/001/TOR). Klein Associates Division, Applied Research Associates, Inc.

Hattie, J., & Timperley, H. (2007). The power of feedback. *Review of Educational Research*, 77(1), 81–112. DOI: 10.3102/003465430298487

Hodges, L. C. (2020). Student engagement in active learning classes. In Mintzes, J. J., & Walter, E. M. (Eds.), *Active learning in college science* (pp. 35–46). Springer., DOI: 10.1007/978-3-030-33600-4_3

Hoffman, R. R., & Militello, L. (2009). *Perspectives on Cognitive Task Analysis: Historical Origins and Modern Communities of Practice*. Psychology Press.

Holleman, G. A., Hooge, I. T. C., Kemner, C., & Hessels, R. S. (2020). The 'Real-World Approach' and Its Problems: A Critique of the Term Ecological Validity. *Frontiers in Psychology*, 11. Advance online publication. DOI: 10.3389/fpsyg.2020.00721

Holuša, V., Vaněk, M., Beneš, F., Švub, J., & Staša, P. (2023). Virtual reality as a tool for sustainable training and education of employees in industrial enterprises. *Sustainability*, 15(17), 12886. DOI: 10.3390/su151712886

Hug, T. (2005). Microlearning: A new pedagogical challenge. In Hug, T., Lindner, M., & Bruck, P. A. (Eds.), *Microlearning: Emerging concepts, practices and technologies after e-learning* (pp. 7–12). Innsbruck University Press.

Hultberg, P., Santandreu Calonge, D., & Safiullin Lee, A. E. (2018). Promoting Long-lasting Learning Through Instructional Design. *The Journal of Scholarship of Teaching and Learning*, 18(3), 26–43. DOI: 10.14434/josotl.v18i3.23179

Hung, S. C.-w., Ho, A. Y.-n., Lai, I. H.-w., Lee, C. S.-w., Pong, A. S.-k., & Lai, F. H.-y. (2020). Meta-analysis on the effectiveness of virtual reality cognitive training (VRCT) and computer-based cognitive training (CBCT) for individuals with mild cognitive impairment (MCI). *Electronics (Basel)*, 9(12), 2185. DOI: 10.3390/electronics9122185

Iarlori, S., Perpetuini, D., Tritto, M., & Monteriù, A. (2024). An overview of approaches and methods for the cognitive workload estimation in human–machine interaction scenarios through wearables sensors. *BioMedInformatics*, 4(2), 1155–1173. DOI: 10.3390/biomedinformatics4020064

Idris, A. Y., Razak, R. A., & Rahman, S. S. A. (2022). Investigating knowledge task difficulties in an e-training system for professional development design. *Docens Series in Education*, 2, 51–65.

IJzerman, R. V., van der Vaart, R., & Breeman, L. D. (2023). An iterative approach to developing a multifaceted implementation strategy for a complex eHealth intervention within clinical practice. *BMC Health Services Research*, 23, 1455. DOI: 10.1186/s12913-023-10439-

Irfadila, M. S., & Noprika, W. (2022). The importance of needs analysis in curriculum development language learning. In *Proceedings of the 2nd International Conference on Language Pedagogy* (pp. 224). Universitas Muhammadiyah Sumatera Barat & Universitas Adzkia.

Ismail, A., Zaharudin, R., Hashim, N., & Ariffin, J. (2020). The Impact of e-NPQEL on the Continuance Intention of Using e-Training among Aspired School Leaders in Malaysia. *International Journal of Interactive Mobile Technologies*, 14(19), 109–123. DOI: 10.3991/IJIM.V14I19.15965

Ismail, S. M., Rahul, D. R., & Patra, I.. (2022). Formative vs. summative assessment: Impacts on academic motivation, attitude toward learning, test anxiety, and self-regulation skill. *Language Testing in Asia*, 12, 40. DOI: 10.1186/s40468-022-00191-4

Jiang, B., Li, X., Yang, S., Kong, Y., Cheng, W., Hao, C., & Lin, Q. (2022). Data-driven personalized learning path planning based on cognitive diagnostic assessments in MOOCs. *Applied Sciences (Basel, Switzerland)*, 12(8), 3982. DOI: 10.3390/app12083982

Jingmiao, Y. (2024). Visual communication design for mobile learning apps: User interface usability and learning engagement. *Comunicar: Media Education Research Journal*, 78(32), 234–248. DOI: 10.58262/V32I78.19

Jönsson, A., & Prins, F. (2019). Editorial: Transparency in Assessment—Exploring the Influence of Explicit Assessment Criteria. *Frontiers in Education*, 3. Advance online publication. DOI: 10.3389/feduc.2018.00119

Kalakoski, V., Selinheimo, S., Valtonen, T., Turunen, J., Käpykangas, S., Ylisassi, H., Toivio, P., Järnefelt, H., Hannonen, H., & Paajanen, T. (2020). Effects of a cognitive ergonomics workplace intervention (CogErg) on cognitive strain and well-being: A cluster-randomized controlled trial. A study protocol. *BMC Psychology*, 8(1), 1. DOI: 10.1186/s40359-019-0349-1

Kasepalu, R., Prieto, L. P., Ley, T., & Chejara, P. (2022). Teacher artificial intelligence-supported pedagogical actions in collaborative learning coregulation: A Wizard-of-Oz study. *Frontiers in Education*, 7. Advance online publication. DOI: 10.3389/feduc.2022.736194

Kaur, D. P., Mantri, A., & Horan, B. (2019). Enhancing student motivation with use of augmented reality for interactive learning in engineering education. *Procedia Computer Science*, 172(3), 881–885. DOI: 10.1016/j.procs.2020.05.127

Kavouras, P., & Charitidis, C. A. (2019). Dual use in modern research. In Iphofen, R. (Ed.), *Handbook of research ethics and scientific integrity* (pp. 1–17). Springer. DOI: 10.1007/978-3-319-76040-7_7-1

Kelemen, W. L., Jhangiani, R. S., Chiang, I.-C. A., Cuttler, C., & Leighton, D. C. (2023). *Research methods in psychology*. The Saylor Foundation. http://www.saylor.org/site/textbooks/

Kennedy, D., Hyland, Á., & Ryan, N. (2007). *Writing and Using Learning Outcomes: A Practical Guide*. University College Cork.

Kim, J., & Maloney, E. J. (2020). *Learning Innovation and the Future of Higher Education*. Johns Hopkins University Press.

Klein, G., & Militello, L. (2001). Some guidelines for conducting a cognitive task analysis. In Salas, E. (Ed.), *Advances in human performance and cognitive engineering research* (pp. 163–199). Elsevier Science/JAI Press. DOI: 10.1016/S1479-3601(01)01006-2

Klepsch, M., & Seufert, T. (2020). Understanding instructional design effects by differentiated measurement of intrinsic, extraneous, and germane cognitive load. *Instructional Science*, 48, 45–77. DOI: 10.1007/s11251-020-09502-9

Knisely, B. M., Joyner, J. S., & Vaughn-Cooke, M. (2021). Cognitive task analysis and workload classification. *MethodsX*, 8, 101235. DOI: 10.1016/j.mex.2021.101235

Kossen, C., & Ooi, C.-Y. (2021). Trialling micro-learning design to increase engagement in online courses. *Asian Association of Open Universities Journal*, 16(3), 299–310. DOI: 10.1108/AAOUJ-09-2021-0107

Kovač, V. B., Nome, D. Ø., Jensen, A. R., & Skreland, L. Lj. (2023). *The why, what and how of deep learning: Critical analysis and additional concerns*. Education Inquiry., DOI: 10.1080/20004508.2023.2194502

Kozanitis, A., & Nenciovici, L. (2023). Effect of active learning versus traditional lecturing on the learning achievement of college students in humanities and social sciences: A meta-analysis. *Higher Education*, 86(6), 1377–1394. DOI: 10.1007/s10734-022-00977-8

Krejtz, K., Duchowski, A. T., Niedzielska, A., Biele, C., & Krejtz, I. (2018). Eye tracking cognitive load using pupil diameter and microsaccades with fixed gaze. *PLoS One*, 13(9), e02036291. DOI: 10.1371/journal.pone.0203629

Krell, M., Xu, K. M., Rey, G. D., & Paas, F. (2021). Editorial: Recent approaches for assessing cognitive load from a validity perspective. *Frontiers in Education*, 6, 838422. Advance online publication. DOI: 10.3389/feduc.2021.838422

Kumar, M. V., & Kumar, A. (2024). Incorporating constructive written feedback to enhance students' writing skills: Strategies and practices. *African Journal of Biological Sciences*, 6(14), 6341–6353. DOI: 10.33472/AFJBS.6.14.2024.6341-6353

Kumar, T., Soozandehfar, S. M. A., & Hashemifardnia, A.. (2023). Self vs. peer assessment activities in EFL-speaking classes: Impacts on students' self-regulated learning, critical thinking, and problem-solving skills. *Language Testing in Asia*, 13(1), 36. DOI: 10.1186/s40468-023-00251-3

Kuyyogsuy, S. (2019). Promoting peer feedback in developing students' English writing ability in L2 writing class. *International Education Studies*, 12(9), 76–87. DOI: 10.5539/ies.v12n9p76

Lang, J. (2020). *Distracted: Why Students Can't Focus and What You Can Do About It*. Basic Books.

Latif, L. A., Subramaniam, T. T., & Khatab, Z. A. (2020). Learner Profiling Towards Improving Learner Success. In Li, K. C., Tsang, E. Y. M., & Wong, B. T. M. (Eds.), *Innovating Education in Technology-Supported Environments*. Springer Education Innovation Series. DOI: 10.1007/978-981-15-6591-5_14

Laurence, P. G., & Macedo, E. C. (2023). Cognitive strategies in matrix-reasoning tasks: State of the art. *Psychonomic Bulletin & Review*, 30, 147–159. DOI: 10.3758/s13423-022-02160-7

Lave, J., & Wenger, E. (1991). *Situated learning: Legitimate peripheral participation*. Cambridge University Press. DOI: 10.1017/CBO9780511815355

Lee, Y. M., Jahnke, I., & Austin, L. (2021). Mobile microlearning design and effects on learning efficacy and learner experience. *Educational Technology Research and Development*, 69, 885–915. DOI: 10.1007/s11423-020-09931-w

Liu, D. (2024). The effects of segmentation on cognitive load, vocabulary learning, retention, and reading comprehension in a multimedia learning environment. *BMC Psychology*, 12(4). Advance online publication. DOI: 10.1186/s40359-023-01489-5

Li, Y., Liu, S., Zeng, A., Wu, J., Zhang, J., Zhang, W., & Li, S. (2024). Interdisciplinary dynamics in COVID-19 research: Examining the role of computer science and collaboration patterns. *Systems*, 12(4), 113. DOI: 10.3390/systems12040113

Loriette, C., Ziane, C., & Ben Hamed, S. (2021). Neurofeedback for cognitive enhancement and intervention and brain plasticity. *Revue Neurologique*, 177(9), 1133–1144. DOI: 10.1016/j.neurol.2021.08.004

Mahmood, M., Aziz, S., & Bibi, M. (2023). Effects of teaching practice in shaping prospective teachers' professional identities: Comparative analysis. *Journal of Education and Educational Development*, 10(1), 72–94. DOI: 10.22555/joeed.v10i1.731

Major, A., & Calandrino, T. (2018). Beyond chunking: Micro-learning secrets for effective online design. *FDLA Journal, 3*(13). https://nsuworks.nova.edu/fdla-journal/vol3/iss1/13

Makhambetova, A., Zhiyenbayeva, N., & Ergesheva, E. (2020). Personalized learning strategy as a tool to improve academic performance and motivation of students. *International Journal of Web-Based Learning and Teaching Technologies*, 16(6), 1–17. DOI: 10.4018/IJWLTT.286743

Malik, A., Iqbal, M., & Sultan, A. (2023). Impact of academic satisfaction on university students' academic achievement. *Journal of Education and Social Studies*, 4(3), 806–812. DOI: 10.52223/jess.2023.4341

Mark, G. (2023). *Attention Span: A Groundbreaking Way to Restore Balance, Happiness, and Productivity*. Harper Collins.

Marshall, D., & Finlayson, M. (2022). Applied cognitive task analysis methodology: Fundamental cognitive skills surgical nurses require to manage patient deterioration. *Nursing Praxis in Aotearoa New Zealand*, 38(1), 25–37. DOI: 10.36951/27034542.2022.04

Martin, F., Chen, Y., & Moore, R. L. (2020). Systematic review of adaptive learning research designs, context, strategies, and technologies from 2009 to 2018. *Educational Technology Research and Development*, 68, 1903–1929. DOI: 10.1007/s11423-020-09793-2

Martini, N., Choong, J. W., Dela Cruz, P. D., Lau, H., Lim, H., Liu, R., & Marshall, D. (2022). Assessing antibiotic prescribing in nurse practitioners: Applied cognitive task analysis. *International Journal of Nursing Studies Advances*, 4. Advance online publication. DOI: 10.1016/j.ijnsa.2022.100101

Mayer, R. E. (2014). Cognitive theory of multimedia learning. In Mayer, R. E. (Ed.), *The Cambridge handbook of multimedia learning* (2nd ed., pp. 43–71). Cambridge University Press. DOI: 10.1017/CBO9781139547369.005

McGee, H. M., & Crowley-Koch, B. (2019). Using behavioral systems analysis to improve large scale change initiatives in autism service organizations. *Perspectives on Behavior Science*, 42(4), 931–954. DOI: 10.1007/s40614-019-00231-0

McNeill, L., & Fitch, D. (2023). Microlearning through the Lens of Gagne's Nine Events of Instruction: A Qualitative Study. *TechTrends*, 67, 521–533. DOI: 10.1007/s11528-022-00805-x

Mergen, M., Graf, N., & Meyerheim, M. (2024). Reviewing the current state of virtual reality integration in medical education: A scoping review. *BMC Medical Education*, 24, 788. DOI: 10.1186/s12909-024-05777-5

Merrill, D. (2002). First principles of instruction. *Educational Technology Research and Development*, 50(3), 43–59. DOI: 10.1007/BF02505024

Merrill, M. D. (2013). *First Principles of Instruction: Identifying and Designing Effective, Efficient, and Engaging Instruction*. Pfeiffer.

Merta, L. W. S., Ratminingsih, N. M., & Budasi, I. G. (2023). The Integration of Technology in English Language Teaching to Stimulate Students' Critical Thinking. *Language Circle: Journal of Language and Literature*, 17(2), 333–341. DOI: 10.15294/lc.v17i2.39097

Mielke, J., Brunkert, T., & Zúñiga, F.. (2022). Methodological approaches to study context in intervention implementation studies: An evidence gap map. *BMC Medical Research Methodology*, 22, 320. DOI: 10.1186/s12874-022-01772-w

Militello, L. G., Hutton, R. J. B., Pliske, R. M., Knight, B. J., Klein, G., & Randel, J. (1997). *Applied Cognitive Task Analysis (ACTA) methodology.*

Militello, , L. GSalwei, , M. EReale, , CSushereba, , CSlagle, , J. MGaba, , DWeinger, , M. BRask, , JFaiman, , JAndreae, , MBurden, , A. RAnders, , S. (2023). Adapting cognitive task analysis methods for use in a large sample simulation study of high-risk healthcare events. *Journal of Cognitive Engineering and Decision Making*, *17*(4), 315-331. DOI: 10.1177/15553434231192283

Militello, L. G., & Hutton, R. J. B. (1998). Applied cognitive task analysis (ACTA): A practitioner's toolkit for understanding cognitive task demands. *Ergonomics*, 41(11), 1618–1641.

Militello, L. G., & Hutton, R. J. B. (1998). Applied cognitive task analysis (acta): A practitioner's toolkit for understanding cognitive task demands. *Ergonomics*, 41(11), 1618–1641. DOI: 10.1080/001401398186108

Miller, B., & Staley, L. (2021). Importance of Feedback, Training and Media Format for Students' Reflective Practice. *International Journal of Knowledge and Learning*, 10, 31–40. DOI: 10.53615/2232-5697.10.31-40

Moges, B. (2018). The implementations and challenges of assessment practices for students' learning in public selected universities, Ethiopia. *Universal Journal of Educational Research*, 6(12), 2789–2806. DOI: 10.13189/ujer.2018.061213

Mohammed, G. S., Wakil, K., & Nawroly, S. S. (2018). The effectiveness of micro-learning to improve students' learning ability. *International Journal of Educational Research Review*, 3(3), 32–38. DOI: 10.24331/ijere.415824

Mohd Yusoff, N. (2015). *A Shared Cognitive User Task Framework for Elearning Storyboard.*

Mokhtar, S. M., Nur, A. A. A., Shazali, N. M., & Ahmad, N. (2020). Why Are Students Involved In E-Learning? A Reasoning Study At Univerisity Kebangsaan Malaysia (UKM). *International Journal of Education and Pedagogy*, 2(1), 152–159. http://myjms.moe.gov.my/index.php/ijeap

Moore, M. L. (2020). *Qualitative Exploration of Instructional Designers' Use of Microlearning for Formal Workplace Training* [Doctoral dissertation, Capella University]. ProQuest Dissertations Publishing

Morgenstern, T., Klichowicz, A., Bengler, P., Todtermuschke, M., & Bocklisch, F. (2024). Towards a cognition-based framework describing interdisciplinary expert team processes for cognitive robotics in Industry 5.0 technologies. *Applied Sciences (Basel, Switzerland)*, 14(10), 4121. DOI: 10.3390/app14104121

Morrison-Smith, S., & Ruiz, J. (2020). Challenges and barriers in virtual teams: A literature review. *SN Applied Sciences*, 2, 1096. DOI: 10.1007/s42452-020-2801-5

Mutezo, A. T., & Maré, S. (2023). Teaching and cognitive presences: The mediating effect of social presence in a developing world context. *Cogent Education*, 10(1). Advance online publication. DOI: 10.1080/2331186X.2023.2171176

Nauert, E., & Gillan, D. (2019). The Cognitive Process of Wildland Fire Chain-saw Troubleshooting: Structure, Content, and Training. *Proceedings of the Human Factors and Ergonomics Society Annual Meeting*, 63(1), 1676–1680. DOI: 10.1177/1071181319631536

Nielsen, E. G., & Minda, J. P. (2019). Problem Solving and Decision Making. *Psychology (Irvine, Calif.)*. Advance online publication. DOI: 10.1093/obo/9780199828340-0246

Nikkhoo, I., Ahmadi, Z., Akbari, M., Imannezhad, S., Anvari Ardekani, S., & Lashgari, H. (2023). Microlearning for today's students: A rapid review of essentials and considerations. *Medical Education Bulletin*, 4(1), 673–685. DOI: 10.22034/meb.2022.355659.1066

Nilsson, R. (2021). *Designing an interactive micro-learning application: A new way to learn thermography* (Examensarbete, 30 hp). Uppsala Universitet.

Noble, H., & Heale, R. (2019). Triangulation in research, with examples. *Evidence-Based Nursing*, 22(3), 89–91. DOI: 10.1136/ebnurs-2019-103145

Nor'ain, M. Y., & Salim, S. S. (2012). Investigating cognitive task difficulties and expert skills in e-learning storyboards using a cognitive task analysis technique. *Computers & Education*, 58, 652–665. DOI: 10.1016/j.compedu.2011.09.009

Nordin, H., Singh, D., & Mansor, Z. (2020). An empirical study of e-learning interface design elements for generation Z. *International Journal of Advanced Computer Science and Applications*, 11(9), 507–515. DOI: 10.14569/IJACSA.2020.0110961

O'Brien, B. C., & Battista, A. (2020). Situated learning theory in health professions education research: A scoping review. *Advances in Health Sciences Education : Theory and Practice*, 25, 483–509. DOI: 10.1007/s10459-019-09900-w

Ordu, U. B.-A. (2021). The Role of Teaching and Learning Aids/Methods in a Changing World. In *New Challenges to Education: Lessons from Around the World* (Vol. 19, pp. 210). Bulgarian Comparative Education Society.

Ozan, C., & Kıncal, R. Y. (2018). The effects of formative assessment on academic achievement, attitudes toward the lesson, and self-regulation skills. *Educational Sciences: Theory & Practice*, 18(1), 85–118. DOI: 10.12738/estp.2018.1.0216

Panke, S., Kohls, C., & Gaiser, B. (2017). Social media strategies for seamless learning: Approaches and metrics. In Şad, S. N., & Ebner, M. (Eds.), *Digital Tools for Seamless Learning* (pp. 278–296). IGI Global. DOI: 10.4018/978-1-5225-1692-7.ch014

Papautsky, E. L., Dominguez, C., Strouse, R., & Moon, B. (2015). Integration of cognitive task analysis and design thinking for autonomous helicopter displays. *Journal of Cognitive Engineering and Decision Making*, 9(4), 283–294. DOI: 10.1177/1555343415602624

Papautsky, E. L., Strouse, R., & Dominguez, C. (2020). Combining cognitive task analysis and participatory design methods to elicit and represent task flows. *Journal of Cognitive Engineering and Decision Making*, 14(4), 288–301. DOI: 10.1177/1555343420976014

Park, Y., & Kim, Y. (2018). A design and development of micro-learning content in e-learning system. *International Journal on Advanced Science, Engineering and Information Technology*, 8(1). Advance online publication. DOI: 10.18517/ijaseit.8.1.2698

Paz, F., Huaracha, A., Paz-Sifuentes, F.-A., Moquillaza, A., Lecaros, A., Falconi, F., Aguirre, J., & Tapia, A. (2023). Comparing two sets of usability heuristics in a user experience evaluation: A case study in an e-commerce mobile application. In *Design, User Experience, and Usability* (pp. xx-xx). Springer. DOI: 10.1007/978-3-031-35702-2_5

Pemberton, I., & Mooney, D. (2024). How cognitive task analysis (CTA) can improve learning from serious accidents. In Golightly, D., Balfe, N., & Charles, R. (Eds.), *Contemporary Ergonomics and Human Factors 2024*. CIEHF.

Penney, G., Launder, D., Cuthbertson, J., & Thompson, M. B. (2022). Threat assessment, sense making, and critical decision-making in police, military, ambulance, and fire services. *Cognition Technology and Work*, 24(3), 423–439. DOI: 10.1007/s10111-022-00694-3

Petersson, L., Larsson, I., Nygren, J. M., Nilsen, P., Neher, M., Reed, J. E., Tyskbo, D., & Svedberg, P. (2022). Challenges to implementing artificial intelligence in healthcare: A qualitative interview study with healthcare leaders in Sweden. *BMC Health Services Research*, 22(1), 850. DOI: 10.1186/s12913-022-08215-8

Pfaff, M. S., Eris, O., Weir, C., Anganes, A., Crotty, T., Rahman, M. T., Ward, M., & Nebeker, J. (2020). Analysis of the cognitive demands of electronic health record use. *Journal of Biomedical Informatics*, 113, 103633. DOI: 10.1016/j.jbi.2020.103633

Pham Thi, T. D., & Duong, N. T. (2024). Investigating learning burnout and academic performance among management students: A longitudinal study in English courses. *BMC Psychology*, 12(1), 219. DOI: 10.1186/s40359-024-01725-6

Piaget, J. (1971). The Theory of Stages in Cognitive Development. In Green, D., Ford, M. P., & Flamer, G. B. (Eds.), *Measurement and Piaget* (pp. 1–11). McGraw-Hill.

Pimenta Rodrigues, , G. AMarques Serrano, , A. LLopes Espiñeira Lemos, , A. NCanedo, , E. DMendonça, , F. L. L. dde Oliveira Albuquerque, , RSandoval Orozco, , A. LGarcía Villalba, , L. J. (2024). Understanding data breach from a global perspective: Incident visualization and data protection law review. *Data, 9*(2), 27. DOI: 10.3390/data9020027

Rafique, S., & Aziz, S. (2019). Relationship of Pedagogical Skills and Content Knowledge with Pedagogical Design Capacity in Higher Education of Pakistan. *Global Regional Review*, IV(IV), 85–94. DOI: 10.31703/grr.2019(iv-iv).10

Rana, K., Poudel, P., & Chimoriya, R. (2023). Qualitative methodology in translational health research: Current practices and future directions. *Health Care*, 11(2665). Advance online publication. DOI: 10.3390/healthcare11192665

Rebelo, S., & Isaías, P. (2020). Gamification As An Engagement Tool In E-Learning Websites. *Journal of Information Technology Education*, 19, 834–854. DOI: 10.28945/4653

Reed, J. E., Kaplan, H. C., & Ismail, S. A. (2018). A new typology for understanding context: Qualitative exploration of the model for understanding success in quality (MUSIQ). *BMC Health Services Research*, 18, 584. DOI: 10.1186/s12913-018-3348-7

Reeves, T. C. (2020). Design research from a technology perspective. In *Educational Design Research* (pp. 64–78). DOI: 10.4324/9780203088364-13

Reniers, J., Mathany, C., Farkas, M., Pollock, H., & Husband, B. C. (2022). Uncharted territory: Curriculum mapping multiple majors simultaneously. *The Canadian Journal for the Scholarship of Teaching and Learning*, 13(1). Advance online publication. DOI: 10.5206/cjsotlrcacea.2022.1.8553

Resnik, D. B., & Hosseini, M. (2024). The ethics of using artificial intelligence in scientific research: New guidance needed for a new tool. *AI and Ethics*. Advance online publication. DOI: 10.1007/s43681-024-00493-8

Riley, R. D., Archer, L., Snell, K. I. E., Ensor, J., Dhiman, P., & Martin, G. P.. (2024). Evaluation of clinical prediction models (part 2): How to undertake an external validation study. *BMJ (Clinical Research Ed.)*, 384, e074820. DOI: 10.1136/bmj-2023-074820

Romanenko, Y. N., Solodovnikova, E., & Maksimenko, N. (2023). Microlearning as a new method of teaching soft skills to university students. *Frontiers in Education*, 8, 1177516. DOI: 10.3389/feduc.2023.1177516

Romero, C., & Ventura, S. (2020). Educational data mining and learning analytics: An updated survey. *Wiley Interdisciplinary Reviews. Data Mining and Knowledge Discovery*, 10(3), e1355. DOI: 10.1002/widm.1355

Ropohl, M., & Rönnebeck, S. (2019). Making learning effective – quantity and quality of pre-service teachers' feedback. *International Journal of Science Education*, 41(15), 2156–2176. DOI: 10.1080/09500693.2019.1663452

Roth, E., O'Hara, J., Dickerson, K., Hughes, N., Fleger, S., & Ki, D. (2020). *Cognitive task analysis: Technical basis and guidance development*. Office of Nuclear Regulatory Research, United States Nuclear Regulatory Commission.

Rubio-Tamayo, J. L., Gertrudix Barrio, M., & García García, F. (2017). Immersive environments and virtual reality: Systematic review and advances in communication, interaction and simulation. *Multimodal Technologies and Interaction*, 1(4), 21. DOI: 10.3390/mti1040021

Russ-Jara, A. L., Elkhadragy, N., Arthur, K. J., DiIulio, J. B., Militello, L. G., Ifeachor, A. P., Glassman, P. A., Zillich, A. J., & Weiner, M. (2023). Cognitive task analysis of clinicians' drug–drug interaction management during patient care and implications for alert design. *BMJ Open*, 13, e075512. DOI: 10.1136/bmjopen-2023-075512

Ryan, R. M., & Deci, E. L. (2000). Self-determination theory and the facilitation of intrinsic motivation, social development, and well-being. *The American Psychologist*, 55(1), 68–78. DOI: 10.1037/0003-066X.55.1.68

Saintika, Y., Astiti, S., Kusuma, D. J. A., & Muhammad, A. W. (2021). Analysis of E-learning readiness level of public and private universities in central Java, Indonesia. *Register: Jurnal Ilmiah Teknologi Sistem Informasi*, 7(1), 15–30. DOI: 10.26594/register.v7i1.2042

Sankaranarayanan, R., Leung, J., Abramenka-Lachheb, V., Seo, G., & Lachheb, A. (2023). Microlearning in diverse contexts: A bibliometric analysis. *TechTrends*, 67, 260–276. DOI: 10.1007/s11528-022-00794-x

Saqr, M., Nouri, J., & Santolini, M. (2019, March). *Towards group-aware learning analytics: Using social network analysis and machine learning to monitor and predict performance in collaborative learning*. Paper presented at the 13th Annual International Technology, Education and Development Conference. DOI: 10.21125/inted.2019.1881

Saunders, M., Lewis, P., & Thornhill, A. (2009). Research methods for business students. *Business* (Fitfh, Vol. 5th). Pearson Education. DOI: 10.1017/CBO9781107415324.004

Seamster, T. L., & Redding, R. E. (2017). *Applied Cognitive Task Analysis in Aviation*. Routledge. DOI: 10.4324/9781315262321

Selgelid, M. J. (2009). Governance of dual-use research: An ethical dilemma. *Bulletin of the World Health Organization*, 87(9), 720–723. DOI: 10.2471/BLT.08.051383

Senadheera, V. V., Muthukumarana, C. K., Ediriweera, D., & Rupasinghe, T. (2024). Impact of microlearning on academic performance of students in higher education: A systematic review and meta-analysis. *Journal of Medical and Technology Research*, 9(1). Advance online publication. DOI: 10.4038/jmtr.v9i1.2

Senge, P. (2006). *The Fifth Discipline: The art and practice of the learning organization*. Random House Books.

Shabadurai, Y., Chua, F. F., & Lim, T. Y. (2022). Investigating the employees' perspectives and experiences of microlearning content design for online training. *International Journal of Information and Education Technology (IJIET)*, 12(8), 786–793. DOI: 10.18178/ijiet.2022.12.8.1685

Shahzad, A., Hassan, R., Aremu, A. Y., Hussain, A., & Lodhi, R. N. (2021). Effects of COVID-19 in E-learning on higher education institution students: The group comparison between male and female. *Quality & Quantity*, 55(3), 805–826. DOI: 10.1007/s11135-020-01028-z

Shail, M. (2019). Using micro-learning on mobile applications to increase knowledge retention and work performance: A review of literature. *Cureus*, 11(8), e5307. DOI: 10.7759/cureus.5307

Shaltry, C. (2020). A new model for organizing curriculum alignment initiatives. *Advances in Physiology Education*, 44, 658–663.

Sharif, A., & Cho, S. (2015). 21st-century instructional designers: Bridging the perceptual gaps between identity, practice, impact and professional development. *RUSC. Universities and Knowledge Society Journal*, 12(3), 72–85. DOI: 10.7238/rusc.v12i3.2176

Sharma, P., Akgun, M., & Li, Q. (2023). Understanding student interaction and cognitive engagement in online discussions using social network and discourse analyses. *Educational Technology Research and Development*. Advance online publication. DOI: 10.1007/s11423-023-10261-w

Sharples, M., Taylor, J., & Vavoula, G. (2007). A theory of learning for the mobile age. In Andrews, R., & Haythornthwaite, C. (Eds.), *The SAGE Handbook of E-Learning Research* (pp. 221–247). Sage Publication.

Shemshack, A., & Spector, J. M. (2020). A systematic literature review of personalized learning terms. *Smart Learning Environments*, 7(1). Advance online publication. DOI: 10.1186/s40561-020-00140-9

Silva, L., Mendes, A. J., & Gomes, A. (2020). Computer-supported collaborative learning in programming education: A systematic literature review. In *IEEE Global Engineering Education Conference, EDUCON* (Vol. 2020-April, pp. 1086–1095). IEEE Computer Society. DOI: 10.1109/EDUCON45650.2020.9125237

Simon, S., & Richards, P. (2022). (Manuscript submitted for publication). Investigating cognitive observational difficulties and expert skills in elite Olympic and Paralympic coaches using an applied cognitive task analysis. *Journal of Expertise*.

Singh, J., Singh, L., & Matthees, B. (2022). Establishing social, cognitive, and teaching presence in online learning—A panacea in COVID-19 pandemic, post vaccine and post pandemic times. *Journal of Educational Technology Systems*, 51(1), 568–585. DOI: 10.1177/00472395221095169

Skulmowski, A., & Xu, K. M. (2022). Understanding cognitive load in digital and online learning: A new perspective on extraneous cognitive load. *Educational Psychology Review*, 34(1), 171–196. DOI: 10.1007/s10648-021-09624-7

So, H. J., Lee, H., & Roh, S. Z. (2020) Examining the design of microlearning for Korean adult learners. *Computer-Based Learning in Context, 2*(1), 40-53. DOI: 10.35542/osf.io/dmu4c

Specht, A., & Crowston, K. (2022). Interdisciplinary collaboration from diverse science teams can produce significant outcomes. *PLoS One*, 17(11), e0278043. Advance online publication. DOI: 10.1371/journal.pone.0278043

Stanik, C., Puhlfürß, T., Mahler, A., Sasu, P. B., Reip, W., & Maalej, W. (2022). Lessons learned from customizing and applying ACTA to design a novel device for emergency medical care. *arXiv*. https://arxiv.org/abs/2108.05622

Stanik, C., Puhlfurs, T., Mahler, A., Sasu, P. B., Reip, W., & Maalej, W. (2021). Lessons Learned from Customizing and Applying ACTA to Design a Novel Device for Emergency Medical Care. *Proceedings of the IEEE International Conference on Requirements Engineering*, (September 2021), 302–311. DOI: 10.1109/RE51729.2021.00034

Stanley, D., Coman, S., Murdoch, D., & Stanley, K. (2020). Writing exceptional (specific, student and criterion-focused) rubrics for nursing studies. *Nurse Education in Practice*, 49, 102851. DOI: 10.1016/j.nepr.2020.102851

Ste-Croix, C., & Angel, H. (2021). *Cognitive task analysis of a complex battle position scenario*. Human Systems Incorporated.

Strom, P. S., Strom, R. D., Sindel-Arrington, T., & Rude, R. V. (2023). Student attention and distraction in community college. *Journal of Educational and Developmental Psychology*, 13(1), 41–49. DOI: 10.5539/jedp.v13n1p41

Stufflebeam, D. L. (2003). The CIPP model for evaluation. In Kellaghan, T., & Stufflebeam, D. L. (Eds.), *Evaluation models: Viewpoints on educational and human services evaluation* (pp. 279–317). Springer Netherlands. DOI: 10.1007/978-94-010-0309-4_4

Su, K.-D., & Chen, H.-Y. (2023). Exploring the learning efficacy of students' STEM education from the process of hands-on practical experience. In *Innovative technologies and learning* (pp. 541–550). Springer. DOI: 10.1007/978-3-031-40113-8_41

Sulaiman, A. H., & Abdullah, S. A. (2024). The use of active learning strategies to foster effective teaching in higher education institutions. *Zanco Journal of Humanity Sciences*, 28(2). Advance online publication. DOI: 10.21271/zjhs.28.2.11

Susilana, R., Dewi, L., Rullyana, G., Hadiapurwa, A., & Khaerunnisa, N. (2022). Can microlearning strategy assist students' online learning? *Cakrawala Pendidikan Jurnal Ilmiah Pendidikan*, 41(2), 437–451. DOI: 10.21831/cp.v41i2.43387

Sutton, R. T., Pincock, D., Baumgart, D. C., Sadowski, D. C., Fedorak, R. N., & Kroeker, K. I. (2020). An overview of clinical decision support systems: Benefits, risks, and strategies for success. *NPJ Digital Medicine*, 3, 17. DOI: 10.1038/s41746-020-0221-y

Swaby, L., Shu, P., Hind, D., & Sutherland, K. (2022). The use of cognitive task analysis in clinical and health services research — A systematic review. *Pilot and Feasibility Studies*, 8, 57. DOI: 10.1186/s40814-022-01002-6

Sweller, J. (2011). Cognitive load theory. In Mestre, J. P., & Ross, B. H. (Eds.), *The psychology of learning and motivation: Cognition in education* (pp. 37–76). Elsevier Academic Press. DOI: 10.1016/B978-0-12-387691-1.00002-8

Tabatabaee, S. S., Jambarsang, S., & Keshmiri, F. (2024). Cognitive load theory in workplace-based learning from the viewpoint of nursing students: Application of a path analysis. *BMC Medical Education*, 24, 678. DOI: 10.1186/s12909-024-05664-z

Tai, J. H. M., Dollinger, M., Ajjawi, R., Jorre de St Jorre, T., Krattli, S., McCarthy, D., & Prezioso, D. (2023). Designing assessment for inclusion: An exploration of diverse students' assessment experiences. *Assessment & Evaluation in Higher Education*, 48(3), 403–417. DOI: 10.1080/02602938.2022.2082373

Tawafak, R. M., Malik, S. I., & Alfarsi, G. (2021). Impact of technologies during the COVID-19 pandemic for improving behavioral intention to use e-learning. *International Journal of Information and Communication Technology Education*, 17(3), 137–150. DOI: 10.4018/IJICTE.20210701.oa9

Taylor, A., & Hung, W. (2022). The effects of microlearning: A scoping review. *Educational Technology Research and Development*, 70, 363–395. DOI: 10.1007/s11423-022-10084-1

Taylor, J. (2019). Reporting research findings to participants is an ethical imperative. *BMJ (Clinical Research Ed.)*, 367, l6324. DOI: 10.1136/bmj.l6324

Taylor, J., Ashford, M., & Jefferson, M. (2023). High performance coach cognition in the wild: Using applied cognitive task analysis for practical insights–cognitive challenges and curriculum knowledge. *Frontiers in Psychology*, 14, 1154168. DOI: 10.3389/fpsyg.2023.1154168

Thakur, S. N. (2023). The human brain, mind, and consciousness: Unveiling the enigma. *Philosophia*, 20(2), 163–179. DOI: 10.13140/RG.2.2.29992.14082

Theodorio, A. O. (2024). Examining the support required by educators for successful technology integration in teacher professional development program. *Cogent Education*, 11(1). Advance online publication. DOI: 10.1080/2331186X.2023.2298607

Trespalacios, J., & Uribe-Florez, L. J. (2020). Developing online sense of community: Graduate students' experiences and perceptions. *Turkish Online Journal of Distance Education-TOJDE*, 21(1), 57–72.

Trowbridge, S., Waterbury, C., & Sudbury, L. (2017). Learning in bursts: Microlearning with social media. *EDUCAUSE Review*, 4.

UK Statistics Authority. (2022, October 21). *Ethical considerations associated with qualitative research methods*. Office for Statistics Regulation. https://uksa.statisticsauthority.gov.uk/publication/ethical-considerations-associated-with-qualitative-research-methods/pages/4/

Valzado, R. B., Subaldo, C. B., Diacor, R. S., & Rellon, J. (2023). Dimensions of pedagogical skills of teachers in the new normal: An exploratory sequential design. *Southeast Asian Journal of Multidisciplinary Studies, 3*(2).

Van Wyk, B. (2023). New Kids on the Block?: Exploring Technological Preferences of a New Generation. *Journal of Higher Education Theory and Practice*, 23(3), 13–19. DOI: 10.33423/jhetp.v23i3.5837

Vats, S., & Joshi, R. (2024). The impact of virtual reality in education: A comprehensive research study. In S. K. Sharma, Y. K. Dwivedi, B. Metri, B. Lal, & A. Elbanna (Eds.), *Transfer, diffusion and adoption of next-generation digital technologies* (Vol. 699, pp. 137-150). IFIP Advances in Information and Communication Technology. Springer. DOI: 10.1007/978-3-031-50204-0_11

Venkat, M. V., O'Sullivan, P. S., Young, J. Q., & Sewell, J. L. (2020). Using Cognitive Load Theory to improve teaching in the clinical workplace. *MedEdPORTAL: the Journal of Teaching and Learning Resources*, 16, 10983. DOI: 10.15766/mep_2374-8265.10983

Volpe, J. A. (2013). *Using cognitive task analysis to inform issues in human systems integration in railroad operations* (Final report). U.S. Department of Transportation, Research and Special Programs Administration, National Transportation Systems Center.

Vu, T., Magis-Weinberg, L., & Jansen, B. R. J. (2022). Motivation-achievement cycles in learning: A literature review and research agenda. *Educational Psychology Review*, 34, 39–71. DOI: 10.1007/s10648-021-09616-7

Vygotsky, L. S. (1978). *Mind in Society: The Development of Higher Psychological Processes*. Harvard University Press.

Wahid, S. A. N., Razak, R. A., & Halili, S. H. (2023). Investigating knowledge task difficulties in designing professional ethics instructional framework for technical educators. *Jurnal Penyelidikan Pendidikan*, 41(1), 45–57.

Wang, L. (2023). The benefits of a well-calibrated micro-learning ecosystem. UNESCO.

Wang, T., Towey, D., Ng, R. Y., & Gill, A. S. (2021). Towards post-pandemic transformative teaching and learning: Case studies of microlearning implementations in two post-secondary educational institutions. *SN Computer Science*, 2(4), 271–279. DOI: 10.1007/s42979-021-00663-z

Wijngaards-de Meij, L., & Merx, S. (2018). Improving curriculum alignment and achieving learning goals by making the curriculum visible. *The International Journal for Academic Development*, 23(3), 219–231. DOI: 10.1080/1360144X.2018.1462187

Wilson, B. G., & Myers, K. M. (2000). Situated cognition in theoretical and practical context. In Jonassen, D. H., & Land, S. M. (Eds.), *Theoretical foundations of learning environments* (pp. 57–88). Erlbaum.

Wilson, R., Joiner, K., & Abbasi, A. (2021). Improving students' performance with time management skills. *Journal of University Teaching & Learning Practice*, 18(4). https://ro.uow.edu.au/jutlp/vol18/iss4/16

Woodward, J., & Ruiz, J. (2022). Analytic review of using augmented reality for situational awareness. *IEEE Transactions on Visualization and Computer Graphics, PP*, (99), 1–1. DOI: 10.1109/TVCG.2022.3141585

Wu, Y., Hu, K., Chen, D. Z., & Wu, J. (2024). AI-enhanced virtual reality in medicine: A comprehensive survey (Version 3). *arXiv*. https://doi.org//arXiv.2402 .03093DOI: 10.48550

Xu, A., Baysari, M. T., Stocker, S. L., Leow, L. J., Day, R. O., & Carland, J. E. (2020). Researchers' views on, and experiences with, the requirement to obtain informed consent in research involving human participants: A qualitative study. *BMC Medical Ethics*, 21(1), 93. DOI: 10.1186/s12910-020-00538-7

Yadav, N. (2024). The impact of digital learning on education. I*nternational Journal of Multidisciplinary Research in Arts. Science and Technology*, 2(1), 2584–0231. DOI: 10.61778/ijmrast.v2i1.34

Yang, S., Yu, K., Lammers, T., & Chen, F. (2021). Artificial intelligence in pilot training and education – Towards a machine learning aided instructor assistant for flight simulators. In Stephanidis, C., Antona, M., & Ntoa, S. (Eds.), *HCI International 2021 - Posters* (Vol. 1420, pp. 689–694). Springer., DOI: 10.1007/978-3-030-78642-7_78

Yin, H., & Chai, C. (2020). Catering for Learner Diversity: Issues and Trends. *ECNU Review of Education*, 3(4), 607–609. DOI: 10.1177/2096531120927620

Yin, J., Goh, T.-T., Yang, B., & Xiaobin, Y. (2020). Conversation technology with micro-learning: The impact of chatbot-based learning on students' learning motivation and performance. *Journal of Educational Computing Research*, 59(1), 073563312095206. DOI: 10.1177/0735633120952067

Younis, M., Imdad, P., & Aneela, A. U. R. (2021). Effects of Constructive and Timely Feedback on Academic Performance of Students. *Pakistan Journal of Educational Research*, 4(4), 670–677. DOI: 10.52337/pjer.v4i4.464

Zajda, J. (2021). Constructivist learning theory and creating effective learning environments. In *Globalisation and education reforms* (Vol. 25, pp. 43–60). Springer. DOI: 10.1007/978-3-030-71575-5_3

Zakeri, Z., Arif, A., Omurtag, A., Breedon, P., & Khalid, A. (2023). Multimodal assessment of cognitive workload using neural, subjective and behavioural measures in smart factory settings. *Sensors (Basel)*, 23(21), 8926. DOI: 10.3390/s23218926

Zarestky, J., Bigler, M., Brazile, M., Lopes, T., & Bangerth, W. (2022). Reflective Writing Supports Metacognition and Self-regulation in Graduate Computational Science and Engineering. *Computers and Education Open*, 3, 100085. DOI: 10.1016/j.caeo.2022.100085

Zhang, L., Wang, X., He, T., & Han, Z. (2022). A data-driven optimized mechanism for improving online collaborative learning: Taking cognitive load into account. *International Journal of Environmental Research and Public Health*, 19(12), 6984. DOI: 10.3390/ijerph19126984

Zhao, L., Zhao, B., & Li, C. (2023). Alignment analysis of teaching–learning-assessment within the classroom: How teachers implement project-based learning under the curriculum standards. *Disciplinary and Interdisciplinary Science Education Research*, 5, 13. DOI: 10.1186/s43031-023-00078-1

Zheng, L., Fan, Y., Chen, B., Huang, Z., Gao, L., & Long, M. (2024). An AI-enabled feedback-feedforward approach to promoting online collaborative learning. *Education and Information Technologies*, 29, 11385–11406. DOI: 10.1007/s10639-023-12292-5

Zhou, Y., Zhao, Y., Xiang, Z., Yan, Z., Shu, L., Xu, X., Zhang, L., & Tian, X. (2023). A dual-task-embedded virtual reality system for intelligent quantitative assessment of cognitive processing speed. *Frontiers in Human Neuroscience*, 17. Advance online publication. DOI: 10.3389/fnhum.2023.1158650

Zimmerman, B. J., & Schunk, D. H. (2011). Self-regulated learning and performance: An introduction and an overview. In Zimmerman, B. J., & Schunk, D. H. (Eds.), *Handbook of self-regulation of learning and performance* (pp. 1–12). Routledge.

About the Authors

Rafiza Abdul Razak graduated with a doctoral degree in Instructional Technology from the International Islamic University of Malaysia. An alumnus of the Universiti Malaya, Malaysia's leading university, she graduated with a Masters of Education (Educational Technology) and Bachelor of Education in Teaching English as a Second Language (B.EdTESL) from the National University of Malaysia. She was the Head of Department of Curriculum and Instructional Technology (CITeD), Faculty of Education from August 2017 till August 2021. An innovative leader who creatively rebrand the department to uplift and boost the departments' wellbeing. Her interest in frugal innovation and creativity inspires her to design and develop several facilities at the Faculty of Education. The Instructional Technology Production-based Learning Studio (InTechPro) was initiated to empower students and clients to create digital production content / e-learning materials for teaching and learning. Inventors Hub is a place for inventors, makers and entrepreneurs to collaborate and innovate. Instructional Technology Hub (InTecH) is in progress of becoming the hub for professional development. Her active marketing strategies for the department has contributed to the growing number of students and networking collaborations at the national and international levels. She initiated industrial collaboration activities with Microsoft thus awarded 32 pre-service teachers with Microsoft Innovative Educator Trainer on top of that 17 of them Microsoft Innovative Educator Expert. Financially the department managed to gain its highest achievement. Thus, she initiated the Curriculum and Instructional Technology Department Research Grant Funding which awarded 10 research grants projects for members of the department. Associate Prof. Dr Rafiza actively organizes international and national seminars. She organized the International Summit of Innovations and Design Exposition (InSIDE) which showcased innovations from primary till tertiary education and also involved industries. She also initiated several MoAs at the national and international level. Her innovations awarded in patents, copyrights and trademarks accumulated to 81 registered products. She recently completed an international research grant funded by Association of Commonwealth Universities focusing on International Innovative Educator Network Program: Microsoft Educator Peer Coaching Universiti Malaya - South Eastern University Of Sri Lanka. She led the project successfully which involved undergraduates and postgraduates students from Asian universities : Universiti Malaya, Malaysia and Southern University of Sri Lanka. Concurrently, she also led the research project with Universitas Jambi Indonesia which also involved pre-service teachers. Apart from international research, she also publishes in high impact journals. She published in a few WoS and Scopus journals. She is currently planning to work with her international collaborators in book publication. Despite academic institutions, her network of collaboration encompasses Ministry of Health. She co-joint in design and development of modules projects and also training. She is currently collaborating with Content Capital Company for sabbatical attachment beginning from 1st December, 2021 to 31 September, 2022. Numerous activities pertaining to training of teacher educators, pre-service teachers and in-service teachers have been considered as part of the collaborative activities. Despite, the company also encourages research and publications activities to further implies high impact outcome and results. Her industrial engagement and expert collaboration also include organizations such as the Division of Allied Health Sciences (Ministry of Health), National Academy of Arts and Culture (Ministry of Tourism & Culture), Centre for Instructor and Advanced Skill Training (Ministry of Human Resources). She has been awarded with prestigious award nationally and internationally. Microsoft has warded her the Microsoft Master Trainer award. She was also awarded the Microsoft Innovative Educator Expert. This commemorates her active involvement in educating teachers in utilizing technological pedagogical knowledge. She was also awarded the Fellowship in Aquaculture Curriculum Development from Universitas Airlangga Indonesia. Consultation services and Professional Development Projects were part of her contributions. Assoc. Prof. Dr Rafiza Abdul Razak was selected as one of the Successful People in Malaysia, with her profile set to appear in 2022 as part of the British Media Publishing House. Her active participation in various activities keeps her motivated in the work that she loves most. Her recent works have been archived in electronic platforms for better sharing and wider view upon variety of medias.

Nurul Fitriah Alias is Nurul Fitriah Alias is a distinguished academic and practitioner in the field of education. She has a Ph.D. in Teacher Training and Education Sciences from Universiti Malaya, where her research focused on innovative pedagogical practices and curriculum development. She also holds a Master's degree in Education from Universiti Putra Malaysia and a Bachelor of Education (Hons) in TESL (Teaching English as a Second Language) from Universiti Teknologi MARA. Throughout her career, she has published extensively in reputable journals, contributing significantly to the fields of educational science, curriculum development, and innovative teaching practices. She has also participated in numerous innovation competitions, earning multiple gold awards for her creative and impactful projects in education. Currently, she serves as a lecturer at the Faculty of Education, Universiti Teknologi MARA, where she has been dedicated to teaching and research for nearly a decade. Her passion lies in advancing education through innovative methodologies, enhancing curriculum design, and contributing to the professional development of educators. Her dedication to excellence and her impressive achievements make her a respected figure in the academic community.

Aizal Yusrina Idris is a highly accomplished academician and experienced educator with over two decades of expertise in teaching Computer Sciences and Information Technology at the university level. She is currently serving as an Assistant Professor at the Kulliyyah Information & Communication Technology, International Islamic University Malaysia, Malaysia. She holds a Ph.D. in Teachers Training and Education Sciences from the Universiti Malaya, Malaysia, a Master's degree in Information Technology (Computer Science), and a Bachelor of Information Technology (Hons.) in Industrial Computing from Universiti Kebangsaan Malaysia. Throughout her career, Dr. Aizal has assumed multiple roles, including lecturer, academic advisor, and course coordinator. She has taught a wide variety of courses ranging from Programming and Software Engineering to Algorithm Analysis and Multimedia Technology, spanning institutions in Malaysia and Saudi Arabia. Notably, she served as the Head of the Computer Science & Engineering Department at Yanbu Industrial College, KSA, where she managed departmental operations and led academic development initiatives. Her academic contributions include authoring and editing books, publishing research articles, and presenting at international conferences. She is also recognized for her consultancy work, having reviewed articles for prestigious journals and contributed as a Subject Matter Expert in courseware development. Dr. Aizal has been honored with accolades such as the Gold and Silver Medals at innovation and invention expos, highlighting her commitment to advancing educational practices and technologies. Dr. Aizal's community involvement and mentorship reflect her dedication to nurturing talent and promoting cultural exchange. Her ongoing efforts in thesis coaching, research supervision, and professional development training continue to inspire and support the next generation of learners and educators.

Index

www.ingramcontent.com/pod-product-compliance
Lightning Source LLC
LaVergne TN
LVHW080341011026
849161LV00009B/2148